George Westinghouse

George Westinghouse
Powering the World

WILLIAM R. HUBER

Foreword by Gary Hoover

McFarland & Company, Inc., Publishers
Jefferson, North Carolina

LIBRARY OF CONGRESS CATALOGUING-IN-PUBLICATION DATA

Names: Huber, William R., 1941– author. |
Hoover, Gary, 1951– writer of foreword.
Title: George Westinghouse : powering the world / William R. Huber ;
foreword by Gary Hoover.
Description: Jefferson, North Carolina : McFarland & Company, Inc.,
publishers, 2022 | Includes bibliographical references and index.
Identifiers: LCCN 2021060942 |
ISBN 9781476686929 (paperback : acid free paper) ∞
ISBN 9781476644141 (ebook)
Subjects: LCSH: Westinghouse, George, 1846-1914. | Westinghouse Electric
& Manufacturing Company—History. | Inventors—United States—Biography. |
Electrical engineers—United States—Biography. | Businessmen—United States—
Biography. | Electric industries—United States—History. | Electrical engineering—
History. | BISAC: BIOGRAPHY & AUTOBIOGRAPHY / Business |
SCIENCE / Physics / Electricity
Classification: LCC T40.W4 H83 2022 | DDC 620.0092 [B]—dc23/eng/20220107
LC record available at https://lccn.loc.gov/2021060942

BRITISH LIBRARY CATALOGUING DATA ARE AVAILABLE

ISBN (print) 978-1-4766-8692-9
ISBN (ebook) 978-1-4766-4414-1

© 2022 William R. Huber. All rights reserved

*No part of this book may be reproduced or transmitted in any form
or by any means, electronic or mechanical, including photocopying
or recording, or by any information storage and retrieval system,
without permission in writing from the publisher.*

On the cover: *inset* George Westinghouse, half-length portrait,
facing front, between 1900 and 1914, photographer
Joseph G. Gessford (Library of Congress); panoramic view of the
World's Columbian Exposition, Chicago, 1893 (Library of Congress)

Printed in the United States of America

*McFarland & Company, Inc., Publishers
Box 611, Jefferson, North Carolina 28640
www.mcfarlandpub.com*

*In memory of my mother,
Roberta Jobe Huber,
who placed this sign in my childhood bedroom,
"Do It Now!"
I never received better advice.*

*"Success is not final; failure is not fatal.
It is the courage to continue that counts."*
—Winston Churchill

Table of Contents

Acknowledgments ix
Foreword by Gary Hoover 1
Preface 3
Introduction 9

1. Origins 11
2. Working on the Railroad 19
3. It's Hard to Stop a Train 22
4. The Smoky City 27
5. Straight Air Brakes 31
6. Automatic Air Brakes 37
7. What's a Wilmerding? 44
8. Two Trains, One Track 53
9. Solitude 60
10. Gas Pains 67
11. More Energy 77
12. AC or DC 81
13. Assembling the Pieces 83
14. The Greatest Inventor Who Ever Lived 90
15. The Greatest Experimenter Who Ever Lived 99
16. Bankers Always Win 113
17. Dying for Electricity 118
18. The Worst of Times 126

19. The White City 135
20. Over a Barrel at Niagara 144
21. New Lands to Conquer 158
22. Rotary Redux 165
23. Trolleys and Trains 171
24. Panic! 181
25. Homes and Family 188
26. Retirement, Honors, and Death 202
27. Memorials 211
28. The Next Century 217

Appendix I—Westinghouse Family Genealogy 223
Appendix II—Automatic Air Brake Operation 225
Appendix III—Electrical Engineering 101 229
Appendix IV—How Does an Induction Motor Work? 235
Appendix V—How Does a Turbine Operate? 239
Appendix VI—Patent Law Primer 241
Chapter Notes 243
Bibliography 259
Index 271

Acknowledgments

In 1987, volunteers and former Westinghouse employees established the George Westinghouse Museum in the former Westinghouse Air Brake General Offices ("The Castle") in Wilmerding, Pennsylvania. In 1998, the Westinghouse Electric Company donated a portion of its corporate records to the George Westinghouse Museum and another part to the Detre Library and Archives at the Senator John Heinz History Center in Pittsburgh.

In 2007, the George Westinghouse Museum closed, and its collection of Westinghouse documents, photographs, and memorabilia moved to the Detre Library and Archives. With 210 boxes (238 linear feet) and oversized materials, the Detre Library holds the most extensive collection of Westinghouse-related material in the world. In short, the Detre Library and Archives is the primary destination for any Westinghouse researcher. Because George Westinghouse himself kept almost no papers or letters, that avenue of inquiry was barren. So the Detre Library became my primary source for Westinghouse documents and photographs.

With the expert aid of Mary Jones, Chief Librarian, and Liz Wright, Librarian, I searched and photographed the relevant portion of the holdings at Detre. As is the case with librarians everywhere, Mary and Liz were patient, helpful, and knowledgeable.

Another librarian provided specific information about George Westinghouse III, son of George and Marguerite. Despite the limitations imposed by the COVID-19 pandemic, Bill Landis, Associate Director for Public Services, Manuscripts and Archives at the Yale University Library, was able to fill in the gaps regarding George III's time at Yale.

The contemporaneous Westinghouse biographies of Leupp (1919)[1] and Prout (1921)[2] provided essential references. They captured the essence of the man but were far too close in time to evaluate his lasting impact.

What is on paper reveals only part of the story. In-person visits to locations relevant to George Westinghouse, his family, and associates uncovered hidden aspects.

John Graf, President and CEO of the Priory Hospitality Group, generously allowed us (my wife and me) to tour "The Castle" in Wilmerding. Giuseppe Provenza, Priory's Director of Facilities, was our "tour guide" and permitted us to go everywhere in the impressive building. Just seeing the street names in Wilmerding (Marguerite, George, Herman, Sprague) recalled its glory days.

After an alfresco lunch, we toured the house and grounds at Clayton, Henry Clay

Breakfast and poker table in Henry Clay Frick's home, Clayton (courtesy Dawn Brean at The Frick Pittsburgh).

Frick's estate. Seeing the breakfast table where Frick, Westinghouse, and their friends played poker breathed life into the verbal descriptions.

A visit to Westinghouse Park, where once stood the Westinghouse home, Solitude, made it clear how close the Pennsylvania Railroad main line was to the house. But it also revealed the transitory nature of man's structures and accomplishments. Without the historical marker proclaiming the Westinghouse Gas Wells, a stranger would have no clue what transpired there.

The Westinghouse Memorial, in a quiet glen adjacent to where I had often played golf in Schenley Park, disclosed the broad scope of Westinghouse's interests and the high esteem in which his employees held him.

Crossing the George Westinghouse Memorial Bridge on Route 30, as I had done so many times in my college years, caused me to realize how industrialized the Turtle Creek valley was—Westinghouse Air Brake, Westinghouse Electric, Westinghouse Machine, and Carnegie's Edgar Thomson steel plant lined the creek and adjacent Monongahela River for miles.

My regular cadre of reviewers, my wife Angie, Bill Reed, Virgil Koning, Carol Cherne, and Carl Lewis Wagner, provided their always insightful comments and critiques.

Then, while searching for some obscure Westinghouse fact, I stumbled on an article by David Bear. It turned out that David, a retired editor for the *Pittsburgh Post-Gazette*, lives adjacent to Westinghouse Park and has a wealth of knowledge about Westinghouse and his family. He kindly agreed to review and edit my manuscript, and, of course, did a professional job. Even more important, he suggested additional lines of research, provided insights that I had missed, and introduced me to other people with specialized knowledge of George Westinghouse, including a few

Westinghouse descendants. David's interest in photography led me to two experts, Harvey Butts and Jim Albright, who contributed beautiful and unique photographs of Pittsburgh.

I have found that the best way to learn about a subject, especially a historical figure, is to write a book. I hope that I have transformed what I learned about the life of George Westinghouse into a book that will educate and entertain you.

"If I have seen further,
it is by standing on the shoulders of Giants."

—Isaac Newton, 1675, based
on Bernard of Chartres, 12th century

Foreword
by Gary Hoover

Entrepreneurs and inventors have driven America's economic growth. These brave men and women were (and are) willing to risk everything, to swim against the tide, and to overcome significant obstacles—with no guarantee of success. Yet many of them remain unknown or forgotten today.

George Westinghouse is a prime example. Westinghouse built a massive industrial empire around his creations that included the air-brake system that stops trains and giant generators for hydroelectric dams. His inventions are still in use today around the world. Unlike some industrialists of his era, Westinghouse treated his tens of thousands of workers uncommonly well. In so many ways, Westinghouse was "larger than life."

Westinghouse was at the center of many historical developments, from harnessing the power of Niagara Falls to lighting the largest and most successful of America's World's Fairs, held in Chicago in 1893. At his peak, he was one of the most famous and revered business leaders in the nation.

On the pages that follow, you will learn the full story of Westinghouse and the times in which he lived. Like other great men, Westinghouse faced ups and downs. His victories were manifold and significant. When Thomas Edison supported the DC power system, Westinghouse instead backed the AC approach and, with the eccentric inventor Nikola Tesla, won the "current wars." On the downside, he lost control of the enterprise nearest to his heart, the Westinghouse Electric Company. Yet, long after that, Westinghouse-branded appliances and light bulbs filled American homes.

William Huber's excellent biography is long overdue. Full biographies of Westinghouse were written about one hundred years ago, in 1918 and 1921. Because they came so soon after Westinghouse's death in 1914, they lacked the benefit of time to achieve adequate hindsight. More recent attempts at telling the Westinghouse story have been incomplete and error-prone. Huber sets the record straight, in immense and fascinating detail. Moreover, the story is told with enthusiasm and includes a wealth of outstanding images that bring George Westinghouse's story to life. Those readers with a technical bent will enjoy the detailed appendices which explain in layman's terms Westinghouse's most important inventions.

I hope that every young, ambitious American will read this story. They might then understand that today's entrepreneurial heroes like Elon Musk stand on the shoulders of such giants as Westinghouse. I also wish that more of our great entrepreneurs and business leaders from the past received such a thoughtful, thorough treatment of their life and work as William Huber has written here.

Entrepreneur, business historian and author Gary Hoover is the cofounder of D&B Hoovers and founder of the American Business History Center (www.americanbusinesshistory.org).

Preface

Why Write About George Westinghouse?

I am a Pittsburgher. An attentive listener can discern that fact after two sentences of random conversation. Our dialect and colloquialisms instantly brand us.

Pittsburgh, at least in the spring, summer, and fall, is a beautiful city. Now that it is no longer the Smoky City, the panoramic view of the Pittsburgh skyline from atop Mount Washington is spectacular. The website *Thrillist*[1] ranks the Pittsburgh skyline as the sixth-best in the U.S.

Pittsburgh has had many heroes, in sports, in politics, in literature, and industry. Names that come to mind include Fred Rogers, August Wilson, Andrew Carnegie, Rachel Carson, Jonas Salk, and H.J. Heinz. *Pittsburgh Magazine* published a list of the 50 greatest Pittsburghers of all time in 2018. They included those names and added George Westinghouse at number 13.[2] In terms of lasting impact on the city, the country, and the world, I would rank him near the top.

As a youngster growing up in Pittsburgh, I remember seeing the name Westinghouse and trademark "Circle-W" everywhere. Westinghouse's Union Switch and Signal factory stood beside the Penn-Lincoln Parkway just east of the Squirrel Hill Tunnels. The Westinghouse Research Center occupied 100 acres in suburban Churchill. My family often drove by the Westinghouse Bettis Atomic Power Laboratory in West Mifflin, near the old Allegheny County Airport. While a student at the University of Pittsburgh (Pitt), I took a course at the Westinghouse Education Center. And, of course, the Westinghouse Memorial Bridge was and is an integral part of U.S. Route 30, the Lincoln Highway.

The Westinghouse name was a daily presence in my life because of KDKA, the 50,000-watt clear-channel AM radio station[a] that was the flagship of the Westinghouse Broadcasting Company. The most popular personality on KDKA was the morning announcer, Rege Cordic. From 1954 to 1965, "Cordic & Company" dominated the drive-time audience with up to 85 percent of radios tuned to KDKA to hear the antics of Rege and his crazy crew.

Cordic's most famous invention was a fictitious beer called "Olde Frothingslosh, the pale, stale ale with the foam on the bottom." In 1955, the joke became a reality when the Pittsburgh Brewing Company started selling Olde Frothingslosh beer and inserting bottles and cans upside down in the case so that the foam would be on the bottom.[3] Cordic's fans and KDKA loved it.

(a) A clear-channel radio station broadcasts on a frequency shared in the U.S. by, at most, two other stations. This restriction allows such stations to be heard without interference over a large geographic area.

Pittsburgh from Mt. Washington; the Allegheny River flows from the north and the Monongahela River joins from the east at the Point to form the Ohio River (courtesy James Albright).

KDKA also featured the games of the Pittsburgh Pirates. One game in particular lives in the hearts of loyal Pittsburghers. On October 13, 1960, Bill Mazeroski, the Pirates' diminutive second baseman, hit a walk-off home run over the left-field wall at Forbes Field to win the seventh and deciding game of the World Series. A little-known fact is that not a single batter struck out in that momentous game, the only time that has happened in World Series history. In a series where the mighty New York Yankees outscored the Pirates 55 runs to 27, only the last run mattered. Many call it "The greatest baseball game ever played."[4] I remember hearing the radio broadcast of the game on KDKA as I sat in Chemistry Lab at Pitt.

As much as the Westinghouse name was prominent in my 20+ years in and around Pittsburgh, I never knew anything about George Westinghouse, the man. My lack of knowledge was, in part, attributable to his personality. Westinghouse was a humble man, not a self-promoter like his rival, Thomas Edison. He hated to have his picture taken, and he hated, even more, giving speeches.

When looking for a suitable subject for this, the second biography I have written, my early exposure to the Westinghouse name surfaced, and I checked to see if other biographies had been written. I was surprised to find that such a significant figure had attracted so little attention. Amazon listed just three biographies, including two written soon after his death and another in 2007. I bought all three and found that the early works lacked the necessary perspective of time, and the recent one was focused more on his financial difficulties and lacked insight into his technical achievements. But all three of the books provided clues about where to find some of the information I would require to write a definitive biography.

I had long thought that the best way to learn a subject is to teach it. Now I know that an even better way is to write a book about it. Through the research for this book, I learned that George Westinghouse converted a childhood propensity for violent outbursts when he did not get his way to a lifelong habit of calm, determined perseverance in the face of adversity. He continually sought to meet his father's expectations and eventually succeeded. Though founding more than 60 companies, Westinghouse remained a loving husband who communicated with his wife every day, wherever he was traveling in the U.S. or overseas. Unlike many of his contemporaries, he sincerely cared about his employees and their welfare. Westinghouse was a man I would have enjoyed knowing.

Until I started writing this book, I had never visited Wilmerding, the company town Westinghouse built, even though I lived within three miles of it in the late 1960s. Nor had I ever seen the stunning Westinghouse Memorial in Schenley Park, even though I attended the University of Pittsburgh, less than one mile away. And I had never been to Westinghouse Park, where his estate and home, Solitude, once stood. Through visiting these sites, I have gained an appreciation for the mutual respect that George and his workers had.

My goal in writing this book is to convey the genius of George Westinghouse, his humanity, his concern for his employees, and the mutual love he and his wife, Marguerite, shared.

He was an imposing man in physical stature, seminal accomplishments, and deep morality, and I have been privileged to learn his story and look forward to sharing it.

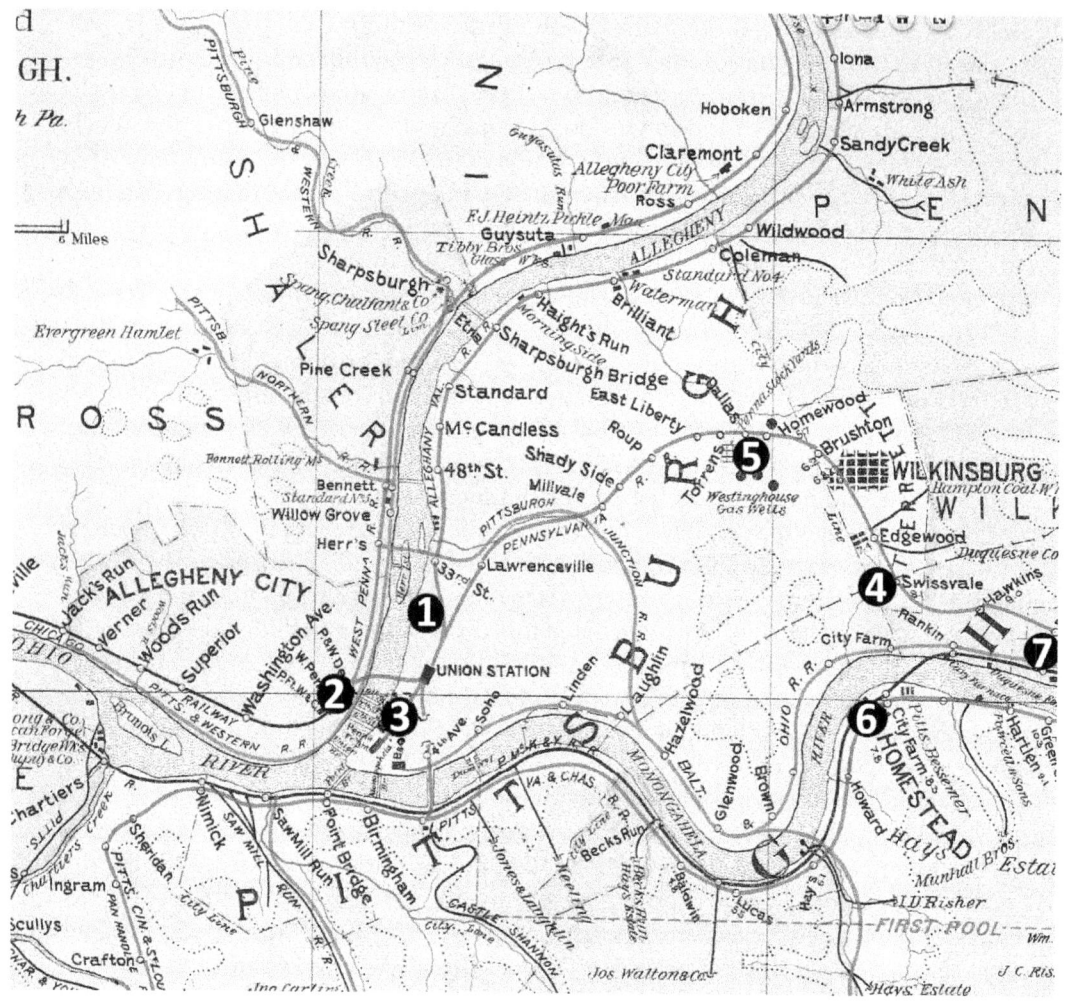

1		Westinghouse Air Brake—25th Street & Liberty Avenue
2		Westinghouse Air Brake—Allegheny City
3		Union Switch & Signal and Westinghouse Electric—Garrison Alley
4		Union Switch & Signal—Swissvale
5		First Westinghouse Home & Gas Wells—"Solitude"
6		Carnegie Steel Company—Homestead
7		Edgar Thomson Works of Carnegie Steel—Braddock

Pittsburgh Area Points of Interest

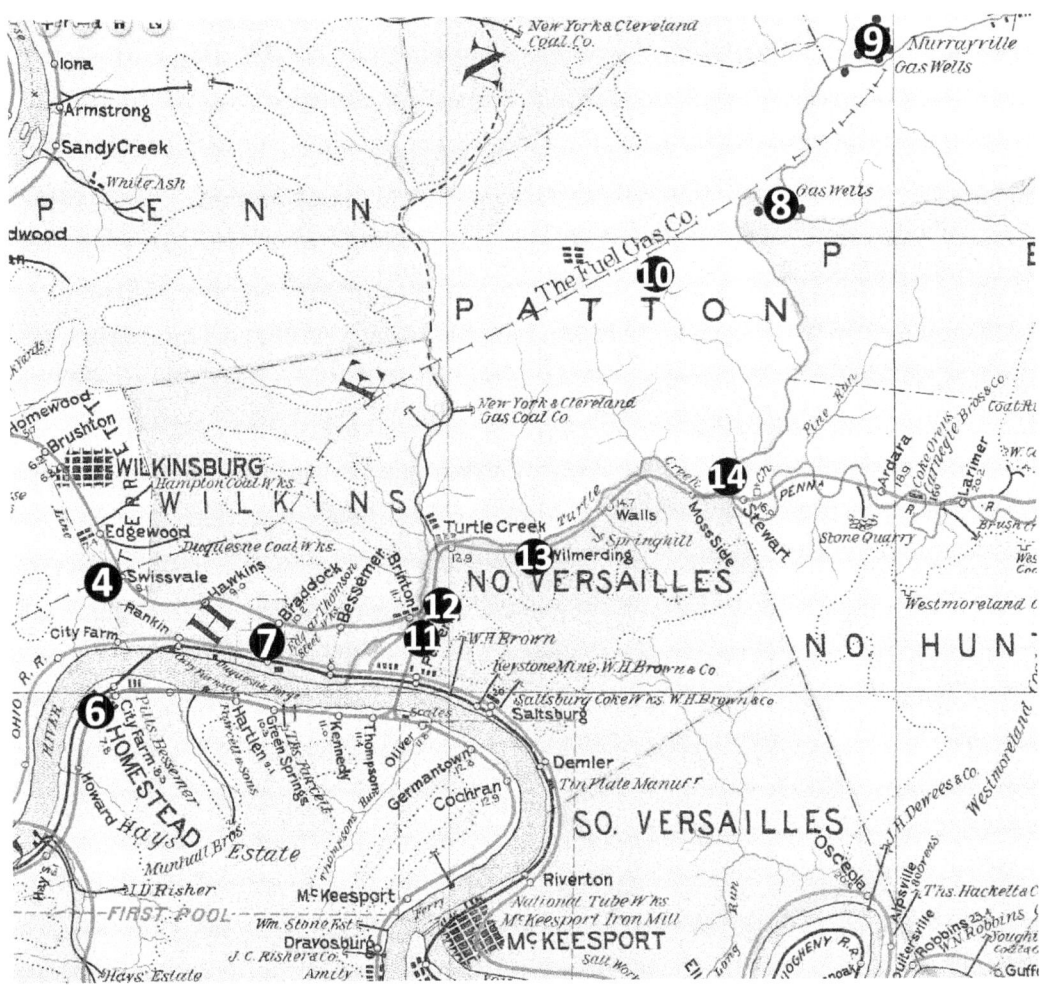

8	Haymaker Gas Wells—Murrysville
9	Other Gas Wells—Murrysville
10	Fuel Gas Line—Murrysville to Pittsburgh
11	Westinghouse Machine Company
12	Westinghouse Electric & Manufacturing—East Pittsburgh
13	Westinghouse Air Brake—Wilmerding
14	Westinghouse Foundries—Trafford

Map from Darlington Digital Library,
University of Pittsburgh Library System

How Do You Spell Pittsburgh?

On November 27, 1758, General John Forbes sent a letter to William Pitt, 1st Earl of Chatham, telling him that his name had been given to the new fort built there, and the surrounding settlement was to be called "Pittsbourgh."

The original version of the city charter, granted on March 18, 1816, called the city "Pittsburgh," and it was known as such for 75 years. In 1891, the United States Board on Geographic Names decided that place names ending in -burgh should drop the final h. Thus, Pittsburgh became Pittsburg. Residents and some organizations refused to drop the -h. The newspaper, *Pittsburgh Post-Gazette*, kept the -h, as did the Pittsburgh Stock Exchange and the University of Pittsburgh. In response to the mounting pressure, the Board reversed its decision, and on July 19, 1911, Pittsburgh again became the official spelling.[5]

Since childhood, I have known only Pittsburgh, so except for direct quotations that use "Pittsburg," I shall use Pittsburgh throughout this book.

Introduction

Have you ever ridden on a train with no concern that it would safely stop at the station? Have you flipped a light switch without wondering if there would be electricity to power the lamp? Did your grandmother have an old but reliable fan or toaster or washing machine with a "Circle-W" on it? Maybe you remember the advertising slogan, "You can be sure if it's Westinghouse." The Westinghouse brand was known around the world, and it still resonates more than a century after the death of George Westinghouse.

The man and the companies he created are responsible for all of these things. But who was George Westinghouse? The goal of this book is to reveal the man whose teachers suspected was mentally disabled, who quit college after one semester, yet founded more than sixty different companies employing well over 100,000 people and was awarded 361 U.S. patents.

After a long career as an electrical engineer, I understand the technologies that Westinghouse developed. For the last 20+ years of my professional life, I was an expert witness in patent litigation cases, and my job was to explain complicated concepts to juries composed of people who were not themselves technically trained.

So I promise you will not be overwhelmed with scientific jargon, arcane acronyms, or complex mathematics. Instead, you will be transported back to the late 19th and early 20th centuries to hear the story of a remarkable man, his wife, their son, and the people around them. More than a biography, this book brings history alive by exploring and illuminating George Westinghouse in the context of his times.

You will read about train wrecks, gas explosions, human electrocutions, labor violence, and a massive and fatal flood. But you will also see kindness, humility, creativity, resilience, and perseverance. Here is the story of a great inventor and successful industrialist, a man who earned the love and respect of generations of his employees. At the same time, his contemporaries were rightly condemned as robber barons.

Westinghouse focused on solving critical real-world problems through the use of his insight and ingenuity. If such solutions resulted in making money, that was a byproduct, not a goal. His inventions applied to railroads, such as air brakes and signaling and control systems, made train travel safe and are still in use today. Without the innovations that Westinghouse and his engineers implemented, we would have smoke-belching, coal-burning power plants spaced a mile apart in cities, while rural areas would have severely limited electrical service. Even the shock absorbers on your car trace back to a Westinghouse invention.

The following chapters reveal not only Westinghouse's technical achievements but the man behind them and the world around him.

Chapter 1

Origins

"We're all immortal, as long as our stories are told."
—Elizabeth Hunter, American author

The great-grandfather of our George Westinghouse came to America in 1755 in the person of John Hendrik Wistinhausen. Fifteen-year-old John and his widowed mother emigrated from England, but previous generations had moved to England from Nordrhein-Westfalen, Prussia. The two Wistinhausen's settled at what is now Pownal, Bennington County, in Vermont's extreme southwest corner. Here John grew to an imposing man of six feet, four inches. He cleared the land and eventually raised two families. With his first wife, Anna Maria Brimmer, he sired a son and a daughter. After Anna Maria died, John Hendrik married Anna Brust, who had four children by two previous husbands. Appendix I—Westinghouse Family Genealogy, provides an Hourglass Family Tree starting with the generation of John Hendrick Wistinhausen and extending for six generations.

John and Anna Maria's son, who would become our George Westinghouse's grandfather, was John Ferdinand Wistenhaus. The spelling of his surname was the first step in anglicizing Wistinhausen. John Ferdinand lived in Pownal for his entire life and, with his wife Catherina Hoenerin, raised a family of twelve children, nine girls and three boys. Their fifth child, George Westinghouse, the father of our George Westinghouse, was born in Pownal on March 20, 1809.

George Westinghouse, Sr., as he would later be known, married Emeline Vedder. Born on September 19, 1810, Emeline was the daughter of Albert Isaac Vedder (1788–1847) and Mary Stilson Vedder (1789–1874).[1] Emeline's ancestors were Dutch and English, and they had resided in Schenectady, New York, for two generations. She was both intelligent and creative and held "clear and definite religious faith,"[2] which she later instilled in her children.

The young Westinghouse couple was married in Minaville, Montgomery County, New York, on July 4, 1830. Soon afterward, they moved to the banks of the Cuyahoga River near Cleveland, Ohio, but the hot summers and accompanying insects did not suit them. They quickly moved back east and settled in Minaville, New York.

Early Settlers

Over a century earlier, the central portion of the British Colony of New York had been settled by immigrants from the Rheinland-Palatinate region in the southwest of

George and Emeline Westinghouse—parents of George Westinghouse, Jr.

what is now Germany. In November 1700, King Charles II of Spain died with no heir, triggering the War of Spanish Succession (1701–1714). The war disrupted life and threatened the balance of power all over Europe, causing many citizens to look elsewhere for peace and stability. England, with massive holdings of thinly-inhabited land in America, encouraged migration of the dissatisfied Europeans to the New World to counter the growing French and Indian influence there.

Colonel Robert Hunter (1666–1734) advanced one specific proposal. Hunter suggested, "settling 3,000 Palatines at New York and Employing them in the Production of Naval stores...." Hunter continued with a description of the land where he proposed to settle the Palatines: "A Tract of Land lying on the Mohaques River, containing about 50 miles in length and four miles in breadth, and a Tract of land lying upon a creek which runs into the said River, containing between 24 and 30 Miles in length. This last-mentioned Land, of which Your Majesty has the possession, is claimed by the Mohaques,[a] but that claim may be satisfied on very easy terms."[3]

The creek mentioned by Hunter is now called Schoharie Creek and flows north through Schoharie and Montgomery Counties to the Mohawk River.[b]

On September 9, 1710, Queen Anne of England approved Hunter's proposal, conveyed 6,000 acres for the settlement, and appointed him (now a Brigadier General) to superintend the transportation of the new settlers to America.[c] Hunter served as the colonial governor of New York from 1710 to 1720.[4]

(a) The Mohaques was one of the Five Nations of Indians present in the region.
(b) Schoharie County was formed from parts of Otsego and Albany counties in 1795. Schoharie is derived from the Mohawk word for driftwood.
(c) My seventh great grandparents, Michael Ittich (later changed to Ickes) and his wife, Elizabetha Von Coblentz, both from Mainz, Rhineland-Palatinate, Germany, were among the 3,000 Palatine settlers who arrived in 1710.

The Wistinhausen family emigrated from Nordrhein-Westfalen and the Palatines came from the southern portion of what is now the German state of Rhineland-Palatinate (Needpix.com).

Beginnings of a Farm Equipment Business

Once resettled among fellow Prussian countrymen in Minaville, New York, George Westinghouse, Sr., continued his occupation as a farmer. But when his neighbor bought a new threshing machine, George could not stop himself from examining and thinking of ways to improve the thresher. Emeline encouraged George, Sr., to change from farming to the manufacturing of farm equipment. They now had three children, and Emeline hoped that manufacturing would better provide for their growing family.

Minaville proved too provincial to support farm equipment manufacturing, so they moved to Central Bridge in Schoharie County, New York. There George established a shop for producing farm equipment, especially threshing machines and other complex mechanical apparatus. In his 1919 biography of George Westinghouse, Jr., Francis E. Leupp wrote of George, Sr.,

> ... a patent he had taken out[d] had begun to bring returns, he made over most of his farm work to hired hands and spent his days at the bench. His mechanical operations gradually outgrew the original shop, and an extension had to be added. This, in its turn, meant more capital and more help, both of which were forthcoming from the neighborhood, where the people had come to recognize in him a man of more than ordinary ability. ... Increased domestic expenses, together with a business competition which was already making itself felt, led Mr. Westinghouse to consider means of reducing the cost of his machines. Though he could make the wooden parts in his shop and do the assembling there, he had to buy all his metal castings in Schenectady and haul them over by wagon—a tedious and expensive process when the roads were out of repair. When, therefore, his business had sufficiently expanded, he decided to remove both factory and family to Schenectady, and in 1856 the change was made.[5]

As Leupp reports, the business was so successful that George decided in 1856 to move it and his family to Schenectady, to be closer to the suppliers of metal parts for his machines. (Ironically, the Westinghouse & Co. factory was eventually demolished to make way for General Electric Building Number 36.)

Amidst all of their moves, George and Emeline had ten children.[6]

Jay Westinghouse, the eldest surviving boy, was 20 years old at the time of the move to Schenectady. He attended the Polytechnic Institute[e] in Troy, New York, but his aptitude was dealing with people rather than machines. Jay worked for his father, meeting customers, hiring workers, and managing the financial accounts.

The next son, John Westinghouse, although gifted with mechanical skills, was also concerned with social work.

Albert was a prodigious reader and loved to debate almost any question. Many expected him to become a prominent attorney.

George Westinghouse, Jr., was a poor student. He was far more interested in mechanical projects in his father's factory and often skipped school to work there. While textbooks bored him, machines fascinated him. He was stubborn and had a pugnacious personality. In his biography, Leupp reported,

> If he felt any especially strong desire, he would not brook the slightest opposition to his efforts to gratify it. When persistent demands were unavailing, he would fly into a rage

---

(**d**) At least seven U.S. Patents were issued to George Westinghouse, Sr. He had, and passed to his son, a strong respect for intellectual property such as patents.

(**e**) Now known as Rensselaer Polytechnic Institute (RPI), a well-respected private research university.

Chapter 1. Origins 15

Children of George and Emeline Westinghouse

Name	Born	Married	Died
Catherine	October 5, 1831; Minaville, NY	Spencer Moore	July 7, 1882 (age 50); Rotterdam, NY
Henry	July 26, 1833; Minaville, NY		February 13, 1835 (age 19 months); Minaville, NY
Jay	July 8, 1836; Minaville, NY	Lovantia Augusta Hall	January 17, 1890 (age 53); Schenectady, NY
Mary	April 15, 1838; Fonda, NY		March 10, 1864 (age 25); Schenectady, NY
John	August 12, 1840; Fonda, NY	Harriet VanVranken Bradt	September 2, 1890 (age 50); New York, NY
Albert	December 9, 1842; Central Bridge, NY		December 10, 1864 (age 22); McLeod's Mill, Mississippi
Herman	March 7, 1845; Central Bridge, NY		September 24, 1849 (age 4); Central Bridge, NY
George, Jr.	October 6, 1846; Central Bridge, NY	Marguerite Erskine Walker; August 8, 1867; Brooklyn, NY	March 12, 1914 (age 67); New York, NY
Elizabeth	February 20, 1849; Central Bridge, NY		January 15, 1865 (age 15); Schenectady, NY
Henry Herman	November 16, 1853; Central Bridge, NY	Clara Louise Saltmarsh; June 20, 1875	November 18, 1933 (age 80); Goshen, Orange Co., NY

House in Central Bridge, New York, where five of the Westinghouse children were born. The site was added to the National Register of Historic Places in 1986 (courtesy of Howard Ohlhous).

which was terrifying to behold. Old neighbors of the family still remember these paroxysms, which took the form first of screaming and stamping, and then throwing himself flat and banging his head against any hard surface that came most convenient—the floor, the wall of a room, the side of a house.... If every one about him remained obdurate, he would keep up the disturbance till his strength was utterly exhausted. Usually, however, some older member of the household, unable to endure the demonstration longer, would yield the point at issue, and his tears, cries, and self-torture would cease as suddenly as they had begun.... Speaking in later years of these outbursts, he remarked with whimsical humor: "I had a fixed notion that what I wanted I must have. Somehow, that idea has not entirely deserted me throughout my life. I have always known what I wanted, and how to get it. As a child, I got it by tantrums; in mature years, by hard work."[7]

In an effort to focus the interest of his son, George Sr. put him on the summertime payroll of Westinghouse & Co. Starting in May 1860, when George Jr. was 13½; his salary was 50¢ a day. He worked through September 1860, and from March through September 1861, he made 75¢ a day. He started again in March 1862 at 75¢

Westinghouse combined clover and grain thresher. Westinghouse traction engine (from 1892 Westinghouse Threshing Steam Engine Catalog, author's collection).

a day, and by April earned 87½¢ a day. School seems to have lost out to work in late 1862, as George Jr. continued working until February 1863, when his rate was raised to $1 per day. In April 1863, he received yet another raise to $1.125 per day until the end of September 1863.

Westinghouse & Co. employed up to 200 workers to manufacture large and complex farm equipment. Two examples of the machines are shown here.

When the U.S. Civil War started in April 1861, George Jr. was 14½ years old. His older brothers, John and Albert, wanted to enlist, but their father convinced them to wait and see if the war would end quickly. George Jr., just as patriotic but less disciplined than his brothers, decided to run away and enlist with one of his buddies. After breakfast one morning, young George took a few essentials in a carpetbag and headed for the train station. Despite his attempts at stealth, a neighbor saw him and promptly notified Emeline. She found George Sr., and he headed for the train station. Just as the conductor was about to release the train, George Sr. boarded and ordered a delay. He confronted George Jr. and said, "George, I guess you'd better come back home."[8]

It was not a suggestion, although George Sr. maintained a calm attitude. Young George returned home and then to school, where his classmates kidded him about his short time in the military.

With the war still raging, Albert enlisted in August 1862 and distinguished himself by his gallantry and efficiency. However, on December 10, 1864, Lieutenant Albert Westinghouse was killed while leading a cavalry charge at the battle of McLeod's Mill, Mississippi.[9]

Brother John Westinghouse had received a commission through his father's partners and served on a Union man-of-war.

Two years later, when George Jr. was 16½,[(f)] the war was still raging, and his parents finally permitted him to enlist. In June 1863, George Jr. committed to thirty days of service in the Twelfth Regiment of the New York National Guard. His term expired in July, and in November, he re-enlisted, this time in the Sixteenth Regiment of the New York Cavalry. He rose to the rank of Corporal before being honorably discharged in November 1863. But his military service was not over. On December 14, 1863, George Jr. passed an examination and was appointed an

George Westinghouse, Jr., in 1864, age 17+.

(f) Prout cites War Department records as showing over 41 percent of enrollments in the Union army were boys eighteen and under, and over 77 percent were twenty-one and under.

Acting Third Assistant Engineer in the United States Navy. He served first on the ship *Muscoota*, then on the *Stars and Stripes*.[10] Both of these ships were used to blockade Southern port cities.

By August 1865, the Civil War was over, Albert Westinghouse had perished, and John and George mustered out and returned to Schenectady. John was almost 25 and went back to work for his father. While skilled at metalworking, he spent his spare time rescuing gang members from a life of crime and guiding them to productive lives. He established a night school and mission and gave time, money, and effort in helping the unfortunate.[11]

George Jr., who would be 19 in October, entered Union College in Schenectady on September 15, 1865. During his first semester, he studied French, German, solid geometry, English rhetoric and essays, and vocal training. Still unsuited to formal education, he lasted just three months at Union. He returned to his father's shop, where he continued his "hands-on" training at the newly negotiated salary of two dollars a day.

On October 31, 1865, three weeks after his 19th birthday, George Jr. received his first patent (U.S. Patent 50,759) for a rotary steam engine. Although never actually used, the engine was the start of a lifelong interest in rotary power. It was also the start of a lifelong series of patents. George Westinghouse, Jr., was the inventor of 361 United States patents, an average of one patent every seven weeks for nearly 49 years. His final patent, for a pneumatic shock absorber for automobiles, was issued four years after his death.[12]

Chapter 2

Working on the Railroad

"Necessity is the mother of invention."
—Attributed to Plato, First English
version by Richard Franck in 1658

On an 1866 trip to Albany for his father's company, George Westinghouse, Jr., met opportunity.

In those early days of railroading, it was common for train cars to derail, leading to the labor- and time-intensive operation of prying and jacking them back onto the track. The last two cars of the train ahead of George's jumped the tracks and resulted in a two-hour delay. Westinghouse watched the re-railing procedure with rapt attention. Once his train was underway, George commented to his traveling companion that, with proper equipment, the cars could have been back on the tracks in fifteen minutes. He then described the concept to his friend, who said, "Why don't you make one and sell it to the railroads?" George replied, "That's a good idea. I'll do it."[1]

The next morning, Westinghouse sketched plans for his "Car Replacer" and started to build a model. When the model was ready, he showed it to his father. As usual, the taciturn George Westinghouse, Sr., expressed little enthusiasm. Without commenting on the functional value of his son's idea, he objected to the cost of pursuing a patent on the invention as well as manufacturing and marketing it. "You'll let yourself in for a pretty penny before you're through, and where's the money coming from?" George Jr. bravely replied, "I thought probably you'd lend it to me." George Sr. relented slightly, saying, "If you are bound to go into this business, I don't suppose I can stop you; but you will have to make my share a very small one."[2]

Undaunted by his father's lukewarm reception, George Jr. scoured potential investors in Schenectady and found two men, Rawls and Wall, who were willing to provide $5,000 each in seed money for his project.[a] The three partners agreed to share the profits equally.

Discussions with railroad men about his new invention revealed an even more serious problem with the burgeoning rail system. Wherever two tracks merged, a custom-made piece of cast iron called a "frog" was required to accomplish the joining. Crossings, where two tracks intersected, also required similar castings.

The frogs were subject to high stress and wear, and they failed frequently. In addition to being costly both in material and labor, replacing them forced a critical stoppage in rail traffic on both tracks.

Westinghouse mulled over the problem and believed he could solve it and also

---

(a) $5,000 in 1866 would be equivalent to about $80,000 in 2019 dollars.

Railroad Frog, allowing two tracks (at bottom of photograph) to merge into one (author's photograph).

improve his car replacer. For the frog, he would replicate the same track pattern on the underside, so it could be turned over when the pattern on the top wore out. For both the frog and the car replacer, he would build them with a new but untested material, cast steel, that promised twenty times longer life without deformation and breakage. Unlike steel, cast iron is brittle, especially at low temperatures. Cast iron frogs could develop undetected cracks and break when a heavy locomotive passed over. Such cracking probably caused one of the worst train accidents in the early history of railroading.

On December 18, 1867, at the same time Westinghouse was developing his cast steel frog, the Buffalo and Erie Railroad's New York Express derailed near Angola, New York. The temperature was below zero, and a frog near the Angola Station had been replaced shortly before the train passed over it. The new cast iron frog failed, causing the wheels on the last car to leave the track. After traveling one-half mile, the train finally stopped, but not before two passenger cars had plunged off the 50-foot-high bridge over Big Sister Creek.

In those two crowded cars, more than 50 people lost their lives, mostly by burning to death when on-board heating stoves set the wooden cars aflame. The disaster was soon labeled "The Angola Horror," and local and national newspapers and news magazines breathlessly reported the carnage, complete with drawings of flaming wreckage.[3]

As he had observed his father doing, George Jr. filed patent applications for his car replacer and frog. In due time, he received U.S. Patents 61,967 for the Car Replacer (issued February 12, 1867) and 76,365 for the Improved Railway Frog (issued April 7, 1868).

The casting of steel was in its infancy, so Westinghouse had to find manufacturers willing to develop the new technology. He identified two foundries; the Bessemer Steel Works in nearby Troy, New York, and another in Pompton, New Jersey, that agreed to work with him.

Returning from a visit to Pompton, George boarded the Hudson River Railroad but found the cars crowded with commuters returning home from work. He passed on some vacant smoking car seats and continued to the last car, where he found an open seat beside an attractive young woman.

Young Westinghouse was self-effacing but not shy, and he soon learned that Marguerite Erskine Walker lived in Roxbury, New York. She had been visiting friends in Brooklyn and was traveling to Kingston, New York, to visit relatives. As Marguerite was preparing to disembark, George asked if he could call upon her. Her hesitant response led him to tear a page from his notebook and write the names and addresses of several responsible friends who could vouch for his family and reputation. She was reassured by this gesture and consented to have him visit. At the Kingston station, George helped her off the train and seemed to float up the steps as he reboarded to continue his journey to Schenectady. Upon arriving home, his first action was to ask his pastor to write to Marguerite, attesting to his character and the standing of the Westinghouse family in the community.

George was happily distracted at dinner that evening, and one of his sisters commented, "You look as if you won a prize in a lottery." He replied, "I am not sure that I have won it yet, but I think I have a good chance." His parents exchanged a startled look, as they disapproved of gambling of any kind. George's next comment caused them even more concern. He said, "I've met the woman I am going to marry." Then followed the usual questions about supporting a wife, living arrangements, and other details, but Westinghouse would not be deterred by trivial issues.

After several visits to see Marguerite at Kingston and Roxbury, George announced to his mother that the couple had decided to be married in Brooklyn on August 8, 1867. And by the way, they would like to live with his parents until they could find a suitable house.[4] Knowing George's determination after having made a decision, his parents could do nothing but acquiesce. George Westinghouse was 20 years old when they married; Marguerite was 25.

Marguerite and George Westinghouse, Jr. (Westinghouse Collection, Detre Library and Archives Division, Senator John Heinz History Center, Pittsburgh, PA).

Chapter 3

It's Hard to Stop a Train

> *"If you assume a train will stop when the engineer sees a car on the tracks, you're right—but trains need at least 18 football fields of track to reach a complete stop."*
> —Warning on NHTSA Website

It seemed that every time George Westinghouse boarded a train, he found another problem to solve. On a short trip in 1866 from Schenectady to Troy, New York, to visit the foundry that was manufacturing his car replacer, the train he was riding stopped between stations for no apparent reason. George and several other passengers disembarked to witness a scene of overturned or damaged cars with their cargo strewn everywhere.

The locomotives that had been pulling two trains collided and were lying on their sides near each other. The weather was clear; the track was straight and in good repair. One or both of the engineers must have been negligent. When Westinghouse suggested this possibility to one of the train crew, he was assured that both engineers had seen the impending disaster but were powerless to avert it because of the slow and cumbersome method of stopping trains at that time.

Stopping a Train with Hand Brakes

When a train engineer desired to stop his train, either for a train station or an emergency, he would signal "brakes down" with the train whistle to alert brakemen who were stationed either in the caboose or on top of the cars behind him. Each brakeman would move to the junction of two cars and turn a wheel attached to a chain.

Turning the wheel would tighten the chain, which in turn would force a brake pad (an iron block) against the moving wheels, thus slowing that single set of wheels. Then the brakeman would move to the next car and repeat the process. Either that brakeman or his companions would perform the same task until brakes were applied to all of the cars. It was an intense and dangerous task, with brakemen leaping from car to car of the moving train to apply brakes until the train halted.

A brakeman's job was incredibly dangerous. His average life expectancy was 27 years.[1] Hazards included falling from moving trains, colliding with nearby structures, bridges or tunnels, and being run over by rolling cars.

On April 30, 1890, Lorenzo Stephen Coffin, former Iowa State Railroad Commissioner, testified before the U.S. Senate Committee on Interstate Commerce. The following distressing statements are extracted from that testimony.

Chapter 3. It's Hard to Stop a Train

One of the deadliest jobs in America was that of brakeman, who worked from the top of moving trains in all weather—engraving by Peckwell, on the cover of *The Railroad Conductor*, vol. 7, no. 15 (Aug. 1, 1890).

- "In this nation there are not less than some thirty-two or thirty-three thousand railway employés either killed or mangled, more or less, for life—crippled in their daily work. The CHAIRMAN. In a year? Mr. COFFIN. Yes, sir; yearly."
- From two causes, hand brakes and link-and-pin couplers (discussed in Chapter 6), over 7,000 railroad workers were killed or crippled for life in 1888.
- In the one-month period between October 19 and November 23, 1889, 100 brakemen were killed and 136 crippled for life.[2]

There Must Be a Better Way

Inventors in the U.S. and Europe had tried to solve the problem of creating a braking system controlled from the engine cab. In England alone before 1870, at least 650 patents were issued for railroad brakes incorporating chains, springs, levers, pulleys, rollers, rods, tubes, and pumps. None worked well enough to solve the problem.[3]

Westinghouse's first idea was to tie together the existing chains that activated the brakes on each car to a separate long chain running the length of the train. Perhaps the long chain could be pulled by some mechanism in the engine, thus giving direct control of braking to the engineer.

While considering the implementation of his idea, business took him to Chicago, where he met with Superintendent A.N. Towne of the Chicago, Burlington, and Quincy Railroad. After transacting their official business, the two talked of railroad safety and the need for an effective braking system. Before parting, Towne said, "Come in tomorrow afternoon, and we'll go down to the yard where they make up our prize train, the *Aurora Accommodation*. We've put a brake on that which seems to do all that can be done in the brake line. I'll have the inventor over to meet you, and we'll inspect the train together. You'll find him an interesting fellow, and he'll talk brake with you from morning till night if you'll let him."[4]

The next afternoon, Westinghouse met Augustine I. Ambler of Milwaukee. Ambler's brake invention, patented in 1862, was surprisingly similar to Westinghouse's concept for a chain brake. When George admitted that he was trying to solve the braking problem, Ambler told him, "You are throwing away your time, young man. I went over all the ground before completing my invention, and my patents are broad enough to cover everything."[5] But Ambler underestimated the determination and perseverance of George Westinghouse, Jr.

In George's view, Ambler's chain brake had several limitations:

- First, the method for tightening the chain was something like a present-day hose reel, driven by a grooved wheel that could be moved to contact the locomotive driving wheel. Westinghouse thought that apparatus was incapable of accurate control and prone to rapid deterioration.
- Second, the long chain passed over rollers to hold it from sagging down to the track. Each roller-chain interface would be the source of wear and possible snagging.
- Third, how was the chain to be lengthened or shortened when cars were added to or removed from the train?

The process of invention is often incremental: addressing the most significant problem to be solved and then refining the new structure to solve less significant issues. The new structure often creates problems of its own, which then must be resolved by further modifications.

Westinghouse tackled the first limitation of Ambler's chain brake, the imprecise method of tightening the chain. He thought that a steam-driven cylinder, fed by steam from the engine, would provide a more precise and controllable operation. But then he considered what would happen as the length of the train increased from the four cars of the *Aurora Accommodation* to fifteen or twenty or more cars. Could a steam-driven cylinder of sufficient size fit on the locomotive? To solve that secondary problem, he thought of distributing the steam-driven cylinder, that is, providing a smaller cylinder on every car instead of a single large cylinder on the locomotive. That potential solution created a new problem: what about condensation of the steam as it traveled further and further from the heat source at the locomotive? Other concerns intervened, and the train brake question went on hold.

Back in Schenectady a few weeks later, George Jr. was eating lunch alone and contemplating some immediate problem in the shop when he became aware of a young woman offering him a brown-paper-covered pamphlet that looked like a magazine. Uncharacteristically, he waved her away, telling her to offer her pamphlet to a group of workers at the next table. She said that she had done so and that they told her to approach Westinghouse. She further stated that she was studying to be a teacher but lacked funds to finish. She hoped that selling magazine subscriptions would earn the required money. By now, George was sympathetic and asked to see the magazine, which was called *Living Age*. He scanned it briefly and saw an article that interested him titled, "In the Mont Cenis Tunnel." Partly from a desire to help and partly from an interest in the article, he asked how long a subscription two dollars would buy. "Three months," she replied. The transaction completed, Westinghouse had to wait to read the article because she could not give him her only sample. He never saw the young woman again, but she had played her part in history.

A few weeks later, the first magazine arrived, but it lay unopened for several days until more pressing matters were resolved. Finally, George returned to the Mont Cenis article.

Mont Cenis is an 11,850 feet high mountain in the Alps between Italy and France.[a] Between 1803 and 1810, Napoleon constructed a road through a pass in the mountain. A rail tunnel through the mountain was begun in August 1857 from the Italian side and four months later from the French side.

The tunnel was initially expected to take 25 years to complete, but the use of pneumatic drilling machines and electrical ignition of explosives cut that time to 14 years. It opened for rail traffic on September 17, 1871. At just under eight miles in length, the Mont Cenis Tunnel was twice as long as any previous tunnel.[6]

Pneumatic was the word that caught George Westinghouse's eye. Unlike most long tunnels, it was impractical to drill shafts at intervals along the Mont Cenis Tunnel route and then bore the tunnel from those shafts. So, at Mont Cenis, tunnel excavation had to be done only from the two ends.[7] Thus, the air (pneumatic) drilling machines were up to four miles from the source of the compressed air that powered them.[b]

Westinghouse reasoned that if the Mont Cenis project could convey compressed air for four miles, carrying it the length of even a massive train would be comparatively simple. Air was free, and, unlike steam, it would not condense into water or freeze. Furthermore, a system using air pressure to activate the brakes would not infringe on Augustine Ambler's patents. Therefore, compressed air would be the ideal source of power to operate the train brakes!

As had become his practice, Westinghouse immediately began preparing detailed drawings of his new air brake system. As the work progressed, he encountered technical and practical problems. If compressed air were to operate the brakes, what would happen if the pressure were lost by cars becoming decoupled from the

---
(a) Due to changes in political boundaries, Mont Cenis is now completely in France.
(b) The article that George Westinghouse read was written when the tunneling equipment was just 3,000 feet from the end of the tunnel, but conveying compressed air 3,000 feet would be more than sufficient for operating train brakes.

train? The pressure in the air hose would be lost and the remaining train would have no brakes, an untenable situation. To avoid that problem, George developed a hose coupling with automatic valves that closed when the hose separated, thus retaining the pressure in both portions of the hose.

To facilitate braking control by the train engineer, he designed a three-way control to serve as the engineer's brake valve. In short, Westinghouse created a complete and functional air brake system, something that no one had done previously. With his drawings completed, he made the now-familiar trip to see his patent attorney, George H. Christy. Together they drafted a patent application and accompanying affidavit attesting to George Westinghouse, Jr., of Schenectady, New York, as the inventor.

On July 10, 1868, Westinghouse and Christy filed the first of many patent applications for an air brake system. George Westinghouse was 21 years old. By 1907, Westinghouse would have filed 103 such air brake patent applications.

With the legalities out of the way, George was free to discuss his invention with railway officials as he traveled the country promoting his railway frog. But in both areas, he faced obstacles. Because the frog was made of durable cast steel and was reversible, replacement frogs were rarely needed, and frog sales declined. And the idea of using air to stop a train encountered great skepticism and even derision among veteran railmen.

The suffering frog sales triggered another problem. George's partners in the frog venture became dissatisfied with the reduced return on their investment and called him to a meeting. Believing they had the upper hand, they presented an ultimatum to George: either buy them out or withdraw from the business and allow them to run it.

Westinghouse was indignant. After all, without his idea for the frog, there would be no business. He said, "You know very well that I am in no position to buy you out, so what's the use of talking about that?" His partner replied, "Well, we left open an alternative." George retorted, "If I retire, what do you propose to pay me for my patents?" "Nothing. You have had the use of our money from the start, in return for your services as salesman. If necessary, we can hire an outside traveling man to take your place, and lay him off when trade is dull."

Westinghouse could no longer suppress his temper. He said, "So you expect me to make you a present of my patent rights? Well, you have missed your guess, for I don't intend to. We'll break up this business here and now, if you say so; but from the moment you and I part company, you make no further use of my patents without paying me as you would a stranger!" As George stormed out, the men were amazed at the defiant attitude of their 22-year-old ex-partner.[8]

With a young wife to support and no income in sight, the future looked dim. But Westinghouse remained confident in the prospects for his air brake, and his travels had planted an idea in his mind. He had heard of a steel-making plant in Pittsburgh that could manufacture his car replacer and frog for far less than he had been paying. Now that his dissatisfied partners were no longer part of the business, he felt free to pursue the Pittsburgh opportunity. Marguerite gave her full support, and George headed to Pittsburgh.

Chapter 4

The Smoky City

> *"New things, new ideas arrived and strutted their stuff and were vilified by some and then lo! that which had been a monster was suddenly totally important to the world."*
> —Terry Pratchett, English Humorist and Author

George Westinghouse arrived at the train station in Pittsburgh with the address for Anderson, Cook & Co., the steelmaker, but no map. He had walked a short distance when he encountered a tall, well-dressed stranger. George asked for directions, and the young man volunteered to walk part of the way with him. As they talked, George found that they were the same age and shared an interest in technical matters. The man, whose name was Ralph Baggaley, was the general manager of a local foundry and a member of one of the most prominent families in Pittsburgh. The men exchanged contact information, and Ralph directed George to his destination.

Upon reaching the offices of Anderson, Cook & Co., Westinghouse met with the senior partner. They quickly agreed that the firm would manufacture the car replacer and frog at their own expense and that George would travel to railroad companies to sell the products. Of course, once he had the ear of his customers, Westinghouse would take the opportunity to present his ideas for a more effective braking system.

He did not find a receptive audience. Responses to his air brake pitch ranged from polite attention to "I have another meeting I have to attend." Commodore Cornelius Vanderbilt, the owner of the New York Central Railroad, proved to be a good listener, but then dismissed the idea as too imaginative for serious consideration.

On one of his periodic trips to Pittsburgh to report on car replacer and frog sales, Westinghouse contacted Ralph Baggaley to discuss his frustration. Over dinner, George revealed his air brake idea, but Baggaley was less than enthusiastic. But as George provided more details, his enthusiasm infected Baggaley, who became convinced that Westinghouse had a concept of great worth.

At the end of their meeting, he said, "Westinghouse, we must lose no time in putting this thing before some of the big men in the railroad world." Of course, that is precisely what George had been trying to do without success. He replied, "I could launch it without much difficulty if I had a little capital. I have seen several railroad men already. They have no way of answering my arguments about the value of the invention if it will work, but I haven't found one yet who was willing to stand the expense of giving it a trial."

Ralph said, "Then a man who has money to risk would be of more use to you just now than one who knows railroading?"

"That's it," said Westinghouse.

"Perhaps your father would help you now, if you put the case before him in that way."[1]

That advice was not what George wanted to hear, but he approached his father anyway. George Sr. quickly made it clear that he had no interest in any such speculative project.[2]

George Jr. was beginning to doubt his own invention. Perhaps there was some obvious flaw in the idea, but the railroad men he had talked to would not risk hurting his feelings by revealing it. He went back to Baggaley with his father's rejection and his doubts. Baggaley had by then agreed to back the idea with a few thousand dollars of his own, but both men wanted some assurance of the utility of George's air brake. Baggaley said, "We are wasting time with so much hesitation. Let me put all the drawings, directions, and claims into the hands of a man I know, the most highly skilled mechanical expert in the city of Pittsburgh, and have him pass on them. It will cost something, for he gets good fees for his opinions, but I think it will pay in the end." Westinghouse, grateful for some progress, agreed.

The expert reviewed all of Westinghouse's papers, and two weeks later, presented a written opinion to Ralph Baggaley. Ralph eagerly read the report but was soon disheartened. The expert condemned the entire idea as unsound and nonsensical. Ralph took the report to Westinghouse, who read it twice before commenting. "How much did your expert charge you for that death sentence?" George asked. "One hundred dollars." "Well, what are you going to do about it?" "Watch me and see," replied Ralph. With that, he tore the report into small pieces and threw them into the fire.

Westinghouse was astounded at Ralph's action, and remarked, "That's a nice way to treat an expert's report. Apparently you don't consider the fellow's opinion worth so much now as you did before you got it?"

"It was worth the hundred dollars I paid for it—every cent; it has taught me a lesson that I could not have bought otherwise for ten times the money. Hereafter I back my own judgment and let outsiders go. George, I'll put up your common sense against the special education of any expert in Christendom. Now let's get to work, so as to be ready for the show that somebody is sure to give us soon."[3]

With Baggaley's financial backing, Westinghouse built the hardware needed for a demonstration. Available funds provided enough to equip one locomotive and one car. But railroad superintendents remained reluctant to commit to an unproven system and would not sponsor an experiment to prove it. This catch-22 situation prevailed until Robert Pitcairn, local superintendent of the Pennsylvania Railroad, called on Westinghouse. Pitcairn listened to George's well-rehearsed pitch and said, "If I can get my people interested, I believe there is enough in the invention to be worth a fair trial."[4] Pitcairn returned soon after, this time with Superintendent Edwin Williams of the Altoona Division and Alexander Cassatt, assistant superintendent of motive power for the PRR. The railroad men agreed to provide the track, train, engineer, and crew free of charge if Baggaley and Westinghouse would manufacture and install all of the braking equipment. The Pennsylvania Railroad, established in Philadelphia in 1846, was on its way to becoming the world's largest railroad, transportation company, and corporation. So, convincing the "Pennsy" to convert to his air brakes would have been a monumental victory for Westinghouse. But Ralph Baggaley and George Westinghouse had no more money to build the necessary equipment.

Chapter 4. The Smoky City

With this missed opportunity haunting them, Westinghouse received an unexpected visit from W.W. Card, Superintendent of the Steubenville Division of the Panhandle Railroad.[a] Card opened by saying, "I understand that you have invented a remarkable brake." Westinghouse was rendered nearly speechless, but he recovered quickly and described the details and advantages of his braking system.

Unlike most other railroad men, Card listened attentively and examined each piece of the mechanism with a critical eye. At the end of the meeting, Card encouraged Westinghouse by stating, "If this will do all it appears capable of, you have opened a gold mine, Mr. Westinghouse. The railroads have been waiting a long time for a really good brake. What we have now will answer only so long as we can find nothing better in the market. When the right one comes along, it will find the roads all ready for it."[5]

A few days later, Card returned, this time with the company's purchasing agent. While the agent was as impressed as Card was by the braking system, he would not commit to sponsoring a test without the approval of the board of directors. He did agree to present a proposal to the board with a strong recommendation that they accept. Predictably the directors refused to sponsor the entire cost but did provide a letter, signed by Thomas Jewett, the president of the company, agreeing to provide a train for the trial.

All that Westinghouse and Baggaley would have to do was equip the train with their braking system and reimburse the railroad for any damages done to the locomotive or cars by the attachment of the system. This offer was less attractive than they had received from the Pennsy. But the young inventor and his financial backer were impatient with the cycle of encouragement and disappointment, so they accepted.

To keep their investment as low as possible, a train of just four cars was specified. As they had already built the equipment for the locomotive and one car, making the additional parts for three more cars was relatively easy. The components of this first air brake system were:

- An air compressor on the locomotive, driven by a steam engine which received its steam from the locomotive boiler;
- An air reservoir, also on the locomotive, to store air compressed at 60 to 70 pounds per square inch;
- A pipe from the air reservoir to a brake valve controlled by the engineer;
- Brake cylinders for the tender and each car of the train; and
- A pipe from the brake valve extending under the tender and cars, connected to each brake cylinder. Between cars, the rigid pipe sections were joined by flexible hoses with couplings that sealed-off the line when disconnected. When the lines were connected, the couplings opened to make a continuous path for the pressurized air.

Westinghouse installed the compressor, air reservoir, and brake valve on the locomotive, the brake cylinders, and shoes on each of the four cars, along with the piping and hoses to connect it all.

On the fateful day in April 1869, the train's last car was occupied by officers of

---

(a) The Panhandle Railroad was so-named because it crossed the narrow portion of West Virginia (the "Panhandle") between the western border of Pennsylvania and the eastern border of Ohio.

the Panhandle Company, railroad men from other lines, and invited guests. Westinghouse provided final instructions to the train's engineer, Daniel Tate, and wished him luck as he left the cab. As soon as Westinghouse had boarded the last car, the train started to move.

The tracks from the Panhandle Station in Pittsburgh entered a one-sixth mile-long tunnel through Grant Hill, emerging at Fourth Avenue, a regular stop to pick up passengers. After two more road crossings, the tracks crossed the Monongahela River and then hit open country on the way to Steubenville. As there was to be no stop at Fourth Avenue, or anywhere else en route, Tate accelerated the train to about thirty miles per hour.

Just as the train emerged from the tunnel, Daniel Tate saw a horse-drawn wagon crossing the tracks at Second Avenue. Simultaneously, the wagon driver heard and saw the approaching train just two blocks away. In a panic, the driver lashed his horses, and they reacted by lunging forward. The sudden movement dislodged the loose crosswise plank the driver was using for a seat, and he plunged to the ground across one of the rails.

Meanwhile, Tate, the train engineer, reacted instinctively by grabbing the brake valve and twisting, just as Westinghouse had instructed him. Pressurized air coursed from the air reservoir through the hoses and to the brake cylinders on each car. The sudden increased pressure jammed the brake shoes against the wheels and, with a loud metal-on-metal screech, brought the train to an abrupt stop, just four feet from the fallen wagon driver.

Tate quickly swung down to examine the driver. Other than being frightened, he was unharmed. Tate then rushed to the rear of the train to check on his riders, including at least three levels of his bosses. All of the passengers were bruised and sore from the sudden stop, some having been thrown the length of the car. But all were elated that the wagon driver was safe, and they were now convinced that the Westinghouse air brake was a rousing success.[6]

After reboarding, the train and passengers continued to Steubenville without further incident, although Tate, pleased as a child with a new toy, demonstrated the stopping power of the air brakes several more times, but with less urgency.

At the end of the return trip to Pittsburgh, a triumphant Westinghouse shook hands with his guests and headed home to tell Marguerite of his victory. Before he got far, he returned to the Panhandle Station and sent the following telegram to his father: "My air brake had practical trial today on passenger train on Panhandle Railroad and proved a great success. George."

George Sr., ever conservative, replied by letter that the brake having already "proved a great success," of course, there would be no further difficulty in procuring all the money needed for manufacturing and marketing it.[7]

George Jr., regardless of his successes, could never please his father.

Chapter 5

Straight Air Brakes

"Everybody says they want to be free. Take the train off the tracks and it's free but it can't go anywhere."

—Zig Ziglar

George Westinghouse's First Company

In July 1869, just three months after the Panhandle Railroad success, 22-year-old George Westinghouse filed a charter under Pennsylvania law for the Westinghouse Air Brake Company. The initial capitalization was set at $500,000. The Board of Directors included, in addition to Westinghouse, Ralph Baggaley, Robert Pitcairn, W.W. Card, Alexander J. Cassatt, Edwin H. Williams, and G.D. Whitcomb.

The company began operations in leased space at Liberty Avenue and 25th Street in Pittsburgh in early 1870. Sixty-one years later, a reviewer of this first factory remarked,

> Our shops, located at the corner of 25th Street and Liberty Avenue, Pittsburgh, were dark and dingy, poorly ventilated, poorly heated and generally unsanitary. Heated by egg stoves, coal or coke-fired, and not too many. Lighted by tallow candles and coal oil lamps; in the offices, the only illumination was kerosene wick fed lamps with glass chimneys; hot, smelly, and poorly adapted for lighting purposes.
>
> Officials wore silk hats and either frock coats or cut-aways with striped trousers. Men dressed pretty elaborately especially if in important positions. Not at all unusual to see H.H. Westinghouse (George's younger brother) visit a plant and witness starting of 60 horsepower engine. Man who started machine usually middle-aged man who had served as journeyman. Wore whiskers and had plenty of dignity. Was well paid, made about 125 a month.[1]

Impetus for Air Brakes

The successful demonstration of air brakes on the Panhandle Railroad did not impress George Westinghouse, Sr., but it did renew the interest of the Pennsylvania Railroad. The PRR, which had first approached Westinghouse and Baggaley about using their air brakes but failed to reach an agreement, outfitted a six-car train with air brakes.

In September 1869, the PRR provided the train to the Association of Master

Mechanics, then meeting in Pittsburgh. The train, with PRR directors and several newspaper reporters on board, traveled to Altoona and used the air brakes to control the speed during the descent of the eastern slope of the Allegheny Mountains. Stops at the steepest parts of the route demonstrated the unprecedented capabilities of the new stopping system. The reporters described the trip, and soon papers across the country extolled the virtues of air brakes.

The PRR added four more air brake-equipped cars and took the ten-car train to Philadelphia in November 1869. A successful demonstration there, followed by similar events in Chicago, St. Louis, and Indianapolis, led to a flood of orders for air brake systems. Both small and large lines, including the Michigan Central, Chicago & North Western, Union Pacific, Old Colony, and Boston & Providence Railroads, placed orders.[2]

Westinghouse focused on standardization when manufacturing critical parts for these systems so that they would be interchangeable and compatible with each other when repairs were necessary.[3]

Cornelius Vanderbilt's massive New York Central Railroad was a prominent holdout against the installation of air brakes. That changed in February 1871 when the Pacific Express, the pride of the New York Central Railroad, headed toward Poughkeepsie, New York. The engineer, Doc Simmons, rounded a bend to see a terrifying sight. A freight train had derailed, and its wreckage was strewn across a drawbridge dead ahead. Because the New York Central had refused to invest in air brakes, Doc had no alternative but to sound the "down brakes" whistle and pray for divine intervention. The brakemen ran to turn their manual brake wheels, but the Pacific Express barely slowed before plowing into the wreck with its oil-filled tank cars. Doc and his

First Westinghouse Air Brake Factory at 25th Street and Liberty Avenue. The building still exists and houses the offices of the Pittsburgh Opera, a group once favored by George and Marguerite Westinghouse (Westinghouse Collection, Detre Library and Archives Division, Senator John Heinz History Center, Pittsburgh, PA). Historical marker photograph by author.

locomotive plunged off the bridge, dragging a baggage car and several passenger cars with it. Back up on the tracks, the tank cars ignited, and flames enveloped most of the remaining Pacific Express cars. Thirty people, including Doc Simmons, died, and dozens were injured.

The crash of the Pacific Express, along with many other fatal train accidents, led an aroused public to demand action. The New York Central quickly began installing Westinghouse air brakes on all of its passenger trains.[4]

Today the U.S.; Tomorrow the World

U.S. railroad managers were initially skeptical about the effectiveness and cost of the Westinghouse air brakes. But the successful demonstrations on the Panhandle and Pennsylvania Railroads convinced them that the improvement over human brakemen justified the investment. George and Marguerite Westinghouse traveled to England in July 1871, hoping that acceptance in the states would translate to success overseas.

It was difficult for the couple to leave, especially because they had just purchased their first house, a beautiful place in the Point Breeze area, on the eastern edge of Pittsburgh. They named it "Solitude" and looked forward to making it home.

George and Marguerite were disappointed with the reception they received in England. The British retained some animosity from the wars of 1776 and 1812 and did not appreciate American encouragement of Irish insurgency. American railroads, in particular, carried the negative connotation of political conniving, stock manipulation, and profiteering. So, the adoption of Westinghouse's air brakes by American railroads convinced no one in England.

George would have preferred to demonstrate his invention on an English railroad, but, not unlike the initial response in the U.S., no one would listen to him. So instead, he approached a professional publication, *Engineering: An Illustrated Weekly Journal*, which was well-known and respected both in England and the U.S. There, he met with one of the editors, Mr. James Dredge, Jr., who was a civil engineer. Their shared interest in engineering eclipsed their political differences, and Westinghouse was soon describing his air brake in detail. Dredge listened politely but said little. At the close of their meeting, Westinghouse gave Dredge a copy of his patent, some additional drawings, and a simplified description of the operation of the air brake. Dredge agreed to examine the documents and, if he found them acceptable, to publish a commentary. As a parting shot, he said, "But I warn you, Mr. Westinghouse, you have put your head into the lion's mouth, and you will have no one but yourself to blame if it is bitten off."[5]

When no article appeared for some time, Westinghouse called on Dredge again. After welcoming him, Dredge said, "I have been favorably impressed with your brake, from the literature about it which you left me. I am keeping that for future use if an occasion offers itself. Just now, however, the thing for you to do is to place your brake on one of our railways and give a public exhibition of its working. The readers of *Engineering* will take far more interest in a statement of what we have seen with our own eyes than in any suggestion we might print, founded on nothing more substantial than your patent and claims."[6]

Dredge also provided a draft of an article that he planned to publish in *Engineering* regarding the need for better brakes on British railroads, and the following list of essential features of those improved brakes.

1. The brakes must be applicable with equal facility by either the locomotive-driver or the guards who might be in various parts of a train;
2. The act of applying the brakes must call for only a slight exertion on the part of the person performing it;
3. The application must be capable of either instantaneous or gradual performance, according to the peculiar character of the exigency;
4. If a part of the train breaks loose from the rest, the brakes must come automatically into play;
5. The system must permit carriages, whether fitted with the brakes or not, to be attached to, or detached from, the train;
6. When a train is divided, the brakes on every division must be capable of working independently;
7. The failure of the brake apparatus on one or more carriages must not interfere with the action of the brakes on the rest of the train; and
8. The brake mechanism must be of simple character, easy to maintain, and not liable to derangement by rough use, or disuse and neglect.[7]

Westinghouse found the list imposing, and his existing air brake did not satisfy all of the requirements. Nevertheless, he embarked on a mission to provide a public exhibition of his air brakes on a British railroad, just as he wanted to do in the first place, and just as Dredge had suggested. Armed with letters of introduction from American railroad executives, Westinghouse visited every British railroad he could find, meeting resistance and denial at each stop. His quest took nine months, during which time he and Marguerite returned to the U.S. and then sailed back to England.

Finally, in March 1872, Westinghouse received permission from the London and North-Western Railway Company to demonstrate his air brakes on their line from Stafford to Crewe. Shortly afterward, the Caledonian Railway between Glasgow, Scotland, and Wemyss Bay permitted him to install air brakes on a locomotive, twelve passenger cars, and two freight cars. A bit later, the South-Eastern Railway allowed a train with a locomotive, tender, and six cars to be equipped and evaluated. All of these demonstrations were a technical success, but still, none of the railroads adopted the Westinghouse brakes as their standard.

In July 1874, the British Board of Trade conducted a series of tests comparing chain, hydraulic, vacuum, and air brakes. A similar set of comparisons was run in 1875 by the British Railways Accident Commission. Both tests revealed the Westinghouse air brake to be superior. Still, railroad managers complained that the cost of conversion was too high. Dredge and his *Engineering Journal* countered that argument by citing evidence gathered by American experts. Those experts concluded that savings on maintenance and repairs more than covered the initial cost of converting to the Westinghouse air brakes.

On some of his trips to England, Westinghouse also ventured to mainland Europe. The reception to his air brakes was warmer there, especially in Belgium. There, a royal commission of engineers compared the Westinghouse brake to all other known brakes and found it to be so superior that it was adopted as standard equipment for all of the state railways. To facilitate manufacturing and sales to the European market, Westinghouse established a British corporation for the production of the air brakes. Executive offices were in London.[8]

Westinghouse Air Brake Patents

The air brake patent application that George Westinghouse, Jr., had filed on July 10, 1868, made its way through the U.S. Patent Office[a] and became U.S. Patent 88,929; issued April 13, 1869. The patent was reissued[b] as Reissue 5,504 on July 29, 1873.

A critical test of the validity of Westinghouse's patents started in 1874. Reissue 5,504 and two related patents were subjects of a lawsuit filed in the Northern District of Ohio. Westinghouse filed suit against Gardner and Ranson Air Brake Company for infringement of the three patents. Because of the importance of this suit, it is worth reviewing the details and conclusions of the court, led by Circuit Justice Swayne and District Judge Martin Welker.

The jurors opened their decision brief as follows: "…the case was argued exhaustively and at great length, by able and eminent counsel. The importance of the case, however, and the large interests involved, as well as the value of the invention itself to the patentee and to the public at large, fully justified the elaborate discussion which the case received, and rendered necessary the careful consideration which we have given to it. The printed record covers about seven hundred and eighty pages; and nearly thirty patents and provisional specifications offered in evidence by the defendant on the issue of novelty and priority of invention, and not included in the printed record, were discussed at the hearing."[9]

The verdict of the two judges can be summarized as follows.

1. The defendant (Gardner and Ranson) infringed each of the three patents in the case. "The right granted under complainant's (Westinghouse) patents is exclusive, and the infringement clearly proven."[10]

2. The reissued patent 5,504 was not materially different than its corresponding original patent. "No material change has been made in the specifications of these patents, and there issued claims are clearly warranted by the specifications of invention, as contained in the original patents."[11]

3. The claims of the 5,504 patent are clear and unambiguous. "…the specification … is sufficiently clear to prevent any misunderstanding of the scope and meaning of the claims in question."[12]

4. Despite the presentation of nearly thirty patents as alleged prior art, the judges concluded that Westinghouse invented new material, and his patents were neither anticipated nor made obvious by the prior literature. "Westinghouse was not the first to conceive the idea of operating railway-brakes by air-pressure, and … he was not the inventor of the larger part of the devices employed for

---

(a) Established in 1790 by President George Washington, the U.S. Patent Office became the U.S. Patent and Trademark Office (USPTO) in 1881.
(b) A patent can be reissued at the request of the holder of the original patent to correct significant mistakes in the original patent or to narrow or broaden the claim coverage. A reissued patent retains the same term of validity as the original patent.

such purposes. But such fact does not detract at all from his merit or rights as a successful inventor. The organisms ... seem to have been entirely new with him; and the incorporation of these elements, together with that of graduating the air pressure in the brake-cylinders ... fully substantiate his pretensions as an original and meritorious inventor, and entitle him as such to the amplest protection of the law. Suggestive as these prior patents and provisional specifications may have been, they do not, any of them, embody that which Westinghouse has invented and claimed; and a prior description of a part cannot invalidate a patent for the whole." "So far as appears from the testimony in this case, none of the alleged prior inventions of air-brake apparatus have ever successfully been applied to practical use; and when we consider the immense importance of the introduction of the air-brake on railroads, and the incalculable benefit which it has conferred on the public in the readiness and certainty with which the trains can thereby be controlled, and comparative immunity from accidents thus secured, ... in connection with the fact that Westinghouse was the first, so far as appears in the record and proofs, to put an air-brake into successful actual use, such considerations only strengthen and confirm the soundness of the conclusions to which a careful examination of these prior patents has led us, that there are substantial and essential differences between these prior patents and the Westinghouse apparatus, and that to these differences we may justly attribute the successful and extensive introduction of the Westinghouse air-brake."[13]

The results of this suit overwhelmingly established the validity of the patents and therefore opened the door to millions of dollars in revenue to George Westinghouse. By 1876, the Westinghouse Air Brake Company had equipped 38 percent of U.S. rolling stock, 2,645 locomotives, and 8,508 cars, with Westinghouse air brakes.[14]

To gain an appreciation of the effectiveness of the Westinghouse straight air brake, consider a typical train on the Pennsylvania Railroad in 1876. That train would consist of a locomotive, a tender, one express car, one baggage car, three coaches, and six Pullman cars.[c] Loaded with fuel, water, and passengers, such a train would have weighed 290 tons. Using hand-operated brakes, activated by brakemen, the stopping distance for such a train traveling at 30 miles per hour was at least 1,600 feet. For a train equipped with Westinghouse's air brakes, that stopping distance was reduced to 500 feet.[15]

Without question, the Westinghouse straight air brake was an enormous improvement. But as George Westinghouse knew, his air brake suffered from critical limitations. James Dredge, Jr's, list of essential features for improved train brakes helped to guide Westinghouse in his search for better solutions.

(c) A Pullman car was a railroad sleeping car built by the Pullman Company (founded by George Pullman) and operated on most U.S. railroads starting in 1867.

Chapter 6

Automatic Air Brakes

"He who stops being better stops being good."
—Oliver Cromwell

What Was Wrong with the Straight Air Brake?

James Dredge, Jr., had identified the critical weakness in the Westinghouse straight air brake without even seeing it. Examining the diagram of the straight air brake system reveals the problem:

A leaking Train Line or disconnected hose would prevent the pressurization of the Train Line and thus would disrupt braking. As an extreme example, consider what would happen if some of the cars became disconnected. The cars that remained attached to the locomotive could be stopped because Westinghouse's patented hose coupling would seal the open end of the Train Line. But the separated cars would become runaways as there would be no way to apply their brakes.

Westinghouse was aware of the problem with his straight air brake before he went to England and met James Dredge, Jr., but Dredge's list of essential brake features probably crystallized his thinking. So he was able, within six months of meeting Dredge, to file two patent applications[a] at the end of 1871, disclosing significant improvements to the straight air brake system. Continuing work by Westinghouse finally resulted in a system that came to be called the automatic air brake, which was disclosed in U.S. Patent 149,901 issued on April 21, 1874. At that time, George Westinghouse was 27 years old.

As with the straight air brake, the automatic air brake included a Compressor, Main Reservoir, and Driver's Brake Valve on the locomotive; and a Brake Pipe (also called a Train Line), Brake Cylinder, and Brake Block on every car. But the automatic air brake added two more components on every car: an Auxiliary Reservoir and a Triple Valve.

The operation of the new Westinghouse automatic air brake was the opposite of the straight air brake. In the straight air brake, increased pressure in the Brake Pipe activated the brakes. In the automatic air brake, a drop in pressure in the Brake Pipe activated the brakes.

The Westinghouse automatic air brake was fail-safe. If part of the train became separated, the pressure in all of the Brake Pipes would drop, and all brakes would be applied, thus stopping the train and preventing runaway cars.

(a) The two patent applications, both filed on December 6, 1871, issued as U.S. Patents 124,404 and 124,405 on March 5, 1872.

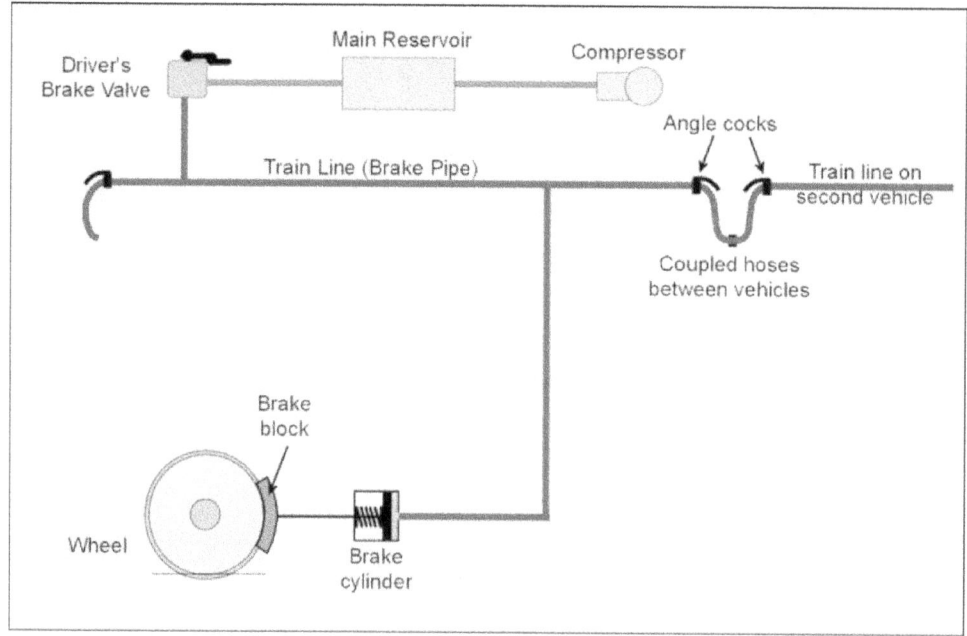

Westinghouse straight air brake system with brakes applied (drawing by P. Connor, www.railway-technical.com).

The detailed operation of the automatic air brake is a bit too technical to include here, so I have placed the explanation in Appendix II—Automatic Air Brake Operation.

Performance of the Automatic Air Brake

A Pennsylvania Railroad train consisting of a locomotive, tender, and seven-passenger cars was equipped with automatic air brakes as later described in U.S. Patent 149,901. On May 20, 1873, the Committee on Science and the Arts of the Franklin Institute evaluated the automatic air brake system. The committee recorded stopping times and distances under various speeds and grades and concluded:

> The committee say that these experiments have demonstrated to them the extraordinary efficiency of this apparatus, and they especially call attention to the value and importance of the arrangement which secures the instant automatic application of the brakes on the engine and on each car of the train independently of the train hand, in certain contingencies which are of common occurrence and are the cause of frequently disastrous accidents.
>
> The committee believe that by contriving and introducing this apparatus, Mr. Westinghouse has become a great public benefactor and deserves the gratitude of the travelling public at least. They believe that his inventions are worthy of and should receive the award of the Scott Legacy Medal.
>
> Resolved, that this Committee (of Science and Arts) recommend to the Board of Managers of the Institute that they make the award of the John Scott's Legacy Premium and Medal to George Westinghouse, Jr., of Pittsburgh, for his improvements in air brakes for railway trains.[1]

First, with the straight air brake, and soon after with the automatic air brake, Westinghouse dominated the market for passenger trains in the U.S. and later in Europe.

But freight trains were an untapped market. Even as late as 1885, almost all freight trains still relied on manual brakes engaged by rooftop brakemen.

Interchangeable Cars

When I was a kid, my closest friend had a massive model train display that he set up at Christmas. His trains were the Lionel brand. My train set was more modest and composed of cheaper Marx rolling stock. While both of our trains used the same three-rail track (O-27 gauge), we could not trade locomotives or cars because the hardware that joined one car to the next (the coupler) was incompatible.

As real trains, particularly freight trains, grew in length and railroad companies expanded operations to more distant locations, the need for interchangeable cars became clear. Rather than transferring cargo from a Canadian Pacific car to a Texas Pacific car, it would be far cheaper to include the Canadian Pacific car in the Texas Pacific train.

Three issues impact car interchangeability: track gauge, car couplers, and braking system. We will discuss each one in turn.

Track Gauge

Track gauge, i.e., the space between rails, had been standardized at 4 feet, 8½ inches in Great Britain since 1845. Because much early U.S. rail equipment was bought from Great Britain, that same gauge found favor here. The Pacific Railroad Act of March 3, 1863, stated, *"Be it enacted by the Senate and House of Representatives of the United States of America in Congress assembled, That the gauge of the Pacific railroad and its branches throughout their whole extent, from the Pacific coast to the Missouri river, shall be, and hereby is, established at four feet eight and one-half inches."*[2] While other gauges remain in use even today, the "standard gauge" was adopted by virtually every commercial railroad in the U.S.

Car Couplers

Car couplers presented a more complex challenge. They must hold together a train weighing hundreds of tons when traveling around curves and up and down hills, and they have to withstand enormous tension and compression stresses as the train starts and stops.

The first couplers were known as "link-and-pin" and were quite simple. A massive iron loop was inserted into the end ("Drawbar") of one car and held in place by an iron pin. Then, as the next car approached, the brakeman inserted the other end of the link into that car's drawbar and secured it with another iron pin. A link and two pins are shown in the photograph.

Coupling cars with the link-and-pin system was the responsibility of the brakeman, the same person who risked his life running on the top of cars to engage the manual brakes. Standing between moving cars holding the end of the link in one hand and the pin in the other was extremely dangerous. If the brakeman's timing were off, he could lose fingers, a hand, or even be crushed between the two cars. The drawing shows a brakeman coupling two cars together.

On April 29, 1873, Eli Janney, a Confederate Civil War veteran from Alexandria, Virginia, received a patent[b] for an improved coupling device. The Janney coupler automatically engaged, so it eliminated the hazardous manual coupling that required a brakeman to stand between the cars. It held train cars with minimal slack, enabling smoother acceleration and more efficiency around curves, but with enough play to allow the train to go up and down hills. It also functioned as a buffer to cushion cars as slack was removed, thus preventing damage to passengers and cargo.[3] The Janney coupler from his patent is shown here.

After testing and comparing the performance of many coupler designs, in 1888, the Master Car Builders Association[c] received a partial waiver of patent rights over the Janney-style coupler, and it became the standard for U.S. railroad cars. The use of the manual link and pin coupler became illegal on mainline railroads in the United States with the passage of the Railroad Safety Appliance Act in 1893.[4]

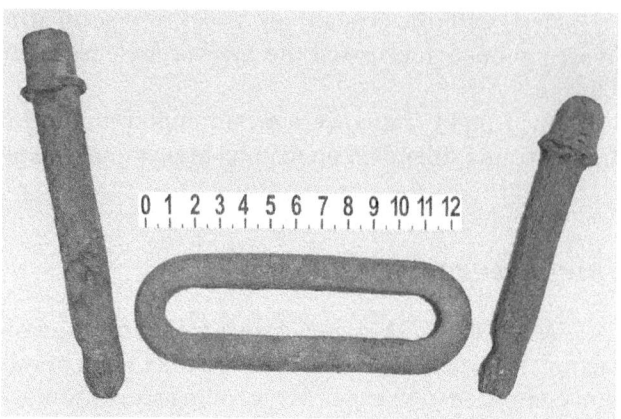

Iron link and pins used for coupling railroad cars together (courtesy Jane Ballard).

Brakeman using the link-and-pin system to couple cars.

(**b**) U.S. Patent 138,405.
(**c**) Organization of railway engineers and executives established on September 18, 1867, in Altoona, Pennsylvania, to standardize railroad procedures, policies, and equipment.

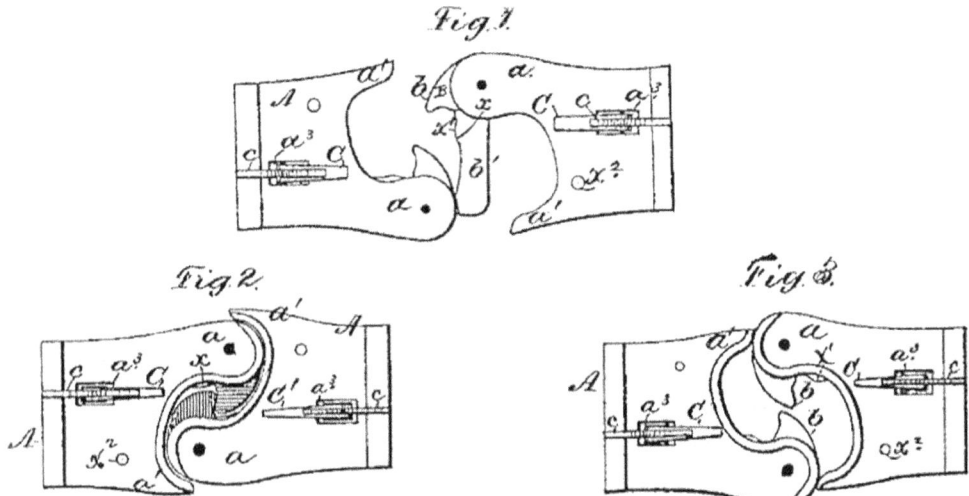

Janney Coupler—Drawing from U.S. Patent 138,405.

Braking System

The Committee on Automatic Freight Car Brakes issued its first report to the Master Car Builders Association in 1885. It recommended a series of comparative trials and tests of then-current braking systems. Two rounds of tests, in 1886 and 1887, were planned on the Chicago, Burlington, and Quincy Railroad at Burlington, Iowa.[5] The location was chosen because of the energy and influence of Lorenzo Stephen Coffin, State Railroad Commissioner of Iowa.

Five brake systems, including the Westinghouse automatic air brake, competed in the 1886 tests, but none was judged suitable for adoption as a standard. The Westinghouse brake performed well in normal use, but emergency stopping produced unacceptable shocks and vibration.

In May 1887, another series of tests commenced. Westinghouse and his engineers had developed a "quick-action" triple valve, which reduced the delay in braking of the 50th car of the test train from twenty seconds to just six seconds. While they had no time to test the improved triple valve, the Westinghouse people were confident that they had solved the problem.

They were wrong. While the stopping time and distance were much shorter, the shocks sustained in the rear cars were even worse than in the 1886 trials. Observers in the last car were thrown the length of the car, and one suffered a broken leg. Once again, none of the candidate braking systems was deemed suitable as a standard.

Westinghouse literally went back to the drawing board. In just three months, he designed and developed a new quick-action triple valve and modified other aspects of the automatic air brake system. The result was a train of fifty cars that stopped from twenty miles per hour in 200 feet without shock or vibration. With this proof of concept, the Westinghouse engineers retrofitted the train used in the 1887 trials with the new automatic air brake system and sent it on a demonstration tour during the last three months of 1887. Godfrey W. Rhodes, Chairman of the Master Car Builders Committee on Automatic Freight Car Brakes, wrote of the results: "Mr.

Westinghouse's wonderful optimism and confidence in the principle of air power, after the early failures were very marked. After the first collapse, in rapid succession, three different triples were invented. Finally came the gem, and then, greatest of all acts, it was exhibited all over the country in operation on a fifty-car train, making stops with no shock in the fiftieth car, not enough jar to upset a glass of water, a marvelous condition when one considers the wreck and disaster that used to take place in the fiftieth car."

At the June 1888 meeting of the Master Car Builders Association, the Committee on Automatic Freight Car Brakes reported: "In our report to the convention last year the main conclusion we arrived at was that the best type of brake for freight service was one operated by air, and in which the valves were actuated by electricity. Since that time your committee has not made any further trial of brakes, but the aspect of the question has been much changed by the remarkable results achieved in non-official trials which have taken place in various parts of the country, and have been witnessed by many of the members of this Association. These trials show that there is now a brake in the market which can be relied on as efficient in any condition of freight service."

The Committee then listed six specific conditions that a freight train brake should fulfill. Although the Committee identified no specific brake, only the Westinghouse automatic air brake with his modified triple valve could meet the requirements.[6]

Through inventiveness and perseverance, Westinghouse won the battle. His automatic air brake system, protected by numerous patents, became the de facto standard brake for freight trains in the U.S.

To assure immediate knowledge of any braking problems, Westinghouse Air Brake (Air Brake) assigned one or more experienced engineers to each of the major railway centers. In the event of an accident, those engineers would immediately proceed to the scene, ascertain the cause of the accident, render any assistance necessary, and report to Air Brake headquarters.

For training purposes, Air Brake constructed a special instruction car equipped with a steam boiler, air compressor, water pump, dynamo to generate electricity for lighting, fifty sets of brake cylinders, and pipes equivalent to the braking system of a freight train. It also had an office and bedroom for the instructors, who lived on board. The car traveled to railroad centers where engineers, firemen, conductors, and other employees learned the construction and operation of train brakes and could practice on a real braking system without the risk and expense of operating a full-sized train. By December 1910, the instruction car had traveled more than 113,000 miles.[7]

Assigning engineers to respond quickly to accidents and taking a special instruction car to the railroad centers were just two ways that Westinghouse implemented his idea of service to present and potential customers. We could look at it as enlightened selfishness, but the result was satisfied air brake users.

Back to England

George and Marguerite had taken the straight air brake to England in 1871 to a less than enthusiastic audience. In 1874, with the spectacular success of his automatic air brake on passenger trains behind him, George Westinghouse was ready to return to London. By then, straight air brakes had been installed on just 148 locomotives and 724 passenger cars. But at least the Westinghouse name was known and respected by railroad men, and the reception for the automatic air brake was positive.

Other inventors had entered the railway brake field, so Westinghouse had competition, especially from vacuum brakes. The most useful of the vacuum brake patents, issued in 1872, belonged to John Y. Smith of Pittsburgh. Smith's work was based on an 1860 patent issued to Nehemiah Hodge of North Adams, Massachusetts.

Westinghouse studied the Hodge patent, took out patents on refinements and improvements to the Hodge apparatus, and eventually purchased the patent. On his 1875 visit to England, Westinghouse was prepared to furnish either vacuum or compressed-air brakes, as required.

Conflicting claims about the performance of various braking systems, often with little evidence, led to a pivotal series of competitive brake tests called the Newark Trials. Held on June 9, 1875, at the Nottingham and Newark Division of England's Midland Railway, trains with thirteen carriages and two vans competed. Two versions of four different braking systems were represented: mechanical, hydraulic, vacuum, and compressed air. The Railway Companies' Association, under the direction of the Royal Commission on Railway Accidents, conducted the tests. The results were conclusive. The best stops, all from an initial speed of 50 miles per hour, were as follows:

- Westinghouse automatic air brake—777 feet
- Clark hydraulic brake—901 feet
- Steel & McInnes compressed air brake—1158 feet
- Smith vacuum brake—1477 feet.

James Dredge, Jr.'s, journal, *Engineering*, commented on the Newark Trials in its June 25, 1875, issue: "Lastly, we come to the Westinghouse automatic arrangement, and this, we think we may safely say, is shown by the recent trials to possess all the requisites of a thoroughly efficient continuous brake. This brake proved more prompt and powerful in its action than any of its competitors. ... Its performances as they stand were far beyond those of any other brake. As regards durability and general reliability in every-day practice also it should be remembered that no brake sent to the trials has been so thoroughly tested as the Westinghouse, and this is a fact which it is well to bear in mind."[8]

The Westinghouse automatic air brake met every one of the essential features of improved railroad brakes delineated by Dredge during Westinghouse's first visit to England in 1871. It was clearly superior in the 1875 Newark Trials. But despite that superiority, the Westinghouse automatic air brake shared the English market with straight air brakes and vacuum brakes for at least forty years after the Newark Trials.[9]

At the time of his death in 1914, the *New York Times* reported about Westinghouse's air brake: "It is said of that invention that it saved more lives than centuries of warfare have destroyed. It has made possible the development of railroad traffic as it is known today—the trains of great length, high speed, large capacity, and increasing frequency. For this contribution to the service of mankind, Westinghouse received great reward and worldwide recognition."[10]

Chapter 7

What's a Wilmerding?

"Every civilization carries the seeds of its own destruction, and the same cycle shows in them all."
—*Mark Twain*

The Westinghouse Air Brake Company was growing like Topsy[a] and outgrew its initial Liberty Avenue and 25th Street facility. In 1881, Westinghouse built a modern plant for his air brake company on the north side of General Robinson Street in Allegheny City, across the Allegheny River from Pittsburgh.[b] The site is adjacent to present-day PNC Park, home of the Pittsburgh Pirates baseball team. In 1889, H.J. Heinz built his Keystone pickle factory near the Westinghouse plant.

Westinghouse built a world-class factory, complete with molding machines imported from Scotland and modern gear and cam driven lathes. Worker comfort was important to Westinghouse, and he included employee locker rooms, washrooms, and showers.[1] On June 1, 1881, he introduced the practice of half-day work on Saturdays at the new factory. Westinghouse had observed that practice in England and was the first to bring it to the Pittsburgh area, and possibly the United States.[2]

But the demand for air brakes continued to grow. Two issues, the plight of the working man and the need for more manufacturing capacity, led Westinghouse to create a "company town" in a farming area fourteen miles east of Pittsburgh. In discussing the decision, an official of Westinghouse Air Brake said, "Mr. Westinghouse said that in his judgment one of the most serious problems in the development of the country along right lines was the proper housing of the masses that were flocking to our industrial centers. He intimated that he had given the question much thought, and, if his affairs permitted, would be glad to attempt its solution along business lines, and yet in the spirit of the highest and most practical philanthropy."[3]

As was often the case, Westinghouse's thoughts soon turned into action. In late 1887 and early 1888, he purchased about 500 acres in the Turtle Creek Valley. A flag stop[c] on the mainline of the Pennsylvania Railroad already served the area. At the suggestion of Robert Pitcairn,[d] Superintendent of the Pittsburgh Division of the

---

(**a**) Topsy was a character in Harriet Beecher Stowe's 1852 novel, *Uncle Tom's Cabin*. "Grow like Topsy" has come to mean growing in an unplanned or uncontrolled way.
(**b**) Allegheny City was the third-largest city in Pennsylvania until it became part of Pittsburgh in 1907. As part of Pittsburgh, it is known as the "North Side."
(**c**) A flag stop is a small railway station between the principal stations or a station where the train stops only on a signal.
(**d**) This is the same Robert Pitcairn who was on the Board of Directors of the Westinghouse Air Brake Company at its founding in 1869.

Chapter 7. What's a Wilmerding?

Wilmerding, 1897, drawn by T. M. Fowler (Wikimedia Commons).

PRR, the train station had been named Wilmerding, from the middle name of Joanna Wilmerding (Bruce) Negley.[(e)] Joanna was the wife of William Backhouse Negley, a prominent attorney and one of the former owners of the farm where the station was located. From that station, the new town took its name. In December 1888, Westinghouse formed a real estate corporation, the East Pittsburgh Improvement Company. On February 25, 1889, he conveyed much of the land which he had purchased to the East Pittsburgh company, excluding that property already allocated to the Westinghouse Air Brake Company and the PRR.

Soon after, construction started on the Air Brake factory. In June 1889, the sale of building lots began, and one of the first purchasers was the Westinghouse Air Brake Company. The Company bought 150 lots and issued contracts for the construction of homes for workers. That construction started in the fall of 1889. The houses were roomy and solidly built frame buildings with all modern (at the time) improvements. Other investors and private individuals bought lots for both dwellings and business locations, and Wilmerding enjoyed a building boom that lasted through 1891. The Borough of Wilmerding was incorporated on March 8, 1890. The Air Brake factory was completed and occupied later that year.[4]

Company Towns

Wilmerding was not the first company town in the United States. George Pullman, who had invented the Pullman sleeping car and was an early advocate

---

(e) The origin and meaning of Wilmerding are unknown, but the family of Henry Augustus Wilmerding and his wife, Nancy Clute, resided in Moscow (now Leicester) in western New York around 1833. See https://pre-prowhiskeymen.blogspot.com/2015/10/jc-wilmerding-went-from-riches-to-rags.html.

for Westinghouse air brakes, built a town focused around his railroad car factory in 1880. Pullman, Illinois, was built on 4,000 acres fourteen miles south of Chicago. The town included a library, theater, hotel, church, market, sewage farm, park, and many residential buildings. In the residential section, 150 acres were dedicated to tenements, flats, and single-family homes with rents from $0.50 to $0.75 per month. The residences featured modern conveniences such as gas, running water, indoor toilets, and regular garbage removal. By July of 1885, the population exceeded 8,600. A town agent was in charge and was responsible for all services and businesses, including street and building maintenance, gas and waterworks, fire protection, the hotel, the sewage farm, and the nursery and greenhouse. Reporting to the town agent were nine department heads and approximately 300 men. Except for the school board, there were no elections. George Pullman selected all of the officials. Pullman, Illinois, was voted the world's most perfect town at the Prague International Hygienic and Pharmaceutical Exposition of 1896.[5]

Pullman differed from Wilmerding in several critical areas. In Pullman, home ownership was discouraged, and the economic model was for rental payments to yield continuing income to the Company. In Wilmerding, employees were encouraged to buy the homes they lived in via monthly installment payments. The November 23, 1904, *Wilmerding News* described the program for Westinghouse employees to purchase their homes: "(T)he purchaser of any property is required to pay about one-fifth of the purchase money in cash upon delivery of deed. He then executes a purchase-money mortgage, payable in five years, with interest payable quarterly at the rate of 5 per cent per annum. While no requirement is made, it is expected that the purchaser shall reduce the principal of the mortgage quarterly by such payments on account as he may be able to make. This plan enables him, during hard times to keep the transaction in good shape by merely paying the interest, while on the other hand, when good wages are earned, he can discharge such part of the principal of his mortgage as he may desire."[6] By 1921, half of the houses in Wilmerding were owned by their residents.

In Wilmerding, as in Pullman, the companies contributed by building schools, churches, and parks. But in Wilmerding, Westinghouse Air Brake refrained from influencing town politics, ignoring the occasional radical socialist or troublesome politician. By contrast, Pullman officials were all selected by George Pullman.

In May 1894, less than a year after the close of the 1893 World's Columbian Exposition (covered in detail in Chapter 19), an event occurred that placed in stark contrast the management philosophies of George Pullman and George Westinghouse. Facing a severe recession and declining demand for his railroad cars, George Pullman reduced wages for his workers by one-third but refused to lower rents on their company houses. The workers called a strike, which eventually spread to the major railroads and stopped most railroad freight and passenger traffic west of Detroit. Massive federal intervention finally ended the strike, but not before 30 strikers died, and $80 million in property was destroyed.

In his 2008 book, Skrabec succinctly captures the differences between Pullman and Wilmerding: "Socialists tended to flock to both towns, but in Pullman City they rioted, while in Wilmerding they joined the democratic elections."[7]

Both towns survive today, Pullman as a National Monument and a portion of

Chicago with a population of 6,500. Wilmerding remains a borough in Allegheny County, Pennsylvania, and has a population of 2,100.

Westinghouse Air Brake and Wilmerding Prosper

Wilmerding grew quickly. In the 1890 U.S. Census, Wilmerding's population was 419 people. Almost as soon as it opened later that year, the Wilmerding Plant of Westinghouse Air Brake had 1,000 employees. With affordable housing nearby, the town mushroomed, and residents moved in as soon as houses were completed.

Retail stores and services were not far behind. The H.G. Croushore Drugstore; the Redfern and Glen Hotels; the post office, railroad station, and bank; Public School No. 1 (Horrocks School) on the south side; Public School No. 2 (Gottwals School) on the north side; First Methodist Church; police and fire departments; and the crowning jewel, the first Westinghouse Air Brake General Office Building, were all established by 1891.

The first house in Wilmerding was headquarters of the East Pittsburgh Improvement Company, the real estate company established to sell property in Wilmerding. The house later was occupied by William Moles, a machinist with WAB from 1884 until his retirement in 1929 (circa 1889, from *Wilmerding and the Westinghouse Air Brake Company*, courtesy Arcadia Publishing).

Good News for Westinghouse and Wilmerding

In the 1880s, on-the-job deaths of railroad workers were second only to those of coal miners.[8] Lorenzo Coffin, the State Railroad Commissioner of Iowa, doggedly lobbied for six years for Federal legislation to correct the railroad industry's abysmal safety record. Finally, on March 2, 1893, Congress passed the Railroad Safety Appliance Act. That act made air brakes and automatic couplers mandatory on all trains in the United States. It took effect in 1900, after a seven-year grace period, and is credited with a sharp drop in accidents on American railroads in the early 20th century. Combined with the recommendations of the Committee on Automatic Freight Car Brakes, the Railroad Safety Appliance Act gave George Westinghouse a virtual monopoly on a multi-billion-dollar market.[f]

The population of Wilmerding exploded; the 419 residents of 1890 grew by a factor of ten, to 4,179 in 1900.

The booming business required more workers, both skilled and unskilled. George Westinghouse made it a priority to educate his new employees in subjects including citizenship, English language, grammar, typewriting, American history, healthy living practices, basic math, electricity, machine design, and air brake operation. Apprentices (almost exclusively men) were paid an hourly rate to attend night school classes. Company managers taught many of the higher-level courses.[9]

The Tonnaleuka Club

The Tonnaleuka Club provided temporary housing and a social structure for the young professional men attracted to the higher-skill-level jobs at Air Brake. In 1904, when local hotels failed to provide adequate facilities, Westinghouse purchased the Glen Hotel at 353 Marguerite Avenue and renovated it to add living and dining rooms and recreation, including billiards and bowling. About thirty men lived at the Club, and associate memberships allowed non-residents to enjoy the recreational amenities.[10]

The picture shows the inside of the Westinghouse Air Brake machine shop. OSHA would have been aghast at the equipment powered by moving belts driven by overhead shafts.

The Relief Department

Of course, OSHA did not exist in the early 1900s, nor did any formalized program of workmen's compensation. But George Westinghouse recognized that his

---

(f) Today, Westinghouse's air brake patents would be deemed "Standards Essential Patents (SEP)," and in return for a standards-setting body including the patent(s) in a standard, the patent holder would be requested to license other manufacturers at a "Fair, Reasonable, and Non-Discriminatory rate (FRAND)."

Opposite: Company-built housing in Wilmerding. Top: Marguerite Avenue Home of A.B. Woods (circa 1902). Bottom: Six-Family Homes (called the "Flats") on Middle Avenue (circa 1903). Both: (from *Wilmerding and the Westinghouse Air Brake Company,* **courtesy Arcadia Publishing).**

Inside the machine shop, circa 1906 (from *Wilmerding and the Westinghouse Air Brake Company*, courtesy Arcadia Publishing).

employees might become unable to work due to accidents or sickness, and their death could leave their families destitute. So, in May 1903, he established a Relief Department at Westinghouse Air Brake. The November 23, 1904, issue of the *Wilmerding News* described the primary features of the program as follows:

> ... to insure a certain income to employees who might become unfitted for work through illness or injury and in the event of death to pay the beneficiary a stipulated sum. Any employee under 50 years of age is entitled to membership, subject to successful physical examination, but membership is not compulsory. Members contribute according to the class in which they belong, there being five, the class being determined by the wages received, varying from $35 to $95 or over per month, the contribution ranging from 50 cents to $1.50. A member may receive benefits for 39 consecutive weeks, in event of disability extending so long a period. The air brake company is the custodian of the funds, but being such does not benefit the company pecuniarily, as it pays four per cent interest on monthly balances to the credit of the relief fund. The company goes further and guarantees payment of all benefits, and if the money received from the monthly contributions be insufficient to meet the requirements the company makes good such deficit.[11]

Benefits and Liabilities of a Company Town

When Westinghouse Air Brake prospered, Wilmerding prospered. But, as with any company town, the dominance of a single employer carried risks. A September 2, 1904, article in *The Wilmerding Times* described the benefits and the problems: "Wilmerding, the Ideal Town," would not be a misleading title for the little industrial center of 5,000 inhabitants fourteen miles from Pittsburg on the Pennsylvania railroad.

No other town in Allegheny county can show such a beautiful park or such broad acres of closely cropped lawn, green as the sward of Old Erin. It is this park and the big expanse of emerald carpet that helps so materially to give the town its ideal appearance."

"There is universal complaint among the Wilmerding business men as to the industrial conditions prevailing there. Of the 4,000 men employed by the air brake company in normal seasons, only one-half find work at present, as the force has been cut down owing to reduced orders for air brakes from the railroads of the country. The town has this misfortune, that it depends almost exclusively on the big works, three-quarters of its bread winners finding employment there, so that naturally when business is slack there is slack in the town, also. The laying off of men, together with the reduction in wages and the increased price of living, have produced this unsatisfactory state of affairs. There are a number of the best business locations in Wilmerding to rent, whereas two years ago room could scarcely be had for love or money. Such periods, however, have occurred ever since the town was laid out, in 1889.[12]

The General Office Building ("The Castle")

No discussion of Westinghouse Air Brake and Wilmerding would be complete without considering the General Office Building, built in 1890. In addition to

Westinghouse Air Brake General Office Building, circa 1960 (Westinghouse Collection, Detre Library and Archives Division, Senator John Heinz History Center, Pittsburgh, PA).

corporate offices, the elegant three-story building contained a bowling alley in the basement, a swimming pool on the first floor, and a restaurant, a library, and two bathrooms on the third floor. In 1894, the Wilmerding YMCA began to use the basement and part of the first floor.

The General Office Building burned almost to the ground on April 7, 1896. Reconstruction began immediately. The new building resembled the old one except that the bowling alley and swimming pool were deleted. The new General Office Building opened in 1897.

The continued growth of Westinghouse Air Brake demanded more space in the General Offices, and an addition was started in 1926. The older portion of the building was remodeled, and rededication of the entire building occurred on October 25, 1927.

Westinghouse Air Brake occupied the 55,000 square foot building until 1985 when it was donated to the American Production and Inventory Control Society. The building was placed on the National Register of Historic Places in 1987.[13]

More recently, the building housed the George Westinghouse Museum. In 2007, the documents and artifacts from the Museum were moved to the Senator John Heinz History Center in Pittsburgh, and a non-profit called *Wilmerding Renewed* acquired the building. Utilities and maintenance proved too expensive, and the building languished. In 2016, *Wilmerding Renewed* sold the building to John Graf and the Priory Hospitality Group for $100,000. Priory Hospitality stabilized the structure and planned to develop it as a boutique hotel.[14] However, the COVID-19 pandemic intervened and those plans had to be scrapped.

Instead, the General Office Building now has become home to the Westinghouse Arts Academy, a public charter high school teaching the normal core courses but adding 30 different arts classes over the four years of school. Areas of specialization include Dance, Digital Arts, Literary Arts, Music, Studio Arts, and Theatre.

In 2019, I was privileged to tour the building. I was impressed by the beauty of the solid wood paneling in many of the rooms and by the excellent state of repair of a century-old structure that has been virtually empty for over 12 years. It is a blessing to see Westinghouse's building restored to beneficial use.

Chapter 8

Two Trains, One Track

"It is a trite axiom, that two bodies cannot occupy the same space at the same time. The duty of the railway signaling engineer may be said to endeavor to prevent two bodies, which are moving at high velocities, from seeking to violate this law of nature."
—Richard Christopher Rapier, Author of
"On the Fixed Signals of Railways" in 1874[1]

For obvious economic reasons, most early railroad lines were constructed with a single track. The single-track configuration suffers from two serious problems. First, what happens when trains travel in opposite directions? Second, what if two trains traveling in the same direction are going at different speeds? If there were two parallel tracks, trains going in opposite directions could travel simultaneously without conflict. But if the two tracks provide for opposite direction travel, the problem of trains moving at different speeds on the same track in the same direction remains. Tracks crossing each other create another train control problem. Some provisions had to be made for trains to pass each other without colliding.

The earliest railroads handled the train control problem by use of passing sidings and a timetable. One train would pull onto the siding based on the timetable and wait for the train traveling in the opposite direction (or faster in the same direction) to pass on the main track at the planned time. Such a system relied on separating trains by *time*. At any given time, only one train would occupy a given space. As long as all trains maintained their schedules, that simple system worked well. But of course, trains rarely ran on schedule, then or now.

The next evolutionary step in train control was telegraphed train orders, first implemented in the U.S. on the Erie Railroad in 1854. Trains still operated on a timetable, but train orders from a dispatcher, sent by telegraph to a train station and thence to the trains, could alter the schedule based on the actual circumstances.

The Long Island Railroad practiced a modification of the time interval system. Flagmen with colored flags were displayed at each station and at set points along the track system. A red flag was displayed as soon as a train passed. After three minutes, a green flag replaced the red flag. After five more minutes, the green flag was removed. A subsequent train had to stop for a red flag but could proceed with caution at a green flag, the engineer knowing that another train was not far ahead.

The predecessor of the modern protocol for controlling trains, called the block system, was first implemented in the U.S. in 1863 by the Pennsylvania Railroad. Under the block system, the railroad line is divided into sections (called blocks), and access of a train to a given section is permitted or denied by a physical signal.[a] The block system separates trains by *space* instead of time. In the first block systems, a railroad employee stationed at each signal would, based on telegraphed instructions, set a rotating arm (often called a semaphore) to different angles to indicate the action required of the train engineer.

The space separation block system is still in use, but the method of detecting trains has changed from human observation to automatic. How the appropriate controlling information is conveyed to train engineers via physical signals has also evolved.

Many inventors tackled various portions of the train control problem, but the contributions of three pioneers to train detection, signaling, and control stand out.

Track Circuit

The first of these pioneers is William Robinson, who recognized the need for a train detection system to automatically detect the presence of a train within a given block. In the 1870s, Robinson developed and improved a track circuit to perform this function.

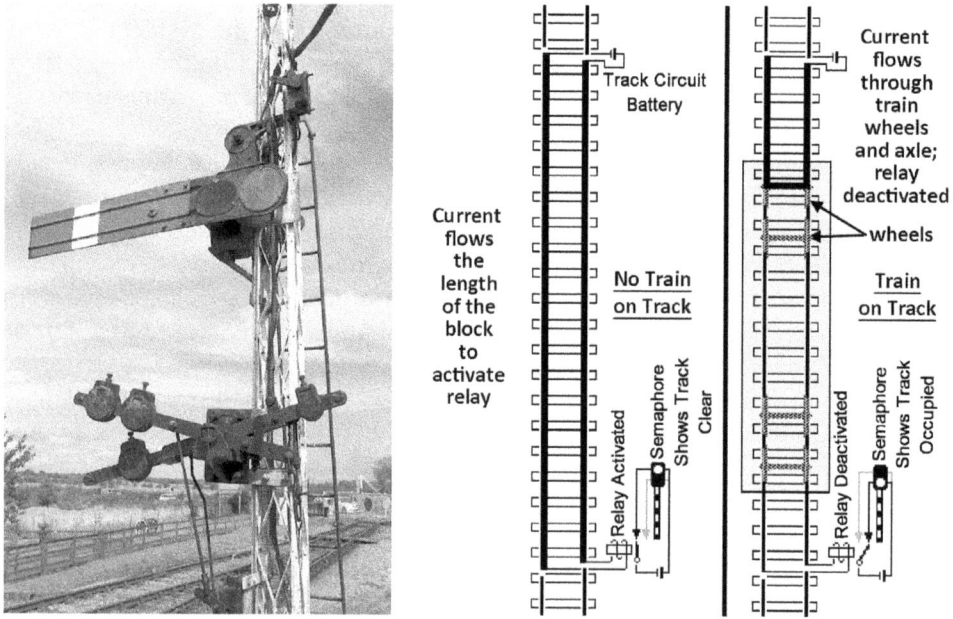

At left: Semaphore for train signaling. At right: Track circuit diagram. Note electrically insulating gap in rails at ends of block (LennartBolks, Wikipedia).

(**a**) An obvious question is, "How long is a block?" The answer is not so obvious. The American Railway Engineering and Maintenance-of-Way Association (AREMA) says, "For block signals, length of block = at least as long as longest normal stopping distance for any train on the route traveling at its maximum authorized speed" and "The longest stopping distance will be used to establish the design block length for the signal system." As with many things in the railroad industry, it is complicated. An oversimplified answer is that on most railroads, blocks are two to three miles long.

The vital requirements of a track circuit are that it must detect the presence of a train within the block without reporting false positives (reports there is a train when there is none) or false negatives (reports no train when there is one). The system must be simple, inexpensive, and easy to maintain. After experimenting with different systems on several railroads, Robinson developed a track circuit that fulfilled these criteria.

Start with the track at the center, showing no train in the block. Robinson connected a low-voltage track circuit battery across the two rails of the track at one end of the block to be monitored. At the other end of the block, he connected the coil of a relay across the rails. With the tracks as conductors, the current from the battery flowed through the relay coil, thereby activating the relay and moving the semaphore (photograph at the left) to the "Track Clear" position.

Move now to the right side of the drawing. When a train entered the block, the wheels and axles would form an electrical connection between the rails, thus shorting out the circuit and diverting the current before it could reach the relay. With no current, the relay would deactivate (drop-out). Electrical contacts on the relay operated the semaphore to indicate "Track Occupied."

Robinson's system required minor modifications to the railroad tracks. First, the rails had to be cut at the ends of each block so that the electrical current would be confined to a single block. Second, the rails within a given block had to be connected electrically so that they would conduct current for the entire length of the block.

On November 22, 1910, the Block Signal and Train Control Board reported to the Interstate Commerce Commission the following comments about Robinson's invention:

> Perhaps no single invention in the history of the development of railway transportation has contributed more towards safety and despatch in that field than the track circuit. By this invention, simple in itself, the foundation was obtained for the development of practically every one of the intricate systems of railway block signaling in use today wherein the train is, under all conditions, continuously active in maintaining its own protection. In other words, the track circuit is today the only medium recognized as fundamentally safe by experts in railway signaling whereby a train or any part thereof may retain continuous and direct control of a block signal while occupying any portion of the track guarded by the signal.
>
> To Mr. William Robinson the Patent Office records concede the honor of having devised the first practical track or "Rail circuit." This comprised what is termed the closed track circuit. Closed track circuits are very reliable, wholly safe in principle, and simple of application and maintenance.[2]

Based on his seminal invention of the track circuit,[b] Robinson formed the Union Electric Signal Company on December 28, 1878.

Interlocking

The second and third of the railroad pioneers were J.M. Toucey, General Superintendent, and William Buchanan, Superintendent of Machinery for the New York Central and Hudson River Railroad (NYC&HRR). They addressed the more complex

---

(b) U.S. Patent 130,661.

problem that occurs when multiple train tracks cross or merge, thus creating the possibility of train collisions.

Toucey and Buchanan developed an interlocking system for train control in 1875. Interlocking is a system of signals and switches to prevent train collisions at intersecting tracks. At a minimum, a successful interlocking system requires:

- Signals must not be operated in a manner that would permit conflicting train movements to take place at the same time on a set route.
- Switches and other appliances in the route must be properly "set" (in position) before a signal may allow train movements to enter that route.
- Once a route is set, and a train is given a signal to proceed over that route, all switches and other movable appliances in the route are locked in position until either:
    ◊ the train passes out of the portion of the route affected, or
    ◊ the signal to proceed is withdrawn, and sufficient time has passed to ensure that a train approaching that route has had the opportunity to come to a stop before passing the signal.[3]

Toucey and Buchanan's system was mechanical. A grid arrangement of steel bars included large, human-actuated levers to operate signals and switches via bars running in one direction. Protrusions on the bars running in the perpendicular direction prevented any operation that could result in trains colliding.

Toucey and Buchanan formed the Toucey and Buchanan Interlocking Switch and Signal Company in Harrisburg, Pennsylvania, in May 1877. The first installations of their mechanism were on the switches and signals of the Manhattan Elevated Railroad Company and the New York Elevated Railroad Company in 1877–78.[4] Manufacturing of the Toucey and Buchanan products was provided by C.H. Jackson at his Jackson Manufacturing Company.

United We Stand

Robinson, Toucey and Buchanan, and Jackson each understood pieces of the train control puzzle and held patents on their respective technologies. But it was left to George Westinghouse to unify the players and assemble the pieces. Westinghouse merged the Union Electric Signal Company and Interlocking Switch and Signal Company on May 1, 1881, to form the Union Switch and Signal Company. C.H. Jackson became General Manager of the combined organization and remained active in the management of the company for many years. All patents held by the former companies and their principals were assigned to Union Switch and Signal. Offices and manufacturing facilities were consolidated in Pittsburgh at the Garrison Alley plant, which had been the Bidwell Plow Works.

The mechanical interlocking system of Toucey and Buchanan proved too unreliable and hard to operate, and under Westinghouse's direction, Union Switch and Signal developed replacements. First was a power interlocking system using air pressure for control and to power the track switches and signaling devices. The first installation of this system was at East St. Louis, Illinois, in 1882.

Electro-pneumatic systems, in which control was electrical, and air pressure

Chapter 8. Two Trains, One Track

Electro-pneumatic interlocking system on the Pennsylvania Railroad at Harrisburg, PA; Installed in 1930 and decommissioned in 1991. The levers and buttons on the lower panel controlled railway switches and signals. The diagram at the top replicates the layout of the tracks being controlled. Built by Union Switch & Signal Company (Niagara, Wikimedia Commons).

remained as the power to operate switches and signals, were first deployed in 1890 at the Pennsylvania Railroad terminal at Jersey City, New Jersey. Electro-pneumatic interlocking remained in use for over 100 years.[5]

In addition to combining the efforts of others in the general field of train control, George Westinghouse contributed fifteen patented inventions of his own. The first of his train control patents was titled "Railway Switch Movement,"[c] and the last, filed and issued after his death, was titled "Automatic Train Control."[d] The most important of Westinghouse's patents in this field was titled "Combined Electric and Fluid Pressure Mechanism," U.S. Patent 245,592, issued August 9, 1881. Not only foundational to the success of electro-pneumatic interlocking and signaling, the structures disclosed in this patent found use in automatic train-stopping devices, drawbridge locks, thermostatic control of ventilators, and anywhere compressed air had to be controlled by electrical signals.[6]

By October 1886, George Westinghouse's focus had turned to electrical apparatus, and he offered to purchase the Garrison Alley plant for his growing Westinghouse Electric Company. On behalf of Union Switch and Signal (US&S),

---

(c) U.S. Patent 237,149; issued on February 1, 1881.
(d) U.S. Patent 1,284,006; issued on November 5, 1918.

The former site of Union Switch and Signal in Swissvale, now the Edgewood Towne Center (Map data: 2020, Google).

Westinghouse had purchased 7½ acres in Swissvale, a Pittsburgh suburb. The US&S officers (led by Westinghouse) agreed to relocate the facilities there. The move was completed in Spring 1887. By 1889, US&S employed 200 people at the Swissvale location and an equal number in the field servicing customers. In 1904, those numbers had grown to 1,000 in-plant employees and 300–600 doing installation and service work.

George Westinghouse remained president of US&S until March 10, 1891. On that day, a "palace revolt" led to the ouster of Westinghouse.[7] The *Railroad Gazette* reported, "At the annual meeting of the Union Switch & Signal Co., held in Pittsburgh, March 10, a new Board of Directors, nominated by A.T. Rowland, the Secretary, who voted the proxies, was elected, replacing all the old directors, who it was expected would be re-elected. The election practically ousts the Westinghouse interest from the management. Mr. Rowland was elected President and Sigourney Butler, of Boston, Vice President…. Mr. Westinghouse was in New York on the day of the election, and says that the new directors owe their election to a betrayal of confidence, and their right to hold office will be contested."[8]

The stockholders soon realized that George Westinghouse and his relationships with customers and bankers were essential to the business of the company, so they called for the immediate retirement of Mr. Rowland. George Westinghouse's candidate, Mr. E.H. Goodman, was installed as president on March 13, 1891.[9]

Union Switch and Signal remained an independent company until 1917, when it became a subsidiary of Westinghouse Air Brake. In 1968, American Standard purchased Westinghouse Air Brake and made Union Switch and Signal a separate division. The US&S Swissvale plant, bordering the Penn-Lincoln Parkway (I-376), closed in 1985 and was demolished in 1986. The site is now a shopping center called Edgewood Towne Center. Everything changes.

Chapter 9

Solitude

"Solitude is pleasant. Loneliness is not."
—Anna Neagle, English Actress

George and Marguerite Westinghouse started their married life in August 1867, living with George's parents, George Sr., and Emeline, in Schenectady, New York. While grateful for the place to live, the newlyweds were joyous to be on their own when they moved to Pittsburgh in July 1868. The young couple settled in a rented flat in Allegheny City, across the Allegheny River from Pittsburgh.

Although she was happy to be with George, the move was difficult for Marguerite. She was close to her family and friends in New York, and Pittsburgh was a dirty, smoky, man's town. Allegheny City, in particular, was home to Irish and Scottish gangs who often clashed. The constant smoke in the air made wearing white or light-colored clothing impractical, and home decorations had to be dark to hide the grimy residue. George had several business interests to occupy him, but he sensed Marguerite's discomfort and started to look for a more enticing permanent home.

The earliest transaction on record for the property that would become the home of the young Westinghouse family is dated September 11, 1785. On that date, Thomas Hutchins sold approximately eleven acres of land in what was then Westmoreland County[a] to John McKee.[1] (This John McKee is probably the same man who founded McKeesport in 1795.[2]) The property changed hands several more times:

- John McKee to Pollard McCormick, 1835[3];
- Pollard McCormick to Peter Schoenberger, 1842[4];
- A singular event affecting the property occurred on December 10, 1852, when the Pennsylvania Railroad ran its first train from Pittsburgh, past the Schoenberger property, to Philadelphia. The completion of the PRR mainline transformed the neighborhood and the state.
- Dr. Peter Schoenberger to Nathaniel Holmes, 1859[5];
- Nathaniel Holmes to Thomas Miller, 1865[6]; and
- Thomas Miller to James Hopkins, 1869.[7] Hopkins added a house to the property.
- In 1871, James Hopkins sold the house and five acres of land to George Westinghouse for $40,000.[8]

(a) Pittsburgh was (and is) in Allegheny County, but at that time, the land east of the city was in Westmoreland County.

Chapter 9. Solitude

Now that he had purchased their new home, it was time to give Marguerite a big present for her birthday. After lunch on Sunday, April 30, 1871, George and Marguerite climbed aboard their carriage for a ride through Pittsburgh. It was a long ride and took Marguerite to the East Liberty Valley, a still-rural area east of Pittsburgh that she had never seen before. When she could contain her curiosity and enthusiasm no longer, the carriage turned onto a gravel driveway lined with mature maple trees. A three-story brick mansion complete with a square tower stood before the couple. George helped his wife out of the carriage and over the lush green lawn to the entrance. On the way, Marguerite commented on the gorgeous roses and hydrangeas.

The couple inspected each room and climbed the tower to view the countryside for miles around. Marguerite said it was the most wonderful birthday present a woman could have. George must have breathed a sigh of relief after making such an extravagant purchase without first consulting his wife.

Marguerite named the house "Solitude." The origin of the name is not clear, especially because the property was bordered by the busy mainline of the Pennsylvania Railroad. Perhaps Marguerite was expressing her happiness about being away from her in-laws and able to create her own home after living with George's parents at the beginning of their marriage.

According to Francis Leupp, Westinghouse's 1918 biographer:

> When they first moved into the house they had not the means to furnish all of it, so the drawing-room was left as it was, and a smaller room on the opposite side of the entrance hall was fitted for social and family purposes. Later, as their circumstances improved, they had the whole house refurnished with some elaborateness, besides extending it to the rear so as to add a spacious and high-ceiled dining room. Westinghouse's favorite place for sitting with his friends during the winter season was a square hall a little back from the main entrance, flanked by an angular staircase and containing an open fireplace. In the warm weather he enjoyed spending his evenings on the porch. He was always a happy host, and rarely a day passed when a few of his friends—most frequently his business associates and their wives—did not dine with him. When some especially perplexing question was occupying his mind, he might slip away from the party after dinner and seek a little library

"Solitude," circa 1900 (Westinghouse Collection, Detre Library and Archives Division, Senator John Heinz History Center, Pittsburgh, PA).

Stable at Solitude, circa 1920. The smokestack serviced a steam boiler (courtesy David Bear).

upstairs where he could be quiet and concentrate his thoughts for a while.[9] ... It was well furnished for such a purpose having a library table, student's lamp, a Remington typewriter No. 6, and both scientific and historical volumes.[10]

In 1879, eight years after he bought the house, Westinghouse purchased the adjacent parcel from James Hopkins for $16,000, increasing the estate area to its present 10.2 acres.

In addition to expanding the house and property, George added a two-story stable with a cellar containing a workroom for his treasured drawing board and a utility room. In a nod to Pittsburgh's snowy winters, he built a brick-lined tunnel stretching 220 feet from the house to the stable.

An underpass of Lang Avenue provided access from the northeast corner of the property to a bridge across the PRR tracks, ending at his private railroad siding at the Homewood train station. Trains from there provided easy access to the Westinghouse plants and to virtually everywhere in the country.

Marguerite proved to be a capable gardener and cultivated vegetables on the side of the house along with flower beds and a winding grape trellis in front.

George Westinghouse had chosen a fortuitous neighborhood for himself and his young wife. Blessed by flat land and mainline railroad service, the area east of Pittsburgh drew wealthy residents to build palatial homes in and around the area called Point Breeze. The neighbors of the Westinghouse family were, or would soon be:

- William Wilkins, whose estate was named "Homewood," was U.S. Minister to Russia and Secretary of War under President John Tyler.

- William Thaw made his fortune in coal, canal boats, and the Pennsylvania Railroad, and built the estate called "Lyndhurst." Thaw's son, Harry Kendall Thaw, married showgirl Evelyn Nesbit on April 4, 1905. Harry Thaw murdered his romantic rival, famous architect Stanford White, in front of hundreds of witnesses on the roof of New York City's Madison Square Garden on June 25, 1906. White, who designed Madison Square Garden and a multitude of other famous public and private buildings, resided in a suite of apartments in the tower of the Garden.[11] A jury acquitted Thaw of the murder by reason of insanity. The Thaw-Nesbit-White love triangle inspired the 1955 movie, *The Girl in the Red Velvet Swing*, and formed a major theme of the 1975 book and derivative musical, *Ragtime*.
- William Nimick Frew, president of the Carnegie Library and Institute, lived at "Beechwood Hall."
- Richard Beatty Mellon was a banker and philanthropist, and the son of Thomas Mellon. His Tudor-style home was so magnificent that it did not need a name.
- Thomas Armstrong founded the Armstrong Cork Company, which manufactured cork and linoleum products. He lived at "Penrose."
- Henry Clay Frick made his initial fortune in coke manufacturing and later became chairman of the Carnegie Steel Company. The Fricks lived at "Clayton," which is the only one of the grand houses still standing. George Westinghouse played poker on Tuesday evenings around Frick's breakfast table with other friends, including Nikola Tesla, Philander Knox, Andrew Mellon, and Robert Pitcairn.
- Henry John (H.J.) Heinz founded the food processing company famous for its "57 varieties" and lived at "Greenlawn," which was adjacent to Solitude.
- Joseph Horne founded the Joseph Horne Company department stores.
- Lillian Russell, world-famous (and infamous) actress and singer, married Alexander Pollock Moore, who later became U.S. ambassador to Spain and Peru.

As Pittsburgh historian Donald Scully, himself descended from a wealthy Point Breeze family, said of the area: "In the comparatively short distance from Beechwood Boulevard and Fifth Avenue to Braddock and Penn avenues, and a few blocks either way, there were more than a score of large estates, most of them surrounded by stone or brick walls, sedate iron gates, or a combination of the three, each one with a stable filled with carriages and horses—maybe a pony or two—luxurious gardener-kept lawns, and accompanying flower and vegetable gardens. Many had the large greenhouses which were in vogue before the craze for swimming pools."[12]

Contrary to its name, Solitude was seldom without guests. Westinghouse often brought his young engineers, including William Stanley, Lewis Stillwell, Oliver Shallenberger, and Benjamin Lamme, home to dinner and discussions about work and the world. In addition to rich and famous neighbors, in later years, George and Marguerite entertained dignitaries from overseas. In October 1896, Prince Mikhail Khilkov, a Russian railroad executive and Minister of Communications, visited Solitude for dinner. He arrived by train, crossed the tracks on the bridge, traveled through the

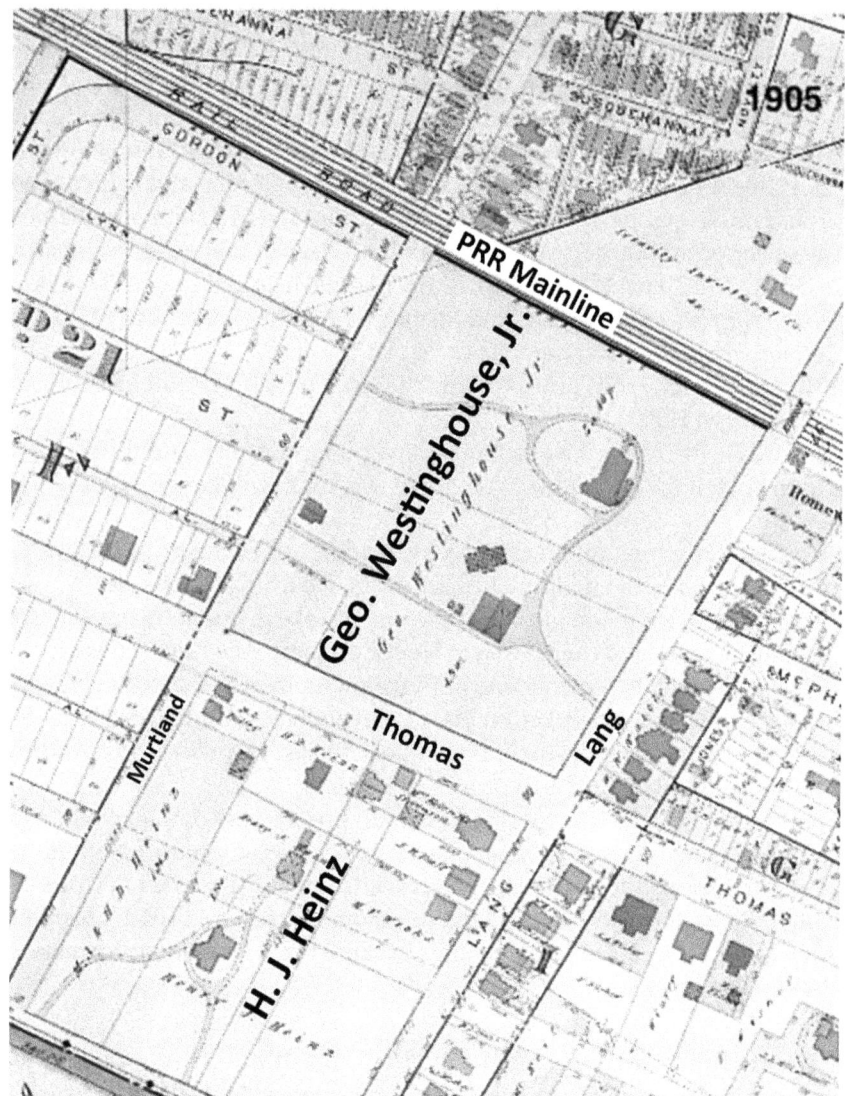

1905 map of area around "Solitude" in Point Breeze—Note 10.2-acre land holding of Westinghouse, and H.J. Heinz property to the southwest.

tunnel, and emerged to a red carpet and canopy stretched from the tunnel to the front entrance.

Marguerite Westinghouse broke out the gold dinner service, and pyramids of orchids adorned the table. Guests included Robert Pitcairn, Andrew Carnegie, and Henry Clay Frick.

Sir William Thomson, who had been knighted by Queen Victoria in 1866 and would become Lord Kelvin in 1892, was a renowned British scientist and inventor. Lord Kelvin became a close friend of George Westinghouse and visited Solitude several times, including a two-day stay in September 1897.[13]

Prince Albert Leopold, later King Albert I of Belgium from 1909 to 1934, received a warm reception when he was a guest at Solitude in May 1898.[14] After he

Monogrammed china of Marguerite Erskine Walker Westinghouse; Note the WWME at top (courtesy Virginia Montanez for *Pittsburgh Magazine*).

visited Pittsburgh, Prince Albert and George Westinghouse traveled together to Niagara Falls. Prince Albert ended his trip to the U.S. with a reception by President William McKinley at the White House.[15]

By 1882, eleven years after moving into Solitude, George and Marguerite were still childless. George was often preoccupied with his inventions and business interests, and Marguerite suffered from several illnesses. But they both desired a family and were delighted when Marguerite became pregnant

George Westinghouse III and his mother, Marguerite, 1883 (Westinghouse Collection, Detre Library and Archives Division, Senator John Heinz History Center, Pittsburgh, PA).

in August 1882. Soon after they learned of the impending birth, the couple moved temporarily to New York City, where the air was far cleaner and medical care more advanced. A son, George Westinghouse III, was born on May 20, 1883, in Manhattan Borough, New York City.[16] He would be the only child of George and Marguerite.

Wherever he was, George Westinghouse had Pittsburgh newspapers sent to him so he could remain current with what was happening at home. The papers were full of news about natural gas—new wells, distribution to homes and businesses, and explosions. He was anxious to return to Pittsburgh with his wife and newborn son to see what all the excitement was about. Several months after George III was born, the newly-expanded Westinghouse family returned to Pittsburgh and Solitude.

Although the family would own two other homes during their lives, as discussed in Chapter 25, Solitude was for forty years the heart of a creative empire that stretched around the world.

Chapter 10

Gas Pains

"Look, natural gas, just like oil, is going to eventually go away. It's not renewable."
—Ed Rendell, former Governor of Pennsylvania

Michael and Obediah Haymaker wanted to get rich quick, but it was taking too long. The brothers had worked as drillers north of Oil City, Pennsylvania, near where Colonel Edwin L. Drake[a] had drilled the first commercially successful oil well in the United States in 1859. The Haymakers sought to duplicate Drake's feat by drilling for oil on the banks of Turtle Creek in Murrysville, Pennsylvania, eleven miles northeast of Solitude, Westinghouse's Pittsburgh home. But the brothers had poor-quality drilling equipment, and it took them a year to pound through the 400-foot-thick Pocono Sandstone. Finally, on November 3, 1878, the brothers struck not oil, but a high-pressure vein of natural gas. The Haymaker well was the first natural gas well in the Pittsburgh area.[1]

Michael Haymaker later described the moment they hit gas: "I'll never forget the day the well came in. We were down 1400 feet. Without the slightest warning, there was a terrific roar and rumble that was heard fifteen miles away. Every piece of rigging went sky high, whirling around like so much paper caught in a gust of wind. But, instead of oil, we had struck gas. It was being shot out under such enormous pressure that it continued to shake the ground and roar for months, rattling windows for miles around. You can't imagine the production at such pressure; we figured the production at 30,000,000 cubic feet/day."[2]

The brothers had no money, no way to cap the well, and certainly no way to pipe the gas to where it might be utilized. For almost three years, that massive vent of gas roared up from the earth. The noisy eruption, nicknamed the "Murrysville Freak," became a tourist attraction.

But the real tourist attraction started on September 18, 1881, when a sightseer's lantern got too close to the gas and ignited it. Haymaker later remembered, "One night, a crowd with a few lanterns got too close. I recall a blinding flash. Perhaps there was an explosion. There must have been. My eardrums were ringing. It was a weird moment. Flames it seemed were everywhere. Over all there was one great flare, reaching high into the air. Then my ears cleared and I heard the familiar roar of the well. I picked myself up. All over the ground others were picking themselves up. Some

---

(a) Drake was not a colonel or even a private. He was a retired conductor from the New York and New Haven Railroad. Letters of introduction sent to businessmen in Titusville prior to his arrival identified him as Colonel Edwin Drake to enhance his authority.

 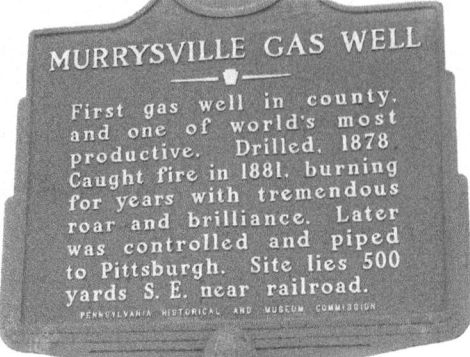

Historical markers commemorating the Haymaker Well in Murrysville—(left photograph courtesy Morgan Downey; photograph at right by author).

remained motionless. After we took stock, we found that there were no very serious injuries.

Gradually, the flame from the well mouth lowered until it settled to an even 100 feet straight up in the air. The original blast had sent the flame hundreds of feet upward, and it was seen in Pittsburgh, 18 miles away.

> "It burned for a year and a half, burning thousands of dollars of potential earnings. All the time we were busy trying to extinguish it. That burning well attracted hundreds of people from all over the country. World travelers told me they had never seen a sight so magnificent. It gave us continuous daylight for miles around."[3]

Murrysville became famous as a place "where there was no night." Tourists came in droves to view what was said to be "one of the greatest wonders of the day." Among them were President and Mrs. Grover Cleveland, with the president calling it "an uncanny picture, a superb spectacle."[4]

The flames were finally extinguished in March 1883. The Haymaker brothers' financial backer, looking for additional capital to commercialize the gas, met Joseph Pew and Edward Emerson.[(b)] Pew and Emerson had incorporated the Keystone Gas Company in 1880, after tapping the Bradford, Pennsylvania, natural gas field. Pew and Emerson incorporated the Penn Fuel Company in 1882 and bought the Haymaker well and surrounding properties in 1883. Penn Fuel then started to pipe natural gas to Pittsburgh's East Liberty and Lawrenceville neighborhoods.[5]

Pew and Emerson went on to form the Peoples Natural Gas Company in 1885, Pennsylvania's first officially chartered natural gas company.[6] They later founded the Sun Oil Company, now commonly known as Sunoco.

George Westinghouse read about the formation of the Penn Fuel Company and another story about a factory that, by converting to natural gas as a source of energy, saved thousands of dollars in a single year. He recognized the possibilities and broached the subject to his wife.

Marguerite replied, "You'd soon get as much absorbed in natural gas as you used to be in brakes when we first married, but the brakes had one advantage over gas—you

---

(b) Emerson was a cousin of Ralph Waldo Emerson, the American essayist, lecturer, philosopher, and poet.

could always work out your problems at home, instead of running off to Murrysville every day." "I can work out my problems at home just the same," he laughed in response; "that is, if you don't mind my boring a well through your flower beds. But don't charge me too much for the privilege. I dare say it will cost me five thousand dollars just to sink the hole and pipe it."[7]

Thus began an adventure and a new business for Westinghouse.

George Drills a Well (or Eight)

As Marguerite had not objected, George got right to work on his gas well at Solitude. On December 29, 1883, he hired the Gillespie Tool Company of Pittsburgh to dig the well, not in the flower beds, but near the stable. George and Marguerite, along with their infant son, George Westinghouse III, had spent the winter in New York City, far from the soot-covered snow of Pittsburgh. When the family returned in the spring of 1884, drilling began in earnest. The Gillespie crew constructed a rough seventy-foot-tall wooden derrick to hold the heavy machinery. All day long, they operated the noisy power drill, much to the chagrin of a skeptical Marguerite and the neighbors. They knew better than to doubt Westinghouse once he embarked on a quest, but his digging for buried treasure was becoming wearisome. The drilling crew struck two small gas veins, but their yields were meager, so Westinghouse urged them to continue digging.

Finally, on May 21, 1884, the foreman reported more signs of gas. When George asked the current depth of the well, the foreman replied, "About 1,560 feet." "Are the signs of gas strong?" "No, sir, weak; but I'm perfectly sure that a good supply is there, or not far away."

Westinghouse concluded, "The only way to find out is to go on. Perhaps by tomorrow we shall get results that amount to something. Only, go slow—feel your way along. Be very careful of the men, and warn them to take no risks."[8]

George slept soundly until he was jolted awake before sunrise by a thunderous explosion. And there was a continuous roaring noise, or had his hearing been damaged by the blast? He dressed quickly and went outside. What had been an organized construction site was now chaos. Gravel, sand, mud, and dirty water covered his usually well-kept lawn. The drilling derrick had been decapitated, and the engine thrown several yards away, landing under the debris. A geyser of mud spewed from the well, along with a hurricane-like roar.

Eventually, workers emerged from behind trees and rattled neighbors and strangers cautiously approached. What had happened?

After carefully extending the well just fifteen feet deeper than the previous day, the workers had heard hissing and rumbling sounds issuing from the hole. They dropped everything and ran for their lives as a great roar pursued them. Water, mud, and gravel spouted high into the air, followed by the continuous roaring noise.

Marguerite retained her usual good humor. George asked, "Are you satisfied with the experiment?" She replied, "Oh, very well. The house still has a roof on it, and the kitchen isn't wrecked."[9]

Westinghouse had a new toy. He and his like-minded friends played with the roaring stream of gas, estimated to flow at about twenty million cubic feet per day,[10] attempting to stop or divert it with large rocks and heavy planks. Whatever they placed in the flow was immediately thrown high in the air or splintered. Then they used the derrick to hold a one-hundred-pound stone on a rope over the opening. As they lowered the rope with its attached stone over the well, the flow shook the stone loose and lifted the line straight up in the air. The roar of the escaping gas continued day and night for almost a week until Westinghouse devised a sturdy stopcock to shut off the flow. Peace and tranquility returned to the neighborhood, at least for a few days.

Then Westinghouse got the idea to see if the gas would burn. He constructed a sixty-foot-high pipe above the mouth of the well with a pulley and wire rope attached to the top. On a fateful evening (as Westinghouse thought any flames would be more impressive at night), he attached oil-soaked rags to one end of the wire rope and turned on the stopcock restraining the gas. George ignited the rags and slowly pulled the other end of the wire rope to hoist the burning rags to the top of the pole where the gas was escaping. A faint bluish flame surrounded the end of the pipe. Then, WHOOSH! Suddenly a one-hundred-foot high pillar of fire emerged from the top of the pipe. Of course, the roar of the escaping gas returned to add another dimension to the display. The flame varied from a cone of sky-blue at its base; a pale yellow at its middle; then a dazzling white; and finally, shades of yellow, orange, and red at the top. People a mile away could read the fine print of a newspaper by the light of the gigantic gas flare.[11]

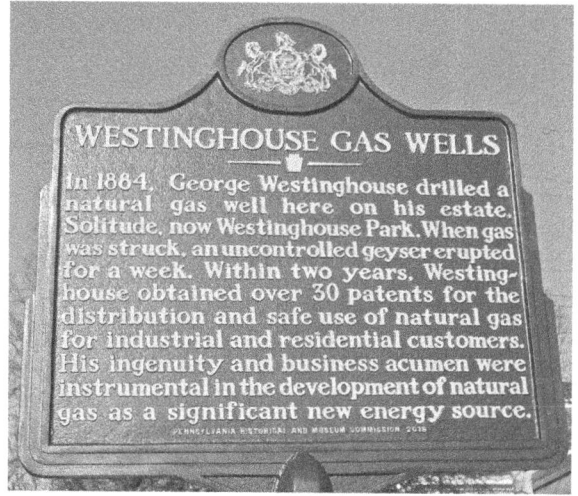

Pennsylvania Historical and Museum Commission marker commemorating Westinghouse's gas wells (author's photograph).

The novelty of perpetual daylight combined with the constant roaring noise wore off quickly. After less than a week, the evening performances ceased, and George announced that he would connect his gas well to a distribution system to supply light, heat, and industrial fuel to the area.

He dug three more gas wells at Solitude, and four other wells in Point Breeze and Homewood.[12] Solitude became one of the first houses in the city to be lighted and heated by natural gas.

Some authors claim that Westinghouse provided gas to his neighbors, Henry Clay Frick and H.J. Heinz, but there is no documentation to support that claim.[13]

For Frick, archival records indicate that gas service to his estate, Clayton, began in 1883 and that the East End Gas Company supplied gas via lines laid by Ernst Axtlhelm.[14] As gas was not discovered at Solitude until May 1884, Clayton's gas most likely came from wells in Murrysville.

Heinz moved into his estate, Greenlawn, on April 6, 1892,[15] long after Westinghouse ended his involvement in natural gas ventures.

Chapter 10. Gas Pains

"Old Number One" flaming gas well at Solitude (*Harper's Weekly*, November 14, 1885, pp. 744–745).

Westinghouse's knowledge and experience dealing with compressed air for train braking gave him an advantage in establishing a gas distribution system. But natural gas has properties that air does not. Both are colorless and odorless,[c] but natural gas is flammable and potentially explosive. Furthermore, gas from a well is at whatever pressure nature determines, and that pressure can fluctuate without notice.

Two gas companies, including Pew and Emerson's Penn Fuel Company, had been serving the Pittsburgh area since late 1882, long enough to identify the hazards of the fuel. Poor design and execution resulted in uneven gas pressure, and sometimes the gas flame went out. If the gas supply was not turned off, as it often was not (the gas was almost odorless, so gave no warning of its accumulation), any spark or flame could cause an explosion.[d] Also, gas leaking from the distribution piping would often seep into buildings, build to an explosive concentration, and blow up. Insurance companies warned of exorbitant rate increases if changes were not made to create safer distribution systems.

Quick to see the business potential, Westinghouse purchased land in the Murrysville area, where there were proven gas wells. He also located a dormant state charter for "The Philadelphia Company," which had been established by Thomas Scott of the Pennsylvania Railroad for railroad lines but could be used for any utility. Westinghouse

---

(c) Since the end of 1976, the chemical mercaptan, which smells like rotten eggs, has been added to natural gas to make the presence of unwanted gas obvious.

(d) An infamous and tragic gas explosion occurred on the North Side of Pittsburgh on November 14, 1927. Twenty-six people died when workmen from Equitable Gas—13 in all, some with acetylene torches—were repairing a leak on top of a massive natural gas storage tank, which had a capacity of 5 million cubic feet. When that tank exploded, it ignited a 4-million-cubic-foot-capacity tank just 200 feet away; then a third tank, this one 500,000 cubic feet, also went up. It was a triple explosion of mammoth, deadly proportions. Video of the spectacular explosion is at https://www.youtube.com/watch?v=1YWVP8w-a8w.

Derricks at Westinghouse's "Solitude" (stable and house visible at right), 1889 (photograph by Ernest Walter Histed).

bought the charter for $35,000 and incorporated the new Philadelphia Company on August 4, 1884.[16] He also aggressively pursued industrial and domestic gas customers and purchased competing firms. By the beginning of 1885, five firms, including Westinghouse's Philadelphia Company,[e] were drilling gas wells in and around Pittsburgh.[17]

Westinghouse recognized and understood the hazards of natural gas and developed a distribution system to avoid them. His first safety-oriented approach was to use double-piping to convey the gas. The smaller-diameter pipe from the well was enclosed within a larger pipe. Gas leaking from the smaller, pressurized pipe was collected inside the larger pipe and released at safe locations, such as at gas street lamps where it was safely burned off. Later, he used a similar but cheaper approach, encasing each joint of the supply pipe in an air-tight enclosure, thus safely trapping any gas leaking from the supply pipe.

Because gas pressure from the well was often high and variable, Westinghouse used a graded-diameter pipe to send the gas to its usage points. Starting at the well, he used an eight-inch diameter line. After four to five miles, the diameter was stepped-up to ten inches then to twelve, twenty, twenty-four, and thirty inches. Later, a thirty-six-inch pipe was used. For a given amount of gas, increased pipe diameter results in decreased pressure. By the time the gas reached its user, standard regulators easily controlled the reduced pressure.

(**e**) Henry Clay Frick, the Pittsburgh industrialist and Westinghouse's neighbor, was a member of the Board of Directors of The Philadelphia Company.

Chapter 10. Gas Pains

To enhance his competitive position, Westinghouse encouraged his Pittsburgh City Council allies to introduce an ordinance requiring the use of Westinghouse-patented inventions, including double-piping of gas lines in the city. The *Pittsburgh Post-Gazette* warned that "the extension of weak and imperfect pipes through the city" meant that its citizens were "living on a powder magazine,"[18] and demanded that the "Westinghouse Ordinance" be passed. City Council passed the "Westinghouse Ordinance" on July 31, 1884. Encouraged by Westinghouse's offer of free gas for the city's firehouses, police stations, markets department, and city property department, Council on November 13, 1884, approved a franchise for the Philadelphia Company to operate within Pittsburgh.[19]

But gas explosions continued, resulting in injuries, deaths, and destruction of property. An editorial in the February 2, 1885, *Pittsburgh Post-Gazette* titled "Death in the Streets" stated, "Save in a state of war we don't believe any large city in the world was ever in a more perilous situation than Pittsburgh is today owing to the dangers of natural gas explosions."[20]

The city council appointed a Natural Gas Commission and, by August 1885, passed an act setting standards for installing and testing pipe under the direction of the city engineer. Again, the act specified several Westinghouse-patented improvements in gas transmission.

One development, in particular, addressed the problem of "flame-out," wherein the flame goes out because of a temporary drop in gas pressure. To prevent explosions from accumulated unburned natural gas, Westinghouse invented the automatic cut-off regulator. If the gas pressure dropped below the required working pressure of four ounces per square inch, the regulator turned off the flow of gas from the street. Only when all gas appliances served by the line were turned off could the regulator be reset, resuming gas flow.

Westinghouse also invented a gas meter so that customers could be charged for actual usage instead of paying a flat monthly fee, as had been the practice. In 1888, the Philadelphia Company installed Westinghouse's meters on all of its natural gas services, thus saving about sixty percent of gas usage.[21] Westinghouse was eventually granted thirty-eight patents related to natural gas.

Jurisdictional disputes between the city of Pittsburgh and the state of Pennsylvania disrupted the regulation of the gas industry. But by 1886, six companies with 107 gas wells were piping gas into Pittsburgh for industrial and residential use.

The largest of these was Westinghouse's Philadelphia Company, which had bought out twenty smaller firms, including the Penn Fuel Company.[22] In 1886, the Philadelphia Company had 58 wells[f] and 184 miles of distribution piping in Pittsburgh, three times as much pipe as the other five gas companies combined. Just two years after its incorporation, the Philadelphia Company was the largest supplier of natural gas in the world.[23]

Growth continued in 1887. That year, the Philadelphia Company expanded to 494 miles of pipe, serving over 12,000 private homes and 582 industrial customers. Natural gas displaced an equivalent of 12,000 tons of smoke-producing coal daily.[24]

In 1888, the Philadelphia Company formed a wholly-owned subsidiary, the Equitable Gas Company. By 1911, Equitable controlled about 1,000 oil and natural gas

---

(f) 48 of these 58 wells were in the Murrysville area.

wells, had leases on more than 440,000 acres, owned 3,000 miles of pipelines, and served 110,000 customers.[25]

Westinghouse resigned as president of the Philadelphia Company in 1889 but remained on the Board of Directors. Meanwhile, as we shall discuss in the next chapters, he became deeply involved in another source of energy, electricity.

Impact of Natural Gas on Pittsburgh

Pittsburgh has a long history of pollution and the struggle to alleviate it. Extractive industries such as coal mining led to some of the most significant and longest-lasting impacts on the environment. The 1762 discovery of a coal seam along the south bank of the Monongahela River resulted in coal becoming the dominant energy source that fueled the industrialization of most of the country. Pittsburgh's environment paid the price. Smoke pollution was the most noticeable effect of coal consumption and gave the city its identity as the "Smoky City."[26] An 1816 visitor described the scene, "*Pittsburgh* was hidden from our view until we descended through the hills within half a mile of the *Allegany River*. Dark, dense smoke was rising from many parts, and a hovering cloud of this vapor, obscuring the prospect, rendered it singularly gloomy."[27]

The discovery and commercialization of clean-burning natural gas provided a respite, however brief, from the pollution. By 1886, estimates of the amount of coal displaced by natural gas ranged from six to twenty million tons per year. Of course, the downside was soaring unemployment in the minefields.

National publications noted the benefits of natural gas and the dramatic reduction of pollution in Pittsburgh. *Harper's Weekly* in February 1892 observed, "A peaceful revolution took place in Pittsburg. The great mills ceased to belch forth huge clouds of smoke, the merchant no longer looked upon soot as the chief enemy of the human race, and the careful housewife put gas burners into her coal stoves and took courage to clean house."[28] As a result, Pittsburgh lost its "Smoky City" title, at least for a time.

Wishful thinking abounded, as local leaders proclaimed the "almost incomprehensible quantities" and "inexhaustible" nature of natural gas supplies. But as early as 1890, fluctuating and dwindling natural gas supplies were already affecting industrial users.

By 1891, Andrew Carnegie's Edgar Thomson[(g)] steelworks became disillusioned with the availability and cost of natural gas and switched back to coke for fuel. George Westinghouse countered by suing Carnegie for a rather large unpaid gas bill amounting to $580,000. The lawsuit was ironic because Carnegie had been instrumental in "securing for Westinghouse a practical monopoly of the natural gas business here" (in Pittsburgh). Also, Carnegie's partner, Henry Clay Frick, was formerly a Director in Westinghouse's Philadelphia Company, which provided the gas to Carnegie.[29] There is no public record of how the millionaires settled their dispute.

In 1892, William Metcalf told the Engineers' Society of Western Pennsylvania,

---

(g) Carnegie named his massive steel plant for his largest customer, Edgar Thomson, President of the Pennsylvania Railroad. The Edgar Thomson Works is known to locals as "ET," a name applied long before the extraterrestrial visited Earth in the 1982 movie.

"Ode to Pittsburgh" by Meda Logan, circa 1907 (Heinz History Center, Detre Library & Archives, GPCC B08.I07).

"We are going back into the smoke. We had four or five years of wonderful cleanliness for Pittsburg, and we have all had a taste of knowing what it is to be clean. We all felt better, we all looked better, we all were better. But we are back into the smoke. It is growing worse every day."[30]

Metcalf was an optimist. Natural gas consumers, both residential and industrial, continued to increase while local gas supplies declined. Gas had to be piped, with the aid of compressor stations, from West Virginia wells. The smoke was back, as described in Meda Logan's "Ode to Pittsburgh."

The smoke and soot remained until after World War II. In 1946, Democrat David L. Lawrence was elected mayor on a platform of cleaning up the city. Lawrence's unlikely alliance with wealthy Republican banker, Richard King Mellon, led to the bipartisan Pittsburgh Renaissance.[31] The recovery of Pittsburgh is a subject covered by several other books, and I recommend *Pittsburgh: The Story of an American City* by Stefan Lorant (1964 and later editions) as a starting point.

Learning by Doing

George Westinghouse's work with compressed air to activate railroad air brakes taught him how to handle compressed gases. Hoses or pipes must be used to carry such gases, and those conveyances leak. The flammability and explosive potential of natural gas added more constraints to the distribution system, as did the need to measure the quantity of gas dispensed. Lessons learned with natural gas would carry over to Westinghouse's next and most challenging venture, the generation and transmission of electricity.

In some ways, electricity is like natural gas. Both are sources of energy, both have to be transported to their users (natural gas by pipes, electricity by wires), both must be metered to recover costs and make a profit, and both can be dangerous or even deadly if improperly used.

So the knowledge that Westinghouse gained from his flirtation with natural gas helped him, but the challenges and personalities ahead dwarfed all that he had faced before.

Chapter 11

More Energy

"Success is almost totally dependent upon drive and persistence. The extra energy required to make another effort or try another approach is the secret of winning."
—Denis Waitley, American Writer

George Westinghouse's interest in electricity started in his childhood when he played with static electricity generated by the movement of drive belts in his father's factory. But, like his father, he was far more interested in mechanical apparatus. In his work with railroad air brakes, George had considered the use of electricity to control the actuation of the brakes, but believed the required wires would be unreliable in the rugged environment of railroads. Railroad signaling and interlocking, which involved stationary equipment, were more amenable to the use of electricity.

Westinghouse's first published activity involving electricity came when he was 34 years old. The work was disclosed in a patent[a] issued on August 2, 1881, just three months after the formation of Union Switch and Signal. From that point on, Westinghouse's patents, excluding those related to natural gas, showed an increasing focus on the generation, transmission, and application of electricity.

In 1881, when Westinghouse started applying electricity to train control, electric lighting was in its infancy. Arc lamps, as their name implies, created a point source of light by forming an electrical discharge (arc) between two carbon electrodes. Because of their extreme and uncontrollable intensity, arc lamps were suitable only for outdoor lighting and eventually became a niche business.

Marguerite Westinghouse played an unwitting role in the development of her husband's electrical business. While the Westinghouse's were traveling in Italy in 1882, Marguerite became dangerously ill and was restored to health by an eminent Italian physician, Dr. Diomede Pantaleoni. George and the doctor became friends, and Pantaleoni introduced his son Guido, who had recently been graduated from the University of Turin and was interested in scientific topics. Westinghouse arranged for Guido Pantaleoni to come to Pittsburgh and work for Union Switch and Signal Company.[1]

In late 1883, Westinghouse started to consider lighting as a business. He assembled a small staff, and they began to study various aspects of such a venture. They focused at first on using DC with incandescent bulbs for residential lighting. An early problem was that incandescent lamps required a narrow range of voltage to operate reliably, and voltage regulation in the face of varying loads on the circuit (more or fewer light bulbs) was poor.

(a) U.S. Patent 245,108; Automatic Electric Current Regulator.

Westinghouse himself had patented a system for voltage regulation with changes in load.[b] His younger brother, Henry Herman Westinghouse, was also an inventor. He had invented a high-speed engine which had self-regulating characteristics.

In early 1884, while exploring the commercialization of these inventions, Herman met a brilliant but eccentric young engineer named William Stanley. Stanley had dropped out of Yale in 1879 after one semester. Rather than the law career that his father advocated, young Stanley wanted to work with electrical equipment.[2] He had recently invented a self-regulating generator and co-invented an incandescent lamp with a filament of carbonized silk. Herman immediately hired Stanley and brought him to Pittsburgh to manufacture his generator and lamp, and to develop a complete electric lighting system.

George Westinghouse was always cognizant of intellectual property rights, so he initiated a study of patents related to William Stanley's inventions. In the area of incandescent lamps, William Sawyer and Albon Man had applied for a patent on a carbonized paper-filament lamp in 1880. An interference proceeding[c] ensued in the U.S. Patent Office because of a similar patent application by Thomas Edison. But Sawyer and Man prevailed, and a patent was issued to them in 1885.[d]

Future developments would show that another feature of the Sawyer-Man lamp was far more critical than the filament structure. Sawyer and Man described a lamp combining two parts—a glass bulb and a separate tightly-fitted stopper. This structure was significantly different than Edison's patented lamp, which had just one piece—a single sealed all-glass bulb. As was his standard practice, in 1887, Westinghouse bought Consolidated Electric Light, the company which held rights to the Sawyer-Man lamp patent.

The patent status of Stanley's electrical generator was less complex. Although a similar patent was issued to Weston in 1883, Stanley's version seemed to have enough unique features to sustain its validity.

Then serendipity struck, as it often does for those who work hard. May 1885, Guido Pantaleoni, whom Westinghouse had brought from Italy in 1882, returned home at the news of his father's death. While in Italy, Guido visited his professor at Turin and was introduced to Lucien Gaulard. Gaulard, a Frenchman, and John Dixon Gibbs, an Englishman, had developed what they called a "secondary generator," which we now call an electrical transformer, or simply, a transformer.[e] The pair patented their invention, established companies, contracted to install lighting, and exhibited their lighting system at the Royal Aquarium in London and at Turin in 1884. Most importantly, they transmitted electrical power from an AC power plant in Tivoli to light lamps in Rome, a distance of twenty miles.

Pantaleoni was so impressed by his encounter with Gaulard that he cabled Westinghouse. George wrote back immediately and asked Pantaleoni to obtain an option for the American rights to the invention. Pantaleoni traveled to London, where he met Gibbs and secured the option.

(b) U.S. Patent 245,591; Automatic Electric Current Regulator, issued August 9, 1881.
(c) A patent interference arises when two or more pending patent applications, or at least one pending patent application and an unexpired patent, contain patent claims covering the same or substantially the same subject matter. See https://www.patentek.com/patent-interference-overview/.
(d) U.S. Patent 317,076; issued May 12, 1885.
(e) For our purposes, a transformer changes (increases or decreases) the voltage of an AC signal.

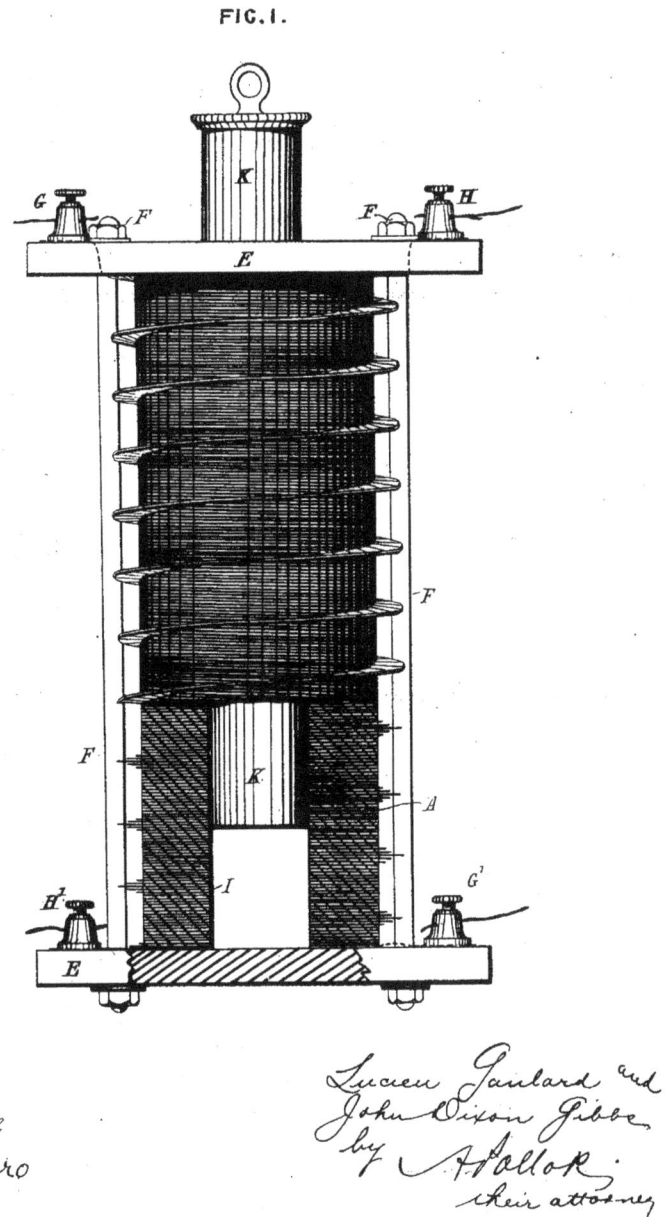

Diagram of Gaulard and Gibbs Transformer from U.S. Patent 316,354; issued April 21, 1885.

Westinghouse instructed one of his patent attorneys, Franklin Pope, to investigate the validity of the Gaulard and Gibbs patents and the practicality of their system. Pope summarized his findings in 1887: "My own impression at first sight was, like that of every one else, an unfavorable one. The knowledge which I had gathered in the ordinary course of my professional experience led me to expect that the loss of energy in conversion would be so great as to render the scheme commercially unprofitable and that this lost energy, appearing in the form of heat, would quickly destroy the apparatus, or at least render it useless; and it was not until I had gone carefully through the published researches of Hopkinson and Ferraris that I found reason to change my opinion. I followed up the matter by personal investigations of the apparatus in operation, and was convinced of its novelty and industrial value. The extracts which I have quoted are but fair samples of the communications and articles which appeared in many of the technical periodicals, and in fact I may say that, so far as I now recollect, all these journals, without exception, whenever they took any notice at all of the work of Gaulard and Gibbs, did so in a spirit of hostile criticism, which continued not only long after the successful installation of the plant in many places, but continues in many quarters up to the present hour."[3]

Westinghouse arranged to have several samples of the Gaulard and Gibbs transformers shipped from England to Pittsburgh. Once the transformers arrived in November 1885, he assigned two of his bright young engineers, Albert Schmid and O.B. Shallenberger, to evaluate their operation. Westinghouse worked closely with his engineers, and within three weeks, determined that the Gaulard and Gibbs transformer was impractical to manufacture. More importantly, they developed an efficient design amenable to rapid manufacturing.

Examination of the original Gaulard and Gibbs transformer quickly revealed its inherent limitations. The transformer consisted of a mahogany base and top with four steel rods enclosing a stack of copper disks separated by waxed paper. Each disk had a projecting metal tag alternatively painted black or red. The black tags were connected together, as were the red tags, thus forming two electrical contacts. There was a central sliding core of iron terminating in a large, nickel-plated knob.[4]

To appreciate the importance of the Gaulard and Gibbs transformer and its successors, we must understand the AC vs. DC issue. Before continuing with the Westinghouse Electric Company story, it will be helpful to understand the competitors in the "Current Wars."

Chapter 12

AC or DC

"Nature may reach the same result in many ways."
—Nikola Tesla

The Current Wars

Scores of books have been written and at least two movies made about the battle to electrify the United States in the 1880s and 1890s. The adversaries were Thomas Edison and George Westinghouse, with Westinghouse aided by the eccentric Croatian-born Nikola Tesla. Because of the massive infrastructure involved, only one system, either Edison's DC or Westinghouse's AC, could win. Westinghouse was just four months older than Edison, but the two men were as different as their preferred electrical systems.

As with virtually all battles for public acceptance, money was the true motivation. Based on their patents, the winner would have the power (no pun intended) to build and operate electrical distribution systems and much of the associated equipment for years to come, in the United States, and eventually the world.

The limited purpose of this chapter is to cut through the highly emotional arguments which culminated in the intentional electrocution of countless dogs and cats, allegedly an elephant, and eventually a man, to examine the underlying technical reasons favoring each side.

DC, Edison's entry in the competition, stands for Direct Current. As the name implies, DC flows as electrons in a steady stream in a single direction. The primary advantage of DC at the beginning of the Current Wars was that electric motors operating on DC were readily available.

AC, favored by Westinghouse, stands for Alternating Current. Again, as the name implies, with alternating current, the electrons flow first in one direction and then in the reverse direction. The rate of reversal is called the frequency.

Many electrical appliances, including heaters, most stoves, and light bulbs, work equally well on either DC or AC. But until Tesla came on the scene in 1886, no practical motors operated on AC.

However, AC has one overwhelming advantage over DC: its voltage can easily be changed either up or down.

Appendix III—Electrical Engineering 101 quantifies the advantages of high voltage for power transmission and provides details of how AC voltage can be changed (increased or decreased) by the use of transformers. Conclusions from that discussion are as follows.

- The critical equation derived in Appendix III shows that the power lost during electrical power transmission is inversely proportional to the transmission voltage squared. For example, if the voltage increases by a factor of 10, the power lost drops by a factor of 100, all other things remaining the same.
- With higher transmission voltages, AC customers can be located much further from the generating station. For example, with the same factor of 10 increase in transmission voltage, customers could be 100 times as far away, again with other factors remaining the same.
- Electrical devices called transformers can efficiently increase or decrease AC voltages. No corresponding device existed in Westinghouse's time to change DC voltages.

Based on these facts, Edison was forced to build his DC generating plant within one mile of his customers. Therefore, he needed many more plants, which were usually coal-fired and smoke-generating. Also, he could not economically serve customers in rural areas.

On the other hand, AC generating plants could be located outside of developed areas and could supply power to many more customers.

Edison based his choice of DC on many factors, among them was the availability of motors operating on DC while no efficient AC-operated motors existed. More important to Edison was the fact that his company manufactured DC generators and stood to profit handsomely if DC power became the standard.

Many experts, including the influential Lord Kelvin, agreed with Edison that DC was preferable. So, despite the clear advantages offered by AC, Edison kept aggressively promoting his DC system.

Before exploring just how aggressive Edison was, we must rejoin Westinghouse in his pursuit of AC power, which required the development of a better transformer and was impeded by the lack of an AC-powered motor.

Chapter 13

Assembling the Pieces

"Once I get on a puzzle, I can't get off."
—Richard P. Feynman, American Physicist

George Westinghouse and his engineers, Albert Schmid and O.B. Shallenberger, had examined the Gaulard and Gibbs transformer and found it to be unmanufacturable. But the function it performed, changing the magnitude of AC voltages, was at the heart of any system of transmitting AC power. So they had to develop an efficient, manufacturable design.

Faced with this well-defined problem, Westinghouse was in his element. Given such a problem, he was able to apply his knowledge, experience, and creativity to find a solution that eluded others. Instead of the stamped copper discs and soldered connections of the Gaulard and Gibbs design, Westinghouse suggested a structure using insulated copper wire that could be machine-wound on a laminated core assembly of thin steel sheets.

Inventing a Manufacturable Transformer

Consider the drawing, which represents a closed core of a transformer. The core is constructed of many laminated layers of steel. To form the windings around one leg of the core, the wire must repeatedly pass through the center of the core. To make this happen, the operator must route the wire through the opening from one side of the core, let go of the end of the wire, and pick it up on the other side of the core. Such an operation would be difficult or impossible for an 1885-era machine to accomplish.

The first step toward a manufacturable transformer taken by the Westinghouse engineers was to cut the core, as shown in the drawing. With this open structure, an operator (or a machine) could easily wind one coil (say the transformer primary) around the top part of the "C" without letting go of the wire. The second coil (the secondary) could likewise be formed around the bottom part of the "C."

William Stanley, whom Westinghouse had added to the transformer team, then made a vital contribution. He suggested that the

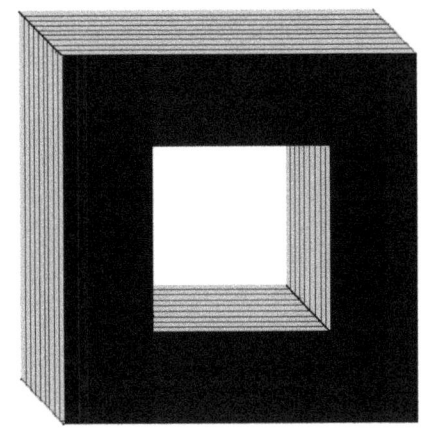

Closed transformer core before coil winding.

primary and secondary coils of the transformer could be wound separately from the core. By winding those coils around a rigid, hollow paper form with a square cross-section, they could be wound with no steel core in place. When the windings were completed, the steel core laminations could then be inserted through the hollow part of the paper form.

Then Stanley suggested a more efficient core structure formed of "E" and "I" shaped laminations, as shown in the drawing. The center bar of the "E" is inserted through the center of the coil form. One layer of the laminated core is inserted from the left and the next layer from the right. The "I" shapes complete the magnetic path.

In just three weeks at the end of 1885, the Westinghouse transformer team derived the concept for a practical transformer. They built and evaluated models in the first months of 1886. The defining patent application for the new transformer structure was filed on February 16, 1886, and was titled "Induction Coil."[a]

"C" shaped portion of transformer core.

In January 1886, Westinghouse sent Guido Pantaleoni and the patent attorney, Franklin Pope, to England to complete negotiations with Gaulard and Gibbs to buy their U.S. patent rights. Upon receiving his instructions, Pantaleoni ventured to ask, "How much are we to pay for the rights?" Westinghouse replied, "They'll tell you their price. Whatever it is, close the bargain, and I'll send the money by cable to you." The pair returned in February with the desired patent rights, which had cost Westinghouse $50,000.[1] In 2019 dollars, that would be equivalent to about $1.3 million. Westinghouse often purchased patents to avoid future litigation, but even for him, this was an expensive acquisition.

"E" and "I" shaped portions of transformer core.

Westinghouse conceived another idea for transformers that would prove to be vitally important in electrical transmission systems.

Transformers are not perfect in their operation. Various factors, such as wire resistance and hysteresis of the core material,[b] result in heating of the structure.

(**a**) U.S. Patent 342,553; Issued May 25, 1886.

(**b**) Hysteresis occurs when the system's output depends not only on its present inputs but also on past inputs. A common example is your air conditioning system. Suppose you set your thermostat to 75 degrees. In a typical system, the AC will turn on when the actual temperature reaches 76 degrees, and turn off when the temperature falls to 74 degrees. The condition of the system at 75 degrees depends on the recent temperature ("past input"). If the recent temperature were 76 degrees or higher, the system would be <u>on</u> at 75 degrees. But if the recent temperature were 74 degrees or lower, the system would be <u>off</u> at 75 degrees. The core of a transformer is a magnetic material, and its present condition of magnetization depends on its past condition.

Chapter 13. Assembling the Pieces

Westinghouse suggested placing the transformer in a sealed container filled with oil to carry away the heat. U.S. Patent 366,362, issued on July 12, 1887, described this invention. Because of its importance, the patent was the subject of extended litigation, all of which it withstood.

In a case regarding this patent called the "Union Carbide Suit," the Court of Appeals for the Second Circuit stated, "The practical result of the invention in suit, as testified to by complainant's experts, was to so increase the capacity of converters (transformers) that, while a dry converter cooled by the natural circulation of air is limited to 10 kilowatts, the oil-insulated converters of the patent in suit are commercially serviceable up to 500 kilowatts."[2]

All of the progress regarding transformers convinced Westinghouse that AC power distribution was not only feasible but far superior to the then-popular DC distribution.

On December 23, 1885, George Westinghouse, Herman Westinghouse, Franklin Pope, Robert Pitcairn, and three others applied for a charter for a corporation to be called the Westinghouse Electric Company (WEC). Among the assets of the new company were 27 patents and applications in the electrical art. The Commonwealth of Pennsylvania granted the charter on January 8, 1886, and formal organization followed on March 8, 1886. George Westinghouse was President, Herman

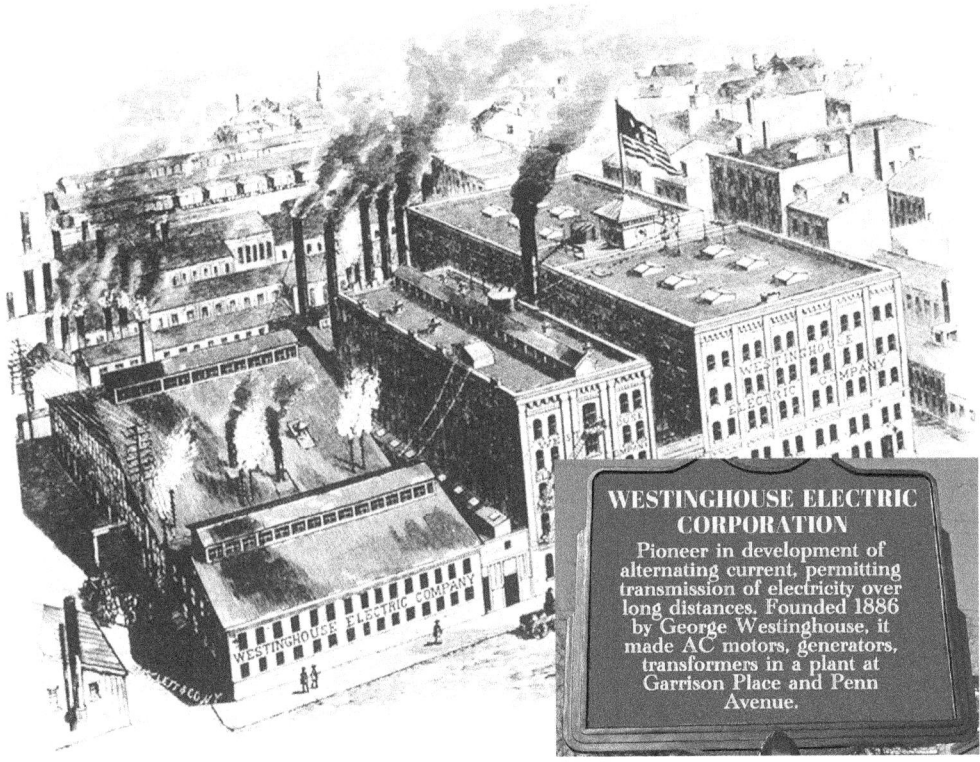

Westinghouse Electric Company facility at Garrison Place, Pittsburgh, Pennsylvania, circa 1893. This facility was initially occupied by Westinghouse's Union Switch and Signal Company (Westinghouse Collection, Detre Library and Archives Division, Senator John Heinz History Center, Pittsburgh, PA. Historical marker photograph by Author).

Westinghouse was Vice-President, and Guido Pantaleoni was General Manager until September 15, 1886, when H.H. Byllesby replaced him.

As mentioned in Chapter 8, "Two Trains, One Track," Westinghouse Electric Company moved into the Union Switch and Signal buildings at Garrison Alley later in 1886.

In 1885, William Stanley, whose already fragile health had deteriorated in the smoke of Pittsburgh, asked Westinghouse how he could get away from the city. He wanted to move Great Barrington, Massachusetts, where as a child, he had often visited his grandparents. Westinghouse agreed to establish a small laboratory in Great Barrington so Stanley could work on commercializing AC distribution systems. While at Great Barrington, Stanley would receive a salary of $4,000 per year ($1,000 less than he had received while working at Pittsburgh) and an additional $200 per month for general laboratory expenses. In return, Westinghouse would have first rights to whatever Stanley invented.

In July 1885, William Stanley, with his new wife, Lila Courtney Wetmore Stanley, moved to South Main Street in Great Barrington. The local newspaper, the *Berkshire Courier*, reported on Stanley's progress: November 4, 1885: "William Stanley Jr. has leased the 'Rubber Factory' (off of Cottage Street) for a term of years, and will occupy it at an early date. It is to be a laboratory for experimental work, in conjunction with Messrs. Westinghouse of Pittsburgh, Pa. A boiler and a 5-horse power engine will be at once set up, and some fine testing machinery, all of which will be under Mr. Stanley's direction. The building will be fixed up inside quite a little, and will make, when finished, a most excellent place for the proposed work."

January 27, 1886: "A Westinghouse Automatic engine, of 25 horse power, was received here on Friday, and will be used by William Stanley Jr., the electrician, in the laboratory he is fitting up in the building known as the rubber mill."

In addition to the Westinghouse steam engine, Stanley installed a steam boiler and a 6,000 watt Siemens alternating current generator. He would generate electricity at 500 volts AC, send it 4,000 feet to the Main Street shops, and use his transformers at each location to step-down the voltage to 100 volts. In total, Stanley wired thirteen stores, two hotels, two doctor's offices, one barber shop, the telephone exchange, and the post office for lighting.[3] The story continued in the *Berkshire Courier*: February 17, 1886: "The introduction of the electric light is soon to be an accomplished fact. Mr. William Stanley Jr., the electrician, has placed globes for incandescent lights in R.I. Taylor's store, and as soon as the wires are ready the lights will be used. The store of C.H. Lillie is to be similarly embellished, and soon we will be a city with all the fixings."

March 10, 1886: "The wires to be used in lighting the electric lights in R.I Taylor's and C.H Lillie's stores, were strung last Saturday."

March 17, 1886: "Last evening the interior and exterior of R.I. Taylor's (Stanley's cousin) store was lighted by three 150 candle power electric lights of the Stanley system. Two of the lights in the store made it as light as noon-day. A large number of business men were present to witness the effect, and were unanimous in their praise thereof."

In 1885, Mary Frances Hopkins, the widow of Central Pacific Railroad magnate Mark Hopkins, had started building a mansion in Great Barrington. Power to light the home was to come from a 100-kW Edison direct current generator. Electricity from that generator was temporarily sent to Main Street in Great Barrington to power eighteen 16-candlepower bulbs in the store of H.A. Brewer.[4] On March 20, Stanley's AC system and Edison's DC system were both energized, and the *Berkshire Courier* reported the results in the March 24, 1886, edition:

THE ELECTRIC LIGHTS

Last Saturday evening an opportunity was afforded those who desire to witness the illuminations by the Stanley and Edison systems of electric lights. Two of the Stanley lights, each of 150 candle power, were placed in C.H. Lillie's drug store, with a thoroughly satisfactory result. A sixteen-candle power light had been placed in Dr. F.P. Whittlesey's office, and proved conclusively that a light of this size would be sufficient for any ordinary sized room. R.I. Taylor's store had undergone a grand transformation. The cases and shelving had been altered to obtain additional room, the ceiling and sidewalls painted a very delicate blue tint, and new show cases are to be added as soon as received, everything was so fresh as paint could make it, while three 150 candle-power Stanley lights, one being outside the door and two in the interior, added the finishing touches. These lights are so powerful, and so perfectly white, that green and blue can be readily distinguished, though they cannot by gas light. A lesser candle power would have the same effect as gas or kerosene light. H.A. Brewer's store was lighted by eighteen 16 candle power Edison lights, making 288 candle power in all. Dividing up the power into so many lights failed to show how strong it would be if concentrated into two or three single lights, so that a comparison of the two systems was difficult, although the majority of those who viewed the results seemed to favor the Stanley lights. With two such enterprising parties as Messrs, Stanley and Bodwell (H.N. Bodwell, superintendent of works at Hopkins mansion) interested, there seems to be little doubt but that, in the near future, the stores and houses, and perhaps the streets of our village will be generally and thoroughly illuminated.[5]

Stanley's Great Barrington system became the first operating AC transformer installation in the United States. For his pioneering work and "...meritorious achievement in invention and development of alternating current systems," Stanley received the Edison Medal in 1912.[6] Ironically, Stanley's demonstration of AC power distribution was a significant factor in defeating Thomas Edison's DC approach. In 2004, Stanley's system was named an IEEE Milestone in Electrical Engineering by the Institute of Electrical and Electronics Engineers (IEEE).[7]

When Edison was informed of the Stanley/Westinghouse AC system in Great Barrington, he initiated what was to become his mantra for the next four years: "Just as certain as death, Westinghouse will kill a customer within six months after he puts in a system of any size. He has got a new thing and it will require a great deal of experimenting to get it working practically. It will never be free from danger."[8]

To build on the success of the Great Barrington installation, in the fall of 1886, Westinghouse Electric Company (WEC) transmitted 1,000 volts AC (later raised to 2,000 volts) from a generator at Garrison Alley to Lawrenceville, about four miles away. There the electricity passed through several of the latest transformers to light four hundred incandescent lamps. The lamps burned continuously for two weeks. This demonstration was the first time AC power had been transmitted for a significant distance in the United States.

After this success, WEC moved the same generator and transformers to Buffalo, New York. There, on November 30, 1886, they were energized to form the first commercial AC transmission system in the U.S.

Because AC power could be transmitted over a much longer distance than DC, it was suitable for use in small towns and rural areas that lacked the population density of large cities. Greensburg, Pennsylvania, about twenty miles from Pittsburgh, became the first town to install a complete municipal AC system. Orders for AC components and systems trickled and then poured in, quickly taxing the manufacturing capability of Westinghouse Electric. The factory was enlarged, and by 1888, WEC employed three thousand people.[9] By August 1889, Westinghouse Electric Company generating stations (DC and AC) powered more than 350,000 incandescent lamps.[10]

Two critical components essential for a complete AC system were still missing, a meter to measure the consumption of electricity and a practical motor to power machinery.

Westinghouse remembered his experiences with providing natural gas to his customers. Before meters were available, users were billed a fixed amount for each gas appliance, and there was no incentive to conserve gas by turning off the device. After gas meters were installed, the gas company was able to charge for the amount actually consumed. Gas usage dropped, and profits rose.

The electric meter problem was quite similar, so Westinghouse knew that he had to solve it. In 1887, he filed two patents for AC electrical meters.[c]

The structure described in those two patents would have reached production and usage except for a fortuitous accident that led to a more accurate and robust meter. In late April 1888, O.B. Shallenberger, then chief electrician of WEC, was examining an AC arc lamp when a tiny coil spring dropped from the lamp to another part of the circuit. Philip Lange, who had designed the arc lamp, was about to retrieve the spring when it started to rotate. Shallenberger exclaimed, "Wait! Let's see what makes that spring revolve."

Upon closer examination, the men found that a shifting magnetic field was influencing the spring. Shallenberger, who had been considering how to construct a meter, said, "There's a meter in that, and perhaps a motor." In less than a month, Shallenberger developed a fully-functioning AC meter, and by August, production models were ready for installation.[11]

Shallenberger's AC electric meter (from Westinghouse Collection, Detre Library and Archives Division, Senator John Heinz History Center, Pittsburgh, PA).

(c) U.S. Patents 383,678 and 383,680; Electric Meter; 383,680 includes Philip Lange as the co-inventor.

Chapter 13. Assembling the Pieces

Shallenberger was right when he said there might be a motor based on the same principle as the meter. But he was not the first to discover such a motor. That honor went to young Croatian-born Nikola Tesla. The name Tesla is familiar today because of the car company founded by visionary Elon Musk. Nikola Tesla was also a visionary and played a dominant role in the development and adoption of AC power.

Chapter 14

The Greatest Inventor Who Ever Lived

"I will stop the motor of the world."
—Ayn Rand, spoken by John Galt in Atlas Shrugged

In Ayn Rand's classic novel, *Atlas Shrugged*, the protagonist John Galt invented a fictional revolutionary motor and then destroyed it. In the real world, Nikola Tesla invented an actual revolutionary motor and changed the world. "Nikola Tesla is the greatest inventor who ever lived." So wrote William Leonard Lawrence, Science News Editor of the *New York Times* in 1945, two years after Tesla's death.[1]

Nikola Tesla was born in 1856 in the village of Smiljan in Croatia. Tesla's 1981 biographer, Margaret Cheney,[2] places his time of birth as "precisely midnight" between July 9 and 10.

His father, Milutin Tesla, was the priest of a Serbian Orthodox Church. His mother, Djuka Mandic, the eldest daughter in a family of seven children, was unschooled but possessed a prodigious memory. She was able to recite entire volumes of European poetry from memory. Nikola, who was the fourth of five children, said that he inherited his photographic memory and inventive ability from his mother.

A tragedy that occurred when Nikola was five years old made a significant and lasting impression on the young boy. His older brother, Daniel, who was brilliant and idolized by his parents, died in an accident. The exact nature of the accident is unknown, but it might have involved either the family's Arabian horse or a fall down the cellar stairs. Regardless of the exact circumstances, Nikola later suffered from nightmares and hallucinations related to his brother's death.

As an adult, Nikola Tesla had an astounding array of phobias, ranging from a violent aversion to pearl earrings on a woman to the smell of camphor to touching the hair of another person. He counted steps when walking and calculated the volume of soup dishes, coffee cups, and pieces of food. If served peas or other countable food items, the number of pieces had to be divisible by three, or he could not eat them.

Tesla's father encouraged his son to become a priest, but the young man was far more interested in science. Starting in 1870, Tesla attended high school in Karlovac, where he completed the four-year program in three years. He later recalled dreaming of harnessing the power of great natural wonders such as waterfalls. In his autobiography, he said, "In the schoolroom, there were a few mechanical models which interested me and turned my attention to water turbines." After his teacher told him about

Chapter 14. The Greatest Inventor Who Ever Lived

Map showing Smiljan, Croatia (Google, Image Landsat/Copernicus, U.S. Department of State Geographer).

Niagara Falls, he said, "I pictured in my imagination a big wheel run by the Falls," and proclaimed to his uncle that one day "he would go to America and carry out this scheme."[3]

Tesla moved back to Smiljan in 1873, but contracted cholera and was seriously ill for nine months and near death several times before he recovered. In 1875, he enrolled at the Austrian Polytechnic Institute (now the Graz University of Technology). He worked tirelessly, later claiming to have studied from 3 a.m. to 11 p.m. seven days a week. He excelled during his first year at Graz, and the dean of the technical faculty wrote to Tesla's father, saying, "Your son is a star of first rank."[4] But by the end of his second year, Tesla had become addicted to gambling and lost his scholarship.[5] In his third year, he continued to gamble, losing but then recovering his tuition money. But he was unprepared for his examinations and never graduated from the university.[6]

Tesla moved to Maribor, Slovenia, where he worked as a draftsman. He continued to gamble, but after losing all of his money and craving another game, he later wrote, "I conquered my passion then and there." "I not only vanquished but tore it from my heart so as not to leave even a trace of desire."[7] He replaced his gambling passion with billiards and became an accomplished player.

His intensity and lack of success resulted in a nervous breakdown around March 1879.[8] During his breakdown, Tesla's senses became incredibly sensitive. "He could hear the ticking of a watch from three rooms away. A fly lighting on a table in his room caused a dull thud in his ear. A train whistle twenty miles distant made the chair on which he sat vibrate so strongly that the pain became unbearable."[9] For the rest of his life, Tesla's senses were far more acute than those of other people.[a]

After his recovery, in 1881, Tesla moved to Budapest, where he worked for the Central Telegraph Office and the Budapest Telephone Exchange. While in Budapest, Tesla conceived the idea for the induction motor. He later recalled the event:

> At that age, I knew entire books by heart, word for word. One of these was Goethe's "Faust." As I uttered Goethe's inspiring words the idea came like a flash of lightening and in an instant the truth was revealed. I drew with a stick on the sand, the diagram shown six years later in my address before the American Institute of Electrical Engineers, and my companion understood them perfectly.
>
> The images I saw were wonderfully sharp and clear and had the solidity of metal and stone, so much so that I told him, "See my motor here; watch me reverse it." I cannot begin to describe my emotions.[10]

Tesla moved to Paris in 1882 to work for the Continental Edison Company and then for the Edison Electric Light Company. He installed indoor incandescent lighting systems, and when his abilities became apparent, he was assigned to designing and building improved DC generators.

In 1884, Tesla's boss, Charles Batchelor, was recalled to the United States to manage the Edison Machine Works in New York City. Batchelor requested that Tesla join him there, and in June 1884, Tesla immigrated to the U.S.

Tesla's time working for Edison in the U.S. was filled with controversy. He was there for just six months, and the ultimate reason for leaving was either an unpaid bonus for improvements to DC generators or for arc-lighting innovations that were never implemented. Whatever the reason, Tesla quit in January 1885 and focused on patenting his arc-lighting and generator (dynamo) improvements.[b]

While discussing his patent applications with Lemuel Serrell, the same patent attorney used by Edison, Serrell suggested that Tesla present his ideas to two businessmen, Robert Lane and Benjamin Vail. The two men agreed to finance an arc-lighting manufacturing and utility company to be called Tesla Electric Light and Manufacturing. But once the lighting system in Rahway, New Jersey, was installed and began operating in 1886, Tesla was eased out of the company and ended up digging ditches in New York City for $2 a day to survive.

But these were not ordinary ditches. These ditches were for communication cables connecting the headquarters of the Western Union Telegraph Company with stock and commodity exchanges. Supervising the work was Alfred S. Brown, a Western Union superintendent. Brown talked with Tesla and realized that he was an extremely intelligent ditch-digger.

Brown introduced Tesla to Charles F. Peck, a New York attorney. Brown and Peck had experience in taking patents from paper to profits and, in April 1887, agreed to back Tesla by forming the Tesla Electric Company. Profits generated by the company would go ⅓ to Tesla, ⅓ to Brown and Peck, and ⅓ to fund future development.

(a) Today, such hypersensitivity is often associated with autism or ADHD.
(b) U.S. Patent 334,823—Commutator for dynamo-electric machines, issued January 26, 1886; U.S. Patents 335,786 and 335,787—Electric-Arc Lamp, issued February 9, 1886; U.S. Patents 336,961 and 336,962—Regulator for Dynamo-Electric Machines, issued March 2, 1886; U.S. Patent 359,748—Dynamo-Electric Machine, issued March 22, 1887.

Chapter 14. The Greatest Inventor Who Ever Lived

The Company laboratory was located at 89 Liberty Street in Manhattan, where Tesla worked on new types of AC motors, generators, and other electrical and mechanical devices.

To protect his inventions, Tesla applied for and was granted patents on his AC components and system. By 1891, he held forty such issued patents.

Before publicizing Tesla's motor, Brown and Peck had it evaluated to verify its performance. William A. Arnold, an American physicist and Cornell University professor, successfully tested the motor. Then he, along with Thomas Commerfield Martin, editor of *Electrical World* magazine, contacted the American Institute of Electrical Engineers (AIEE) to arrange a demonstration.

On May 16, 1888, Tesla revealed his AC induction motor to the world in a lecture that is viewed as a classic. In his presentation titled, "A New System of Alternating Current Motors and Transformers,"[c] he described the technology and its applications. One attendee, Dr. Bernard A. Behrend, a noted electrical engineer and inventor, wrote, "Not since the appearance of Faraday's 'Experimental Researches in Electricity' has a great experimental truth been voiced so simply and so clearly."[11]

George Westinghouse's engineers heard of Tesla's presentation and told their boss of the new AC motor. Westinghouse immediately arranged to meet Tesla at his Liberty Street laboratory. The two men understood each other and soon became friends. Unlike Edison, who would not even listen to Tesla's ideas about AC, Westinghouse was a fellow inventor in the field and worthy of respect. Westinghouse was anxious to license Tesla's AC patents and started negotiations.[12]

Some previous biographers have clouded the results of those negotiations. For example, the classic Tesla biography by John J. O'Neill, written in 1944, says of the first meeting between Tesla and Westinghouse: "So favorably impressed was Westinghouse that he decided to act quickly. The story was related to the author by Tesla.

'I will give you one million dollars cash for your alternating-current patents, plus royalty,' Westinghouse blurted at the startled Tesla. This tall, suave gentleman, however, gave no outward sign that he had almost been bowled over by surprise. 'If you will make the royalty one dollar per horsepower, I will accept the offer,' Tesla replied. 'A million cash, a dollar a horsepower royalty,' Westinghouse repeated. 'That is acceptable,' said Tesla. 'Sold,' said Westinghouse, 'you will receive a check and a contract in a few days.'"[13]

O'Neill is not to blame; Tesla was known to exaggerate at times.

Based on Westinghouse documents from the Detre Library and Archives Division of the John Heinz History Center in Pittsburgh, this is what actually transpired.

- Monday, May 21, 1888: Just five days after Tesla's landmark presentation, Brown and Peck met with H.H. Byllesby, General Manager of Westinghouse Electric. The group then met with Tesla at his laboratory, and Byllesby found him to be "straight-forward, enthusiastic." The motors worked as expected and could be reversed without short-circuiting. In discussions without Tesla present, Brown and Peck claimed that a Mr. Butterworth from San Francisco had proposed a bid of $200,000 plus a royalty of $2.50 per horsepower. Byllesby wrote to George Westinghouse as follows:

---

[c] The lecture has been preserved and is available as a booklet on Amazon, https://www.amazon.com/System-Alternating-Current-Motors-Transformers/dp/1934451789.

"The terms, of course, are monstrous; and I so told them; and they replied that they could not possibly hold the matter over longer than the date mentioned" (10:00 a.m. Friday, May 25). "I told them I thought there was no possibility of our considering the matter seriously, but that I would let them know before Friday."[14]

- Undated, but written before July 5, 1888: In a draft agreement apparently prepared by Peck, he proposed the following terms.
  - Sixty days option (no price mentioned), during which Westinghouse Electric Company will construct one ten horsepower and one twenty horsepower motor under plans and specifications to be furnished by the Tesla Electric Company and with Nikola Tesla supervising the testing of the motors
  - At option period expiration, if WEC elects to purchase, WEC shall pay
    - $10,000 in cash
    - $60,000 total in three payments spaced six months apart
    - A royalty of $2.50 per horsepower sold during the life of the patents
- Thursday, July 5, 1888: In a letter from Westinghouse to one of his attorneys, T.B. Kerr, he wrote, "If the Tesla patents are broad enough to control the alternating motor business, then the figures that have been named by Mr. Byllesby, viz, Five Thousand Dollars for a sixty days option, and if the option is accepted, Ten Thousand Dollars in cash and Sixty Thousand Dollars in deferred payments, are not unreasonable; in fact, The Westinghouse Electric Company cannot afford to have others own the patents that are necessary to enable it to make motors to work on the alternating current system. And believing as I do, that the Tesla patents will be a valuable addition to our property, I sincerely hope that you and Mr. Byllesby will come to a definite understanding Friday that will enable us to announce that we have secured a successful alternating current motor, which with our meter will materially advance our interests."[15]
- Friday, July 6, 1888: In a letter to George Westinghouse, most likely from Byllesby, the writer states, "As per telegram of this evening we have substantially closed the agreement of option of purchase of the Tesla matter." Then he lists changes from the previous terms:
  - Option for ninety days instead of sixty days
  - Cash payment upon acceptance of option to buy increased to $20,000 from $10,000
- The writer added, "(W)e brought Mr. Shallenberger with us and after an interview which he had with Mr. Tesla he was strong in his advice to secure the patents including the other ideas and devices which Mr. Tesla revealed to him."[16]
- Saturday, July 7, 1888: Apparently, the parties had not fully agreed on the terms of an agreement. On Saturday, they came to a meeting with slightly different versions, as summarized in the table.
- Monday, July 9, 1888: Kerr wrote to Westinghouse and said, "Enclosed herewith I send your copies of the papers executed between the Tesla Electric Co. and The Westinghouse Electric Co. Besides these papers it was agreed that Mr. Tesla should come to Pittsburgh and superintend the construction of the motors here. I understand from Mr. Byllesby and Mr. Shallenberger that

Proposed Terms for Sale of Tesla Patents as of July 7, 1888

	Tesla Proposal	Westinghouse Proposal
Option		
Duration of Option Period	None	90 days
Price for Option	N/A	$5,000
Payment for Patents		
Initial Cash	$25,000	$20,000
Installments at six month intervals	3 of $16,667	3 of $16,667
Per horsepower royalty	$2.50	$2.50

these motors can be built ready for testing within forty days, so we have ample time in the ninety days provided, to settle all questions of efficiency, cost, title to the patents, and other matters of importance.
- Mr. Tesla agrees to come as an expert for the company if the purchase is completed, when called upon from time to time, during the contract, at the rate of $10 a day and expenses."[17]
- Other letters and agreements followed regarding Tesla's stock in the Tesla Motor Company[18] and the Tesla Light and Manufacturing Company.[19]
- Friday, August 2, 1889: Westinghouse Electric signed an agreement to purchase patent rights to past and future work of Tesla during his time at Westinghouse Electric in Pittsburgh for 150 shares of Westinghouse Electric stock.[20]

As a result of selling his patent rights, Tesla became a rich man, at least for a while. The royalty agreement ($2.50 per horsepower) proved to be an albatross around the neck[d] of Westinghouse Electric. O'Neill reported that Tesla voluntarily tore up the hard-won contract to save the Westinghouse Electric Company. Here is a portion of what O'Neill wrote, again based on what Tesla told him: "Mr. Westinghouse," said Tesla, drawing himself up to his full height of six feet two inches and beaming down on the Pittsburgh magnate who was himself a big man, "you have been my friend, you believed in me when others had no faith; you were brave enough to go ahead and pay me a million dollars when others lacked courage; you supported me when even your own engineers lacked vision to see the big things ahead that you and I saw; you have stood by me as a friend. The benefits that will come to civilization from my polyphase system mean more to me than the money involved. Mr. Westinghouse, you will save your company so that you can develop my inventions. Here is your contract and here is my contract—I will tear both of them to pieces and you will no longer have any troubles from my royalties. Is that sufficient?"[21]

Documents in Westinghouse company files tell a different story. "We have selected from our Westinghouse Historical collection information on Nikola Tesla and find that we have the copy of the Agreement between George Westinghouse and Mr. Tesla concerning the patent rights for his invention. There is no indication from these papers that it was destroyed as Mr. O'Neill in his book, *Prodigal Genius*, has indicated."[22]

(d) From "*The Rime of the Ancient Mariner*" by Samuel Taylor Coleridge.

What Did He Buy?

What is an induction motor, and why was it so valuable? To answer those questions, consider motors in general.

The function of any motor is to move one thing relative to another. The vast majority of motors move that one thing in a circular path relative to the stationary surroundings. The portion of the motor that is stationary is called the stator (as in "stationary"). The part that moves is called the rotor (as in "rotate"). In all motors before Tesla's work, both the stator and the rotor had to be supplied with electricity. Because the rotor is moving (rotating), providing it with electricity required either slip rings (continuous bands of metal contacted by metal or carbon brushes) or a commutator (segmented band of metal usually contacted by carbon brushes).

The brushes had to be in electrical and physical contact with the rotating slip rings or commutator, so wear and associated maintenance were constant problems. Also, especially with the commutator, sparking at the contact point was a problem, both accelerating wear and creating a fire or explosion hazard.

Many people tried to develop an AC motor that would eliminate the need for supplying electricity to the rotor by physical contact. Among these researchers were Galileo Ferraris, Guido Pantaleoni's professor at the University of Turin; O.B. Shallenberger, Westinghouse's electrical engineer; and Nikola Tesla.

Tesla had conceived his design in Budapest in 1882; filed a patent on the idea on October 12, 1887; received that patent on May 1, 1888; and described and demonstrated the invention to the AIEE on May 16, 1888. In legal terms, Tesla's invention had priority over all other inventors.

As specified in the agreement transferring his patents to Westinghouse, in late 1889, Nikola Tesla moved temporarily to Pittsburgh. He and George Westinghouse now greatly respected each other, so George invited Nikola to live at Solitude. That arrangement provided an opportunity for the two electrical geniuses to discuss the possible future of the electrical industry, as well as play poker either at Solitude or at Frick's Clayton.

But after a few months, the Westinghouse servants grew weary and suspicious of Tesla's strange, compulsive behavior, and Marguerite had to find him an apartment nearby.[23]

Tesla was to work with the Westinghouse engineers on commercializing his

Left to right: Slip rings, commutator, carbon brushes (photographs by author).

induction motors, but they had vastly different working styles. Where the engineers drew precise plans containing all details required to build their conceptions, Tesla scrawled a few lines on a scrap of paper and expected the machinists to produce a functioning result. He could envision in his mind all of the details and expected others to match his gift.

Furthermore, the Westinghouse engineers had designed their electrical equipment (generators and transformers) to operate at a frequency of 133 cycles per second. But Tesla's motors were designed for 60 cycle operation. The engineers resented this foreign dandy with a strong accent, telling them to redesign what they had worked so hard to create. Tesla prevailed, and when the frequency was reduced to 60 cycles, the induction motors started to fulfill their promise.

The induction motor, as invented by Tesla and improved by him working with Westinghouse's engineers, proved superior to other motor designs in several areas:

- Simplicity—The only moving part is the rotor, and nothing is in contact with the rotor except the supporting structure. There are no brushes, slip rings, or commutator to wear out.
- Low cost—Because of their simplicity, induction motors are less expensive to manufacture and maintain.
- No sparks—Because there are no brushes to cause sparks, induction motors could be used in hazardous or explosive environments.

Tesla's induction motor, 1888 (Westinghouse Collection, Detre Library and Archives, John Heinz History Center, Pittsburgh, PA).

- Self-starting—Three-phase induction motors self-start when power is applied.
- Reversible—Most induction motors can be reversed by changing wire connections.
- Variable-speed—The speed of an induction motor is determined by its structure and the frequency of the incoming power. Until the advent of solid-state circuits, varying the frequency was quite difficult, so lack of speed control was viewed as a disadvantage. But frequency, and therefore motor speed, can now be easily controlled.[24]

Because of these advantages, about 70 percent of machines in industry are driven by three-phase induction motors.[25]

How Does It Work?

If the rotor is not supplied with electricity by slip rings or a commutator (i.e., by physical contact), how can the motor work? The answer goes back to the transformer we discussed earlier in conjunction with increasing or decreasing AC voltage. Just as current flowing in the primary of a transformer induces a voltage in the secondary of the transformer, current flow in the stator windings induces a voltage in the rotor windings. The conceptual details are described in Appendix IV—How Does an Induction Motor Work?

More Battles in the Current Wars

With rights secured to Tesla's inventions, especially his induction motor, and with Shallenberger's AC meter, Westinghouse now had all the pieces he needed to build, sell, and install his AC transmission systems. But Thomas Edison had other plans.

Chapter 15

The Greatest Experimenter Who Ever Lived

"I have not failed. I've just found 10,000 ways that won't work."
— *Thomas Edison*

George Westinghouse invented things because he needed them for his businesses. Nikola Tesla invented things for the joy of inventing. Thomas Edison was an inveterate tinkerer whose experiments led to 1,093 patents, about one for every eleven days of his adult life. At the time of his death in 1931, Edison was the most prolific inventor in the world,[a] although he viewed himself not as an inventor but as a "perfector." He did not look for problems in need of solutions, but rather solutions in need of improvements.[1]

His parents, Samuel Ogden Edison, Jr., and Nancy Elliott, were married in Vienna, Ontario, Canada in 1828, when Sam was 24, and Nancy was 18. Sam was a hotel proprietor in Vienna when he joined the Mackenzie insurgents in the Rebellion of 1837.[b]

When the rebels lost, Sam and Nancy and their four children fled to the United States, where they settled in Milan, Ohio, in 1842. Three more children were born in Milan, the last of them was Thomas Alva Edison on February 11, 1847.[2]

"Al," as he was known until he was 39, was seven years old when the family moved to Port Huron, Michigan. In Port Huron, Edison attended school briefly but, although bright, did poorly under the traditional teaching methods. Modern-day historians and medical professionals have surmised that Edison may have had Attention-Deficiency/Hyperactivity Disorder or ADHD.[3] A simpler explanation is that Edison suffered from poor hearing from an early age, likely caused by scarlet fever and untreated middle-ear infections. Edison's mother, Nancy, was a schoolteacher and removed him from public schools so she could home-school him. He also learned by reading, which became his lifelong passion.

Edison, the Entrepreneur

In 1859, the Grand Trunk Railroad added a station in Port Huron, connecting the town to Detroit, 63 miles away. Twelve-year-old Al got a job selling newspapers

---

(a) In more recent times, many inventors have accumulated more U.S patents than Edison. For example, Shunpei Yamazaki holds at least 2,591 U.S. patents and over 9,700 worldwide patents.
(b) See https://www.thecanadianencyclopedia.ca/en/article/rebellions-of-1837.

and magazines on the train, traveling the four hours to Detroit starting at 7 a.m. each morning. To fill his time in Detroit while waiting for the evening return trip, he joined the Detroit Young Men's Society and read science books in the Society library.[4] Edison decided to print his own newspaper, so he bought a portable letterpress and published the two pages of his *Grand Trunk Herald* from the baggage car of the train. He then hawked his papers to passengers as he walked through the train cars.

One day in August 1862, in Mt. Clemens, Michigan, Al saw Jimmie, the two-year-old son of the stationmaster, James Mackenzie, playing on the train tracks. Edison managed to save the boy from being struck by a train, and the grateful father taught him Morse code as a reward.[5] By 1863, sixteen-year-old Thomas was proficient enough to become an apprentice telegrapher.

Edison worked for the Western Union Telegraph Company and traveled throughout the Midwest before moving to Boston in 1868. His particular skill was sending and receiving news reports, and he made friends with reporters who would prove useful in promoting him and his inventions in later years.

Edison, the Inventor

On June 1, 1869, Edison received his first patent[c] for an electronic vote-tallying machine. He soon left his job as a telegrapher to move to New York and devote himself full-time to his inventing career. In October 1871, 24-year-old Edison met Mary Jane Stillwell, a 16-year-old employee at his News Reporting Telegraph Company. The couple wed on Christmas Day, 1871, in Newark, New Jersey.[6] Their first child, daughter Marion Estelle, was born soon afterward.[d] Edison gave her a nickname, "Dot," from the Morse code character.

Based on his knowledge of telegraphy, in 1874, Edison invented the "quadruplex" system that permitted four telegraph signals, two in each direction, to be sent simultaneously on a single telegraph line.

His former employer, Western Union, bid for the rights to his invention, but they were slow in finalizing their offer. Jay Gould, the ruthless financier, wanted to challenge Western Union's control of long-distance telegraphy, and Edison's quadruplex system was vital to his plan. Late in 1874, Gould offered $30,000,[e] and Edison quickly agreed.[7] Edison used the money to buy about 34 acres in rural New Jersey from the family of one of his employees, William Carman. A former real estate office became home for Thomas, his wife Mary, their daughter Marion, and their infant son, Thomas Jr., who had been born on January 10, 1876, in Newark, New Jersey. Thomas Jr. was soon called "Dash," from the other Morse code character.

Edison brought his now-widowed father[f] to build a 2½ story laboratory and workshop, glass blowing shop, carpenter's shop, and blacksmith's shop on the property that became known as Menlo Park.[8] In March 1876, Edison moved into the new facility, along with two key associates, Charles Batchelor and John Kruesi.

(**c**) U.S. Patent 90,646; "Improvement in electrographic vote-recorder."
(**d**) Various sources list dates from February 18, 1872, to February 18, 1873.
(**e**) $30,000 in 1874 would have the purchasing power of $645,000 in 2019. See https://www.in2013dollars.com/us/inflation/1874?amount=30000.
(**f**) Edison's mother, Nancy, died on April 9, 1871, in Port Huron, Michigan.

Chapter 15. The Greatest Experimenter Who Ever Lived

Unlike Westinghouse, who created his own detailed drawings of his ideas, Edison could not reduce his concepts to paper. Charles Batchelor performed that critical step so that Swiss-born master-machinist Kruesi could build physical models. Ten years later, Batchelor would become Nikola Tesla's supervisor at the Edison Electric Light Company in Paris and would bring Tesla to the United States.

Menlo Park became Edison's Invention Factory. It was the first facility established for the express purpose of producing constant technological innovation and improvement. The top floor was for chemistry experiments, and the main floor was a fully-equipped machine shop. Edison said he wanted the lab to have "a stock of almost every conceivable material."[9]

In his 1999 study of technology, Seth Shulman cites an 1887 newspaper article revealing the seriousness of Edison's desire, stating the lab contained "eight thousand kinds of chemicals, every kind of screw made, every size of needle, every kind of cord or wire, hair of humans, horses, hogs, cows, rabbits, goats, minx, camels ... silk in every texture, cocoons, various kinds of hoofs, shark's teeth, deer horns, tortoise shell ... cork, resin, varnish and oil, ostrich feathers, a peacock's tail, jet, amber, rubber, all ores...."[10]

Soon after moving into Menlo Park, Edison focused on improving the microphone for telephones. Alexander Graham Bell had patented the telephone on March 7, 1876,(g) and many inventors sought to improve Bell's disclosure. Edison (or his workers) succeeded in enhancing Bell's loose-contact ground carbon microphone by roasting the carbon. Bell Telephone Company introduced a microphone using Edison's roasted carbon invention, and it was the standard transmitter in all telephones from 1890 until 1980.[11]

Edison's next significant invention, and the one that earned him the title, "Wizard of Menlo Park," resulted from his search for a machine to inscribe telegraphic messages as indentations on a paper tape for later sending. He soon

Edison (age 31) and his phonograph—photograph taken in Mathew Brady's studio in Washington, D.C., in April 1878.

(g) U.S. Patent 174,465; "Improvement in telegraphy." Disputes over inventorship of the telephone arose immediately, with Elisha Gray in the U.S. being the primary opponent.

realized that a similar device could record sound. His first version of what would become the phonograph used a diaphragm attached to a needle that traced sound waves onto tinfoil wrapped around a wooden cylinder. When another needle retraced the wave pattern, the original sounds were reproduced, albeit with poor fidelity. Upon seeing Edison's invention, Joseph Henry,[h] the first Secretary of the Smithsonian Institution and President of the National Academy of Sciences, called Edison "the most ingenious inventor in this country ... or in any other."[12]

Edison obtained a patent for the phonograph in 1878,[i] but, contrary to his usual practice of improving existing inventions, he initially did little to develop it. Even in its original and primitive form, the phonograph amazed all who saw and heard it. When Edison demonstrated the machine to President Rutherford B. Hayes in April 1878, Hayes and his wife recorded and played back their voices until 3 a.m.[13]

Alexander Graham Bell and his cousin, Chichester Bell, along with Charles Tainter, made significant improvements to Edison's phonograph, in particular, switching to a wax cylinder in place of Edison's fragile tinfoil, which could be played only a few times. The work of Bell and Tainter culminated in a patent issued on May 4, 1886.[j] After patenting his incandescent lamp in 1887, Edison resumed work on the phonograph based on the work of the Bells and Tainter. On October 8, 1887, he formed the Edison Phonograph Company to market his machine and wax cylinders.[14]

Life Inside the Invention Factory

From his days as a telegrapher, Edison developed a rapport with newspapermen that served him well in telling the world of the inventions emerging from Menlo Park. A newspaper report also gives us a view inside the Invention Factory. A January 17, 1879, article in the *New York Herald*, which was cited in Edison's Electric Light: The Art of Invention,[15] provides this description:

> The ordinary rules of industry seem to be reversed at Menlo Park. Edison and his numerous assistants turn night into day and day into night. At six o'clock in the evening the machinists and electricians assemble in the laboratory. Edison is already present, attired in a suit of blue flannel, with hair uncombed and straggling over his eyes, a silk handkerchief around his neck, his hands and face somewhat begrimed and his whole air that of a man with a purpose and indifferent to everything save that purpose. By a quarter past six the quiet laboratory has transformed into a hive of industry. The hum of machinery drowns all other sounds and each man is at his particular post." ... "Every man seems to be engaged at something different from that occupying the attention of his fellow workman. Edison himself flits about, first to one bench, then to another, examining here, instructing there: at one place drawing out new fancied designs, at another earnestly watching the progress of some experiment. Sometimes he hastily leaves the busy throng of workmen and for an hour or more is seen by no one. Where he is the general body of assistants do not know or ask, but his few principal men are aware that in a quiet corner upstairs in the old workshop, with a single light to dispel the darkness around, sits the inventor, with pencil and paper, drawing, figuring, pondering. In these moments he is rarely disturbed.

(**h**) The unit of inductance is named the Henry, in his honor.
(**i**) U.S. Patent 203,018; "Telephones or speaking-telegraphs," Filed December 13, 1877; Issued April 30, 1878.
(**j**) U.S. Patent 341,214; "Recording and reproducing speech and other sounds," Issued May 4, 1886, to C. A. Bell and S. Tainter.

The Incandescent Lamp

Who invented the light bulb? In 1838, Marcellin Jobard invented the first incandescent light bulb, using a carbon filament with a vacuum atmosphere. Or was it Frederick de Moleyns of England, who was granted the first patent for an incandescent lamp in 1841, with a design using platinum wires contained within a vacuum bulb? He also used carbon as a filament. Or maybe it was Moses Farmer, who built a platinum filament incandescent light in 1859. Later that year, while living in Salem, Massachusetts, he lit the parlor of his home at 11 Pearl Street with incandescent lamps, the first house in the world to be lit by electricity.[16] Friedel and Israel[17] list 23 inventors of incandescent lamps before Thomas Edison, including Joseph Swan, a British physicist and chemist whose work paralleled Edison's in time and content.

Edison joined the quest for a practical incandescent light bulb on September 15, 1878. On that day, he announced to the usual group of reporters visiting Menlo Park that he had solved the problem that had baffled all other investigators—a practical incandescent light. Oh, and by the way, he would also develop a new industry to distribute electric power to run machines, trains, and all of those unique light bulbs. He claimed that he would do all of this in six weeks! Despite having nothing to demonstrate, the Wizard of Menlo Park had spoken.

On September 16, the *New York Sun* proclaimed, "Mr. Edison says that he has discovered how to make electricity a cheap and practicable substitute for illuminating gas. Many scientific men have worked assiduously in that direction, but with little success." "It has been reserved for Mr. Edison to solve the difficult problem desired. This, he says, he has done within a few days. His experience with the telephone, however, has taught him to be cautious, and he is exerting himself to protect the new scientific marvel, which, he says, will make the use of gas for illumination a thing of the past."

Edison was quoted as saying, "When the brilliancy and cheapness of the lights are made known to be public—which will be in a few weeks, or just as soon as I can thoroughly protect the process—illumination by carbureted hydrogen gas will be discarded."[18]

The next day, stocks in natural gas companies, the purveyors of ubiquitous and dangerous gas lights, tumbled. Investors, including the influential J. Pierpont Morgan, rushed to invest in Edison's ideas.[19]

By 1878, other researchers had determined that a vacuum or neutral gas environment was essential to prevent oxidation of the filament material. A sealed and evacuated glass bulb was the obvious structure to provide a transparent, gas-tight carrier. Direct current from batteries was readily available as the source of power. The remaining unknowns were the filament material and how to attach the filament to the wires providing the electricity. Edison and his workers tried some 3,000 different filament candidates of both organic and inorganic origin, such as carbon and platinum.

Edison filed his first patent in the field on October 14, 1878.[k] The description disclosed a platinum filament and a shunt that bypassed the lamp when the filament burned out. Arc lamps, which were common at that time, were connected in series. When one lamp burned out, the entire string went dark. Edison's initial patent

---

(**k**) U.S. Patent 214,636; "Improvement in electric lights;" Issued April 22, 1879.

assumed this same series connection, but the shunt allowed other lamps in the string to remain lit when one burned out. The same shunt idea is used in Christmas light strings today.

The day after filing his first lighting patent, on October 15, 1878, Thomas Edison began operation of the Edison Electric Light Company. The company was to provide financial support for electric light research and experiments in return for control of the resulting patents.[20] Capitalized at $300,000, investors in the Edison Electric Light Company included William H. Vanderbilt and J. Pierpont Morgan. There was just one problem—Edison did not have a workable incandescent light. The quest for that holy grail would take another year, a year that stretched the patience of Edison's investors to the breaking point.[21]

Eleven days after forming the Edison Electric Light Company, on October 26, 1878, Edison's wife, Mary, presented him with their third child, William Leslie Edison. Thomas continued his practice of remaining distant from his children and had little direct contact with any of them. He preferred to work in his laboratory and did so for long and late hours.

Platinum was too expensive for use in filaments, so Edison returned to carbon as his filament material of choice. But what kind of carbon was the best? Different materials were oxidized to form carbon and tried as filaments. The first success was on October 22, 1879, when a filament remained illuminated for 13.5 hours. Edison filed a patent application[(I)] on November 4, describing "a carbon filament or strip coiled and connected … to platina contact wires." The disclosure said that the filament could be made of "cotton and linen thread, wood splints, papers coiled in various ways." As this patent plays a significant role in the upcoming battles with Westinghouse, the specific claims were critical. The patent had four claims; the first two were relevant:

    1. An electric lamp for giving light by incandescence, consisting of a filament of carbon of high resistance, made as described, and secured to metallic wires, as set forth.

    2. The combination of carbon filaments with a receiver made entirely of glass and conductors passing through the glass, and from which receiver the air is exhausted, for the purposes set forth.[22]

The "high resistance" specified in Claim 1 makes it clear that this light was intended for a parallel connection, so the failure of one light would have no effect on other lights in the circuit. The Claim 2 requirement, "entirely of glass," would provide the difference that Westinghouse would later rely on so he could produce his two-piece "stopper" lamps based on the Sawyer-Man patents.

On the last day of 1879, Edison invited the public to a demonstration of his light bulbs which would illuminate Menlo Park. A crowd of 3,000 people arrived by carriage, train, and on foot. They were amazed to see the buildings and streets remain lighted despite wind and rain, and the lights could be turned on or off without having to visit each fixture, as was the case with gas lamps.

One interested visitor to the lighting display at Menlo Park was Henry Villard, president of the Oregon Railroad and Navigation Company. Villard wanted Edison to install electric lights on his company's new steamship, the *Columbia*. Edison agreed,

---

(I) U.S. Patent 223,898; "Electric lamp"; Issued January 27, 1880.

and he and his workers installed the lighting system in April 1880, thereby creating the first commercial application for his new lights. Soon after, continuing work at Menlo Park pointed to the use of carbonized bamboo for lamp filaments, resulting in bulb lifetimes of 1,200 hours.

On December 17, 1880, Edison formed an electric utility, the Edison Electric Illuminating Company (or Edison Illuminating Company), to compete with existing gas lighting utilities. Oregon Railroad and Navigation Company president Henry Villard and J. Pierpont Morgan were significant investors. Edison retained a large number of shares for himself and named his personal attorney, Sherbourne Eaton, as president. Eaton and Villard shared tactical control, but it was Morgan who made the critical decisions.[23]

Pearl Street Station

In 1881, the Edison Illuminating Company started construction of the first commercial power plant in the U.S. at 255–257 Pearl Street in New York City to serve nearby residents and businesses. Edison chose the Pearl Street location "because I thought the property could be purchased for about $10,000 a lot as it was in a slum district. I wanted a plot 50 ft. × 100 ft. and was stunned when they quoted $75,000 a lot," Edison told L.W.W. Morrow in a 1922 interview.[24] It was no accident that the area Edison chose to serve included the offices of several influential newspapers, including the *New York Times*. Edison sent surveyors to the area around the proposed plant location to determine the number of gaslights burning every hour on each customer's premises. They found about 1,500 potential customers using 18,043 gas jets.[25] Based on this information, Edison could predict the demand for electricity from his plant. He also examined the financial reports for the gas companies, and "(K)new I was up against stiff competition, so I used all available labor-saving devices for handling coal and ashes at the plant."[26]

Edison decided on underground distribution for the wires, as shown in the drawing. He later said, "I adopted the underground system of distribution because I knew that the public never would approve of a wilderness of overhead wires and also

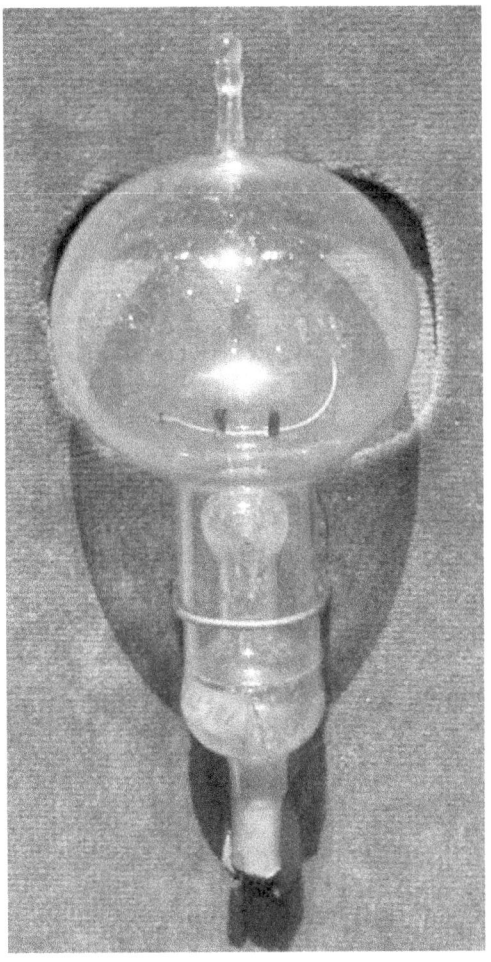

An original carbon filament light bulb from 1879 from Thomas Edison's shop in Menlo Park (Terren, Wikimedia Commons).

because the electrical system, so similar to gas lighting, could only be made reliable through underground distribution."²⁷

Edison personally supervised the installation of the underground cables and had to overcome obstacles erected by the arc lighting and gas companies, his competitors. Of course, the government added impediments in the form of city inspectors. Edison said he had to pay five inspectors $5 per day each, but "None of those inspectors ever appeared to inspect, but they all appeared, however, at the office each Saturday to get their pay."

Finally, it was time. Edison's anxious board of directors met in J.P. Morgan's office at 3:00 p.m. on September 4, 1882, to watch the throw of a switch that would start the Pearl Street Station. Someone bet Edison $100 that the system would fail. "Taken," said Edison as he pulled the lever.

The system worked as Edison said it would.²⁸ Pearl Street came on-line and supplied 110-volt DC electricity to 59 customers with 400 light bulbs. By the end of 1882, the Pearl Street Station supported 231 customers with 2,323 lamps. It is interesting to note that one of those attending the activation of Pearl Street was H.H. Byllesby,²⁹ who four years later would become the general manager of the Westinghouse Electric Company.

The *New York Times* reported, "Edison's central station, at No. 257 Pearl street,

Illustration from the June 21, 1882, *Harper's Weekly* titled "The Electric Light In Houses: Laying The Tubes For Wires In The Streets Of New York." The illustration records the construction of Thomas Edison's first domestic incandescent lamp electric utility in New York City with the central station at Pearl Street. Edison's company chose to bury its power lines under the streets instead of mounting them on poles (Wikimedia Commons).

Edison's 1,500 light, 110-volt "Jumbo" steam dynamo with 150 horsepower Porter-Allen engine at Pearl Street Station (Fae, Wikimedia Commons). Plaque Commemorating the Pearl Street Station (courtesy of Bethel Bilezikian Charkouduian).

was yesterday one of the busiest places down town, and Mr. Edison was by far the busiest man in the station. The giant dynamos were started up at 3 o'clock in the afternoon, and, according to Mr. Edison, they will go on forever unless stopped by an earthquake." ... "Yesterday for the first time THE TIMES Building was illuminated by electricity. Mr. Edison had at last perfected his incandescent light, had put his machinery in order, and had started up his engines, and last evening his company lighted up about one-third of the lower City district in which THE TIMES building stands."

"It was about 5 o'clock yesterday afternoon when the lights were put in operation. It was then broad daylight, and the light looked dim. It was not till about 7 o'clock, when it began to grow dark, that the electric light really made itself known and showed how bright and steady it is. Then the 27 electric lamps in the editorial rooms and the 25 lamps in the counting rooms made those departments as bright as day, but without an unpleasant glare. It was a light that a man could sit down under and write for hours without the consciousness of having any artificial light about him."[30]

As discussed previously, because they produced DC, central generating stations such as Pearl Street could serve only customers within a half-mile radius. For a city like New York, that would require a checkerboard of polluting power stations. For sparsely populated areas, such stations would be uneconomical.

Lighting Morgan's House

In the late spring of 1882, as work on the Pearl Street Station was underway, J.P. Morgan decided to invest more than money in Edison's inventions. He would have Edison install electric lights in his 219 Madison Avenue mansion, which was then undergoing a complete renovation. Morgan would show the world that he embraced innovation while promoting what he expected would become a hugely profitable investment. In addition to wiring the entire house and installing lighting fixtures in every room, jobs comparable to what they had done on Henry Villard's steamship, Edison and his crews would have to build a self-contained generating station on Morgan's property. The station, complete with a coal-fired boiler, steam engine, and two electric generators, was to go under the horse stable. One crew excavated a dirt-lined cellar to hold the power equipment while other teams ran wires inside the walls of the house and out through holes where yet other workers installed fixtures. Wires in a shallow, bricked-over trench connected the power source under the stable to a distribution panel in the house. Morgan's elegant mansion must have looked like an anthill, with workers scurrying everywhere.

The grand experiment culminated on June 7, 1882. Surprisingly, neither Thomas Edison nor Edison Electric Light Company president Sherbourne Eaton attended, but the illumination of J.P. Morgan's home was a great success. Morgan and his wife basked in the glow of 385 amber-hued incandescent lights.

The problems of being a pioneer soon surfaced. The generator required the attention of an expert engineer who worked from 3:00 p.m. to 11:00 p.m. At the start of his shift, he lighted the coal-fired burner to create steam to run the steam engine. About an hour later, the residents of the house could enjoy their electric lights. But at 11 p.m., the lights would dim and go out as the generators wound down, and servants had to scramble to find not-so-modern candles to provide emergency lighting.

Morgan's neighbors were displeased with the loud rattle and clang of the machinery beneath the stable, which destroyed the previous silence of the genteel neighborhood. Next-door neighbor Mrs. James Brown claimed that the machines made her entire house vibrate, and the smoke and fumes from the burning coal caused her precious silver utensils to tarnish. Morgan, not often cowed, had to apologize and demanded that Edison solve the problems. Eventually, Edison's men, by extensive application of soundproofing and installation of resilient motor mounts, tamed the noise. The smoke was carried to the existing chimney on the house by way of another brick-covered trench across the yard.[31] Those on the cutting edge sometimes bleed.

But the future was here, and respected investors such as Darius Ogden Mills, California banker, and William H. Vanderbilt, son of Cornelius Vanderbilt and president of the New York Central Railroad, became major investors in Edison companies.

Patent Problems

On October 8, 1883, the U.S. Patent Office ruled that Edison's basic electric lamp patent (U.S. Patent 223,898; discussed above) was invalid because of prior work by William Sawyer. After six more years of litigation, a judge finally ruled that Claim 1, the one specifying "a filament of carbon of high resistance," was valid.

Joseph Swan, whose work on incandescent lamps was contemporaneous with Edison's, sued Edison for patent infringement and won. As a settlement, Edison agreed to form a joint company, Ediswan, to manufacture and market lamps in Great Britain.

More Central Power Stations

Roselle, New Jersey, was the first community served by overhead electrical wires. Because it was so expensive and disruptive to install underground wiring to serve the small population of Roselle, Edison chose wires on poles. The system began service on January 19, 1883, and remained in service for ten years. The Roselle installation was important because it illustrated that Edison's DC system could supply power to a relatively small town, but no similar facilities followed.

On August 9, 1884, Thomas Edison suffered a devastating blow. His wife, Mary Jane, died at the age of 29. Various sources speculate that she died of a brain tumor, a morphine overdose, or typhoid fever. Regardless of the cause, Edison, who had neglected his wife and often forgot birthdays and anniversaries, went into a deep depression. He could never return to Menlo Park, where he had lived with Mary. He had three children to raise, but he realized that he was not up to the task.

Fortunately, a friend, telephone switchboard inventor Ezra Gilliland, knew Lewis Miller, an inventor of farm machinery. Lewis Miller was instrumental in establishing Chautauqua, the retreat in western New York, that drew cultured, educated visitors. Also, Miller had a daughter, Mina, who had mingled with the Chautauqua crowd and was poised and educated herself. Gilliland introduced Edison to Lewis Miller, and Edison thus met Miller's daughter, Mina. Despite their age difference (Edison was 39, and Mina was 19), they were attracted to each other. Edison's hearing had declined so much that he was virtually deaf, so he taught Mina Morse code so they could communicate more easily.

He asked Mina to marry him by sending a coded message by tapping on her foot, and she responded by coding, "Yes." They married on February 24, 1886, in Mina's home town of Akron, Ohio. Mina suddenly became the mistress of a 13-acre estate, including a 23-room mansion called Glenmont in West Orange, New Jersey, and step-mother to Edison's three children.[32] Mina was just seven years older than the eldest child, Marion. Marion said Mina was "too young to be a mother but too old to be a chum."[33]

Adjacent to Glenmont was a new laboratory for Edison and his associates to continue their tinkering and experimenting. Preserved today as the Thomas Edison National Historical Park, the 300,000 artifacts and five-million-page document archive make it the third-largest museum collection in the National Park Service.[m] The documents include Edison's personal and business correspondence, laboratory notebooks, legal files, patent records, engineering drawings, manufacturing and financial records, advertising and sales material, payroll records, historical photographs, trade catalogs, sheet music, and the papers of Edison associates and family members.[34]

The West Orange laboratories were like their predecessor facility at Menlo Park, but far more extensive and better planned. Products developed at the West Orange labs included the motion picture camera, silent and sound movies, improved phonographs, sound recordings (the iconic wax cylinders), and the nickel-iron alkaline storage battery.[35] Edison worked at these labs from 1887 until his death on October 18, 1931, but in later years, he transitioned from inventor to business manager and industrialist.

Despite the distractions of a new wife and new laboratory, work on installing central power stations had to continue. By 1885, Edison had installed between fifty and sixty DC central power stations, which operated 24 hours per day year-round.[36]

Competition for Edison

By 1887, Edison had 121 DC central power stations installed in the U.S. Westinghouse had 68 AC stations built or under construction. But there was another company in the business—the Thomson-Houston Electric Company, under license from Westinghouse, had 22 AC stations completed or under construction.

Edwin Houston graduated from Central High School of Philadelphia (a degree-granting institution rather than an ordinary high school) in 1864 and then became a professor there.[37] One of his students, Elihu Thomson, followed the same career path, graduating from Central High in 1870 and also becoming a professor and then Chair of the Chemistry Department.[38] Together, Houston and Thomson developed a generator for arc-lighting applications and, with capital from a group of investors led by businessman Charles Coffin, formed the Thomson-Houston Electric Company in 1882. Many of the company's original investors were shoe manufacturers from Lynn, Massachusetts, so Thomson-Houston established headquarters there. With the growing importance of electricity in arc-lighting and industrial applications, Thomson-Houston flourished and reached sales of $10 million and employed 4,000 people in 1892.[39]

Edison was installing DC central power systems, and Westinghouse and Thomson-Houston were installing AC central power systems, but everyone knew that only one system could win. Westinghouse was confident that his AC system was

---

(m) Excellent pictures of the interior and exterior of Edison's West Orange laboratories are at https://artsandculture.google.com/exhibit/edison-s-west-orange-laboratory%C2%A0-thomas-edison-national-historical-park/agJyIGxTlBg6JA?hl=en.

economically and technically superior, so he was comfortable in allowing those factors to determine the victor. Edison, dealt an inferior hand but worshipped by the press and public, refused to concede and instead launched constant attacks on the AC system, using the ancient weapons of F-U-D, Fear, Uncertainty, and Doubt, to win the war.

Edison, Westinghouse, and Tesla

With 1,093 U.S. patents, Edison far surpasses Westinghouse's 361 and Tesla's 112. But, of course, the numbers do not tell the whole story.

Edison ran his laboratories as a micro-manager with lots of workers. Although he invented some brand-new technology, his focus was on improving existing products. Almost all patents that resulted from work at Menlo Park, West Orange, or any other Edison facility carried the boss's name, Thomas Edison, in addition to the names of the workers who made the invention.

Westinghouse did it differently. He encouraged his employees to develop creative solutions to the problems they encountered in their everyday work. Patents resulting from their efforts were issued in their own name, without including Westinghouse as one of the inventors.

Westinghouse had great respect for intellectual property and bought patent rights at a high cost, even when their utility was questionable. Tesla once said of Westinghouse, "He is one of those few men who conscientiously respect intellectual property, and who acquire their right to use inventions by fair and equitable means…. Had other industrial firms and manufacturers been as just and liberal as Mr. Westinghouse, I should have had many more of my inventions in use than I now have."[40]

Westinghouse also encouraged his employees to develop companies based on their inventions that were unrelated to Westinghouse businesses. A prime example is Hugh Rodman, founder and head of the Rodman Chemical Company of East Pittsburgh. Rodman relates what happened to him:

> For several years I was research engineer for the Machine Company, making such investigations as Mr. Westinghouse or the management directed, and, as a matter of course, turning over the results to the company. One investigation carried me to the case-hardening department, where, after considerable work, I developed patentable processes and materials which apparently had commercial value apart from the company's ordinary activities. These I reported as usual, and the question was raised as to who properly owned them. I held that, as the company was not interested in chemical manufacturing, it should retain only a working right to the processes, leaving me to patent them for my own benefit in other respects. The company argued that, its money and equipment having been used, the processes belonged to it. We appealed to Mr. Westinghouse as arbitrator. His decision was that, though the company might legally maintain its right to the inventions, he would make no move to do so, and he not only turned over to me the entire rights in the inventions, but offered me enough capital to erect and run a small factory, of which he left me in full control. I feel great satisfaction in adding that the investment proved worthwhile, and in bearing this witness to his fine generosity![41]

In contrast to both Edison and Westinghouse, Nikola Tesla was a lone-wolf inventor. His staff consisted of one or two technicians to help turn his concepts into hardware.

J. Pierpont Morgan

Edison was a partial owner in most of the power stations he constructed, and he also supplied their equipment. J. Pierpont Morgan wanted more of a share in this burgeoning business and invested almost two million dollars in Edison's manufacturing arm. Edison and Morgan renamed it the Edison General Electric Company, and Henry Villard became president.[42] Among all of Edison's investors, J. Pierpont Morgan held the largest share and carried the most influence, both with Edison and with other investors. As the only non-technical player in the War of the Currents, he is often neglected. But Morgan had a significant impact not only on Edison but on Westinghouse and Tesla as well. We shall consider him next.

CHAPTER 16

Bankers Always Win

"Giving debt relief to people that really need it, that's what foreclosure is."
—J. Pierpont Morgan

When J.P. Morgan died in 1913, John D. Rockefeller learned that Morgan's fortune was $80 million. Rockefeller, whose own fortune was nearly one billion dollars, commented, "And to think, he wasn't even a rich man."[1] Whatever your definition of rich, Morgan wielded an outsized influence on American commerce and government in the Gilded Age.[a]

John Pierpont (J.P.) Morgan was not born with a silver spoon in his mouth; it was a golden spoon. His parents were Junius Spencer Morgan, American financier and banker, and his wife, Juliet Pierpont, daughter of the poet, teacher, lawyer, merchant, and Unitarian minister, John Pierpont. Born in Hartford, Connecticut, on April 17, 1837, J.P. Morgan was educated at public and private schools in New England. In 1852, rheumatic fever interrupted his high school studies, and he convalesced in the Azores for almost a year. Morgan returned to the English High School of Boston, where he was graduated in 1854. His father then sent him to the Swiss village of La Tour-de-Peilz on Lake Geneva, where he attended Bellerive, an exclusive boy's school established in 1836. The next stop for Morgan's continental education was the University of Gottingen, Germany, where he studied the German language and art history. He would later apply his knowledge of art in acquiring a personal collection valued at $50 million at his death. The collection is now housed in the J.P. Morgan Library and Art Museum on Madison Avenue in New York City.

His formal education completed, Pierpont, as he preferred to be called, moved to London in 1857, and entered banking at his father's firm, Peabody, Morgan & Company. Returning to New York in 1858, he worked first as an unpaid intern at the merchant banking firm of Duncan, Sherman & Company, the American representatives of Peabody, Morgan.

Personal Life

Young Morgan was an ambitious Christian gentleman and made friends among the best New York families. He spent Sunday evenings at their homes, especially if

---

(a) The Gilded Age was the era during the late 19th century, from the 1870s to about 1900. The term for this period became popular in the 1920s and 1930s and was derived from writer Mark Twain's and Charles Dudley Warner's 1873 novel *The Gilded Age: A Tale of Today*, which satirized an era of serious social problems masked by a thin gold gilding.

J. Pierpont Morgan encountering a photographer.

they had attractive daughters, and sang hymns with them in front of the fire. He often wrote to his family, joined St. George's Church, and was a serious and capable worker at Duncan, Sherman.[2]

Morgan was physically imposing at 6'2" tall[(b)] with a muscular build and piercing eyes. As a young man, he was considered attractive, although he was shy. He suffered from acne rosacea, which caused redness in his face and ruptured blood vessels on his nose. The condition worsened when he reached middle age and caused rhinophyma, resulting in growths, lesions, and pockmarks on his nose.[3] In his comprehensive 2010 biography, Ron Chernow says, "The nose certainly contributed to an insecurity and lack of social ease that were thinly masked by a barking voice and tyrannical manner."[4]

Morgan carefully controlled photographs of himself and forced photographers to retouch their pictures to make his nose look less deformed. When walking in public, if someone had the temerity to try to take his picture, he would strike them with his cane and force them to cease.

The Love of His Life

Soon after arriving back in New York, Morgan met Amelia "Memie" Sturges, daughter of Jonathan Sturges, a well-known merchant and patron of the arts. In 1859, the Sturges family embarked on a grand tour of Europe, for which Morgan planned

---

(b) The average male height in the United States at that time was 5'7". See https://www.todayifoundout.com/index.php/2016/07/j-p-morgan-giant-nobbly-purple-nose/.

the itinerary. At the end of the Sturges' journey, Morgan met them in London and toured the city with Memie every day for two weeks. He joined the family on their cruise back across the Atlantic to New York.

In the spring of 1860, the couple agreed to marry, but by winter, Memie developed a severe and persistent cough. Morgan and Memie married on October 7, 1861, and embarked on a Mediterranean honeymoon. In Paris, a doctor diagnosed Memie's ailment as tuberculosis. She wrote to her mother, "I wish you could see his loving devoted care of me, he spares nothing for my comfort and improvement." Despite Pierpont's efforts and doctor's care, Memie died in Nice, France, on February 17, 1862, a little over four months after their wedding. At the age of 24, Pierpont was a crushed and grieving widower.[5]

Business Comes First

In 1861, Morgan formed his own company, and, with support from his London associates at Peabody, Morgan, he opened a small office. By 1864, he formed J. Pierpont Morgan & Co. His father, Junius, thought some of Pierpont's investments were too speculative, but he had raised an independent-thinking son.

Another Wife

On May 31, 1865, less than two months after the end of the Civil War and the subsequent assassination of Abraham Lincoln, Morgan married Frances Louisa "Fanny" Tracy. While Frances and Morgan traveled in the same social circles, they were mismatched from the start. Morgan loved New York City, hard work, a busy social life, travel, and luxury in art, houses, clothing, and yachts. Fanny preferred a quiet domestic life with her children and a few friends.[6] The workaholic Morgan never spoke to Fanny about business. Despite their differences, the couple had four children:

- Louisa Pierpont Morgan, who was devoted to her father and often traveled with him, was born on March 10, 1866;
- J. Pierpont Morgan, Jr., was born September 7, 1867. After his father died in 1913, he took over the business interests, including J.P. Morgan & Co.;
- Juliet Pierpont Morgan, born July 19, 1870; and
- Anne Tracy Morgan, who was born July 25, 1873, and became a philanthropist and financed relief efforts during and after the World Wars.

The Morgan's marriage faltered in the late 1870s. J.P. Morgan became a notorious womanizer, but they never divorced. Pierpont died on March 31, 1913, at the age of 75, and Fanny was 79 when she died on November 16, 1924.

Devoted to Business

As noted earlier, Morgan used the contacts he had made in London at Peabody, Morgan & Co. to bring British capital to cash-hungry U.S. firms such as railroads. In the process, he often gained control of the railroads which received the foreign funds by a process that became known as "Morganization." Because he used this tactic on Edison and Westinghouse, it deserves closer examination here.

What is "Morganization?"

Morganization was based on Morgan's core belief, shared with other robber barons such as Cornelius Vanderbilt and John D. Rockefeller, that cutthroat competition was wasteful, and combinations and size would reduce competition and increase efficiency. Morgan developed a strategy for creating monopolies without the attention-getting merger of large competitors. Morganization involved some or all of the following steps:

- Take control of small, underfinanced companies in the target industry, often by loaning them needed capital and then foreclosing when the loans could not be repaid;
- Buy other small competitors;
- Lower prices for goods and services until other competitors go bankrupt trying to compete;
- Buy the bankrupt former competitors, thus forming a monopoly; and
- Slash the workforce and lower wages, thus maximizing the monopoly's profits.[7]

Morgan first focused on the railroad industry, which was highly capital-intensive and competitive, two of the characteristics necessary for Morganization to work.

One of his first targets was the small Albany and Susquehanna Railroad in upstate New York, which passed through Schoharie Junction, near the birthplace of George Westinghouse. In 1869, rival financier Jay Gould was attempting to seize control of the line when Morgan stepped in on the side of the existing management. In a proxy battle, Morgan emerged as vice-president and Director of the A&S.

Through the 1880s, Morgan, financed by European money, guided the reorganization of the railroad industry to achieve improved efficiency and develop an integrated transportation system, i.e., one with less internal competition.

In 1885, he reorganized the New York, West Shore & Buffalo Railroad and leased it to the New York Central. Also, in 1885, Morgan met on his yacht, the *Corsair*, with the feuding directors of two of the largest railroads in the country, the New York Central and the Pennsylvania. The group sailed up and down the Hudson River, with Morgan refusing to dock until the executives reached a compromise that reduced wasteful competition. The agreement came to be called the *Corsair Compact*.[8]

In the last half of the 1880s, he re-financed or reorganized a multitude of railroads, including the Philadelphia & Reading in 1886, the Chesapeake & Ohio in 1888, and the Baltimore & Ohio in 1890. With each of these maneuvers, Morgan obtained stock in the railroads and membership on their boards of directors, thus greatly expanding his influence.

Henry Ford, never shy about expressing his opinions, once said the following about bankers and businesses: "And that is the danger of having bankers in business. They think solely in terms of money. They think of a factory as making money, not goods. They want to watch the money, not the efficiency of production.... Bankers play far too great a part in the conduct of industry. Most businessmen will privately admit that fact. They will seldom publicly admit it because they are afraid of their bankers. It required less skill to make a fortune dealing in money than dealing in production. The average successful banker is by no means so intelligent and resourceful a man as is the average successful businessman. Yet the banker through his control of credit practically controls the average businessman.... The banker is, as I have noted, by training and because of his position, totally unsuited to the conduct of industry."[9]

Later in their lives, Henry Ford and Thomas Edison became close personal friends and neighbors in Fort Myers, Florida. Ford and Edison, along with Harvey Firestone and other friends, often went on elaborate "picnics" together. Edison had encouraged Henry Ford to build his first gasoline-powered car, and Ford repaid this gesture by giving Edison "the first car off each Ford assembly line—the first Model T, the first Model A, the first V-8, the first Lincoln."[10] While Ford's animosity toward bankers came from his own experiences, it was likely intensified by Edison's interactions with J.P. Morgan.

New Industries to Control

By the end of 1887, Morgan was Edison's primary banker and had invested almost two million dollars in his electrical businesses. Three companies competed in the rapidly emerging electrical equipment and distribution arena: Edison General Electric Company; Westinghouse Electric Company; and the Thomson-Houston Electric Company. Morgan virtually controlled Edison's company but had no influence over the other two. Such competition ran counter to his instincts and made him quite uncomfortable. The next five years would see momentous changes to the electrical industry and to Morgan's comfort level.

Chapter 17

Dying for Electricity

"I am an expert on electricity. My father occupied the chair of applied electricity at the state prison."

—W.C. Fields

Edison Fights Back

Despite his head start in building central power stations, by 1889, Edison was losing ground to Westinghouse and Thomson-Houston. He had to do something, so he chose to attack the safety of AC power distribution. That November, he wrote an article that was published in *The North American Review*.

"There is no plea which will justify the use of high-tension and alternating currents, either in a scientific or a commercial sense. They are employed solely to reduce investment in copper wire and real estate." Edison argued against the prevailing wisdom that burying high-voltage wires would diminish their danger. He continued, "The public may rest absolutely assured that safety will not be secured by burying these wires. The condensation of moisture, the ingress of water, the dissolving influence of coal gas and air oxidation upon the various insulating compounds will result only in the transfer of deaths to manholes, houses, stores, and offices, through the agency of the telephone, the low-pressure systems, apparatus of the high-tension current itself." He concluded, "I have always consistently opposed high-tension and alternating systems of electric lighting."[1]

Edison held a losing hand with his DC system, and he knew it. But he had invested too much time, energy, and emotion to concede. Some writers suggest that Edison, who had almost no formal education, could not understand the complexities of AC. While that might be true, the combination of emotional attachment and future profits likely motivated him to defend his position, even to the point of violating his fundamental principles.

AC is Dangerous!

Unknown to Edison, George Lemuel Smith, a 30-year-old dockworker from Buffalo, New York, contributed his life to Edison's anti–AC campaign.

The Brush Electric Light Company had installed a dozen brilliant white arc lights in an industrial area of Buffalo and illuminated them every evening starting on July

13, 1881. Residents, amazed at the transformation of the black night into artificial daylight, flocked to see the spectacle. To gain publicity, the company allowed visitors to enter the power plant containing the steam engines and AC generators.

Smith and three friends viewed the lights and visited the power plant on the evening of August 7, 1881. They, like other spectators before them, linked hands, and those on the end of the human chain grabbed the steel railing separating them from the machinery. They giggled as a harmless induced current from the generator flowed through the group. Then it was off to a local bar where Smith consumed far more beer than he should have.

Just after 10 p.m., Smith returned, alone and very drunk, to the power plant. Seeing his condition, the plant manager, G.W. Chafee, ordered him to leave. Smith soon returned, and the eviction was repeated. Chafee was otherwise occupied and did not see Smith return yet again.

After yelling that he was going to stop the generator, Smith placed his left hand near the brush of a 4,800-pound arc-light power generator and his right hand on the frame of the machine.[a] Chafee and others who were trying to reach Smith to evict him watched in horror as his body went rigid and collapsed. Only by stopping the generator could they release Smith's dead body from its grip on the brush.[2]

Dr. Alfred Porter Southwick, a dentist and former steamboat engineer, was interested in all things scientific and especially electricity. He and a local physician, Dr. George E. Fell, discussed the results of Smith's autopsy and concluded that electrocution appeared to be a quick and painless way of death. When sharing their thoughts with Colonel Rockwell, head of the Buffalo Society for the Prevention of Cruelty to Animals, Rockwell said that the BSPCA had been looking for a more humane method of executing the hundreds of stray dogs and cats collected by the city of Buffalo each year.

Southwick thought he could solve the problem of strays and started by learning all that he could about generating and handling electricity, conductors and insulators, and the effects of electricity on living things. In 1882, he and a few friends assembled the apparatus they needed and began their experiments in killing. Many questions confronted them, such as where and how to attach the electrodes, how much voltage and current would be lethal, and how long it would take for the animals to die. Their small generator proved inadequate, so they moved to a former police station where they would have access to the high-voltage wires that fed the city's arc-lights. Eventually, with an almost unlimited supply of stray dogs and cats to experiment on, they achieved consistent results. Southwick wrote articles calling electrocution "the safest and kindest method of killing."

Not surprisingly, Southwick and his friends soon talked of applying their successful dog-killing technology as a more humane method of capital punishment. Public sentiment against capital punishment was growing, primarily because of the gruesome stories reporting on the brutality of hanging, which was by far the prevalent method of execution at the time. But if a better, more humane approach could be found, that opposition would subside.

(a) By touching the brush with one hand and the generator frame with the other, George Smith was creating an electrical path through his body from a high-voltage point to ground. It would be similar to touching both bare wires from an electrical receptacle today, except that Smith was dealing with a much higher voltage.

Southwick published papers in scientific journals in 1882 and 1883 and began to think seriously about the details of human execution. Based on body weight alone, he could predict the required voltage and current. As a dentist, he quickly decided that a chair would be the appropriate setting, as straps could be easily applied, and a dead body was unlikely to fall off.[3] New York had endured a series of botched hangings, so politicians in that state were receptive to Southwick's articles.

In 1886, Governor David B. Hill appointed a three-man commission to study the issue. The commission chairman was Elbridge Gerry, philanthropist and animal rights activist. The other members were Alfred Southwick and Matthew Hale, an Albany attorney and politician. The commission studied the history of capital punishment with emphasis on the thirty-four methods of execution used over the centuries. All of these previous methods were deemed inappropriate, so newer methods were sought. The commission identified two new methods: lethal injection and electrocution. Because lethal injection would require participation by a physician, it was rejected, leaving only electrocution. Already in favor of that option, Southwick sought outside support to bolster his position by writing to Thomas Edison.

Southwick's letter dated November 8, 1887, posed a dilemma for Edison. While he knew he could advance his fight against AC by judicious suggestions, he was morally opposed to capital punishment. Widely quoted as saying, "The dove is my emblem. I want to save and advance human life, not destroy it,"[4] Edison had also said that he would "join heartily in an effort to totally abolish capital punishment."[5] In considering his answer to Southwick's letter, Edison's moral convictions triumphed, and he replied accordingly.

But Southwick was not going to allow five years of work go to waste when he was so close to achieving his goal. He wrote back to Edison on December 5, 1887, stating, "Science and civilization demand some more humane method than the rope. The rope is a relic of barbarism and should be relegated to the past." He appealed to Edison's vanity, saying, "(Your) reputation as an electrician would help much with the legislature" in the crusade for a more humane form of punishment.[6]

Edison replied to Southwick's second letter just four days later, on December 9, 1887. After assuring Southwick that electricity was reliable as a lethal force, he wrote as follows: "The best appliance in this connection is, to my mind, the one which will perform its work in the shortest space of time, and inflict the least amount of suffering upon its victim. This, I believe, can be accomplished by the use of electricity, and the most suitable apparatus for the purpose is that class of dynamo-electric machinery which employs intermittent currents. The most effective of these are known as 'alternating machines,' manufactured principally in this country by Geo. Westinghouse.... The passage of the current from these machines through the human body even by the slightest contacts, produces instantaneous death."[7]

Edison had relented and became an active participant in supporting the adoption of the electric chair as the state-approved means of execution. If Edison's frequent statements about saving and advancing human life reflected his true beliefs, he must have suffered great inner turmoil.

Edison's reply exerted significant influence on the commission chairman, Elbridge Gerry. His commission on modes of capital punishment reported to the New York legislature in January 1888, noting that they had consulted with experts in electricity, including Edison. The recommended method of execution was electrocution, but there were no details of AC or DC, the current, voltage, or method of attachment to the victim. A bill adopting the commission's recommendations passed the New York legislature and was signed by Governor Hill on June 8, 1888. The provisions of the law were to become effective on January 1, 1889.

Because some critical details were not specified, Edison was asked for his opinion. He replied that high-voltage alternating current applied through a pair of handcuffs attached to the arms would be most effective.

Ironically, decades later, during World War I, Edison headed the Naval Consulting Board, where he worked on several projects, including submarine detectors and gun-location techniques. However, due to his moral indignation toward violence, he specified that he would work only on defensive weapons, later noting, "I am proud of the fact that I never invented weapons to kill."[8]

Determination of the details of execution by electricity was turned over to the state's Medico-Legal Society.[9] That organization that had been formed in 1867 to bridge the often-conflicting objectives of law and psychiatry.[10] The Medico-Legal Society appointed a committee to study the issues, and that study led them to Harold P. Brown.

Self-Appointed Expert

Harold Brown had entered the fray with a lengthy letter published in the June 5, 1888, edition of the *New York Evening Post* titled, "Death in the Wires"—Technicians attempting to service overhead wires in New York City encountered a spiderweb of electrical and telephone wires. Brown, who claimed to be an electrical engineer, was motivated to write his letter by three recent deaths by electrocution in the city. The essence of Brown's diatribe was that alternating current was far more dangerous, even fatal, than direct current. He wrote,

> ... (S)everal companies who have more regard for the almighty dollar than for the safety of the public have adopted "alternating" current for incandescent service.
> If the pulsating current [used for arc lights] is dangerous, then the alternating current can be described by no adjective less forcible than damnable.
> The only excuse for the use of the fatal alternating current is that it saves the company operating it from spending a larger sum of money for the heavier copper wires, which are required by the safe incandescent systems. That is, the public must submit to constant danger from sudden death, in order that a corporation may pay a little larger dividend.

Brown's letter concluded with his proposed "RULES FOR STATION LIGHTING," which focused first on arc lighting. But his final rule was aimed squarely at the AC systems of Westinghouse and Thomson-Houston: "No alternating current with a higher electro-motive force [voltage] than 300 volts shall be used."[11] Of course, the critical advantage of AC was its ability to be transmitted at high voltage for long distances and then reduced in voltage at the customer's location. As he intended, Brown's proposed rule would nullify that advantage.

Researchers have found no link between Brown and Edison at the time of Brown's initial letter, but a close relationship was established soon afterward. Brown was criticized for his unsupported claim that AC was more dangerous than DC, so he approached Edison for equipment to perform experiments to prove his hypothesis.

Edison quickly saw an ally in his fight against Westinghouse, so he supplied not only the needed equipment but also an expert assistant in Arthur Kennelly. Kennelly had over a decade of experience as a telegraph engineer in Great Britain before immigrating to the U.S. in 1887. He would go on to author or co-author 28 books and 350 papers and predicted the existence of the ionosphere.[12]

In mid–July 1888, Brown and Kennelly conducted experiments at Edison's laboratory in West Orange, New Jersey. They executed dogs and cats obtained from neighborhood children for a bounty of 25¢ per animal. Their results were consistent with Brown's earlier statements that AC was more dangerous than DC at comparable voltage and power levels.

Laboratory studies and even published papers on the dangers of AC were one thing; public demonstrations were quite another. On July 30, 1888, Brown and Kennelly took their apparatus, provided by Edison, to the Columbia School of Mines. There, with the assistance of Dr. Frederick Peterson of the Medico-Legal Society, Brown led what he claimed was a vicious 76-pound Newfoundland dog to the stage in front of hundreds of onlookers. Brown subjected the dog to direct current at voltages up to 1,000 volts, but the dog survived. Brown then changed to alternating current at 330 volts, and the dog died. He then killed three more dogs with AC before the American Society for the Prevention of Cruelty to Animals put an end to the carnage.[13]

Soon after the Columbia demonstration, Peterson was named chairman of the Medico-Legal Society committee charged with determining details of the human execution protocol. That committee issued a preliminary report on November 15, 1888, recommending that either DC or AC could be used for executions, "but preferably the latter." A final recommendation was to be issued in December.[14]

Brown was optimistic but wanted to be sure that AC would be the final recommendation. The only remaining issue was whether the results on dogs would translate to larger animals (such as humans). So he went back to Edison to obtain permission to use equipment and facilities at the West Orange labs to electrocute larger animals.

Edison consented, and on December 5, 1888, two healthy calves and a horse were successfully electrocuted with alternating current.[b] Present at the spectacle were Elbridge Gerry, the chairman of the commission which had written the bill on human electrocution, the members of the Medico-Legal Society committee, Brown, Kennelly, and Edison. Less than one week later, the Medico-Legal Society committee unanimously recommended that AC be used for legal human electrocutions.

Implementation of the Killing Machine

With the technical details decided, the New York superintendent of prisons, Austin Lathrop, was tasked with implementing the system at three prison locations. Lathrop contacted Harold Brown and asked him to procure the necessary apparatus

---

(**b**) Many writers attribute the killing of a much larger animal, "Topsy" the elephant, to Brown or Edison. Topsy was indeed killed by a combination of poison, strangulation, and electrocution, but the execution occurred on January 4, 1903, more than a decade after the War of the Currents was resolved. Neither Harold Brown nor Thomas Edison had anything to do with Topsy's demise. However, a movie crew from the Edison Manufacturing Company filmed the event. See https://en.wikipedia.org/wiki/Topsy_(elephant).

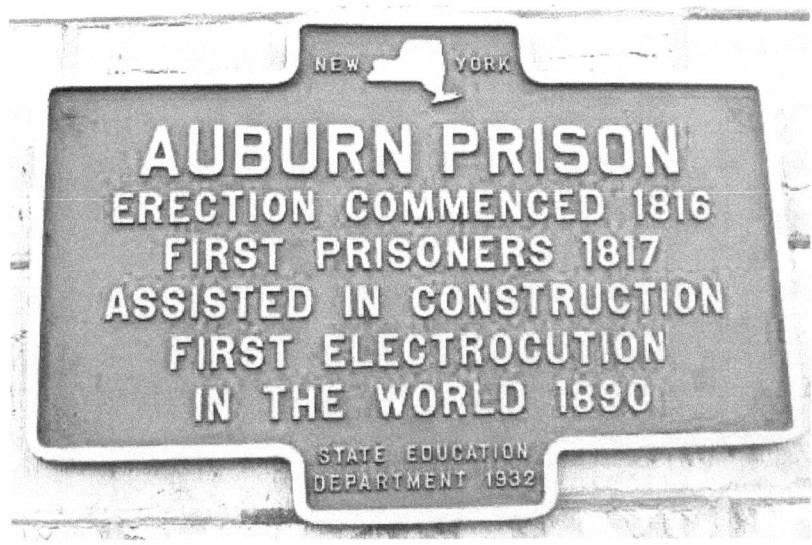

Historical marker at Auburn Prison (courtesy Anton Schwarzmueller, *The Historical Marker Database*).

and design a means of applying the fatal current. Brown did not have the knowledge or skill to create the "electric chair," so two physicians, Drs. Carlos MacDonald and A.D. Rockwell designed it and had it constructed in the Auburn Prison woodshop.[15] Brown's job was to procure the Westinghouse generators.

As Westinghouse was trying to prevent the use of his equipment for killing people, he refused to sell anything to Brown or the New York prison system. With help from the Thomson-Houston Electric Company, which installed Westinghouse equipment under a license agreement, Brown located and purchased three used Westinghouse AC generators and had them delivered to the prisons.

All We Need Is a Victim

William Kemmler was an ax murderer (more precisely, a hatchet murderer). When he was sober, which was not often, the 28-year-old Kemmler was also a fruit huckster in Buffalo, New York. He lived under the name John Hort with his common-law wife, Matilda "Tillie" Ziegler, and her 4-year-old daughter, Ella.

On Friday, March 29, 1889, after drinking all night, Kemmler and Tillie argued over another man, John DeBella, who happened to be an employee of Kemmler's. In a drunken rage, Kemmler struck Tillie twenty-six times with a hatchet. Although short of Lizzie Borden's reputed forty whacks, Tillie's wounds were fatal, and she died the next day. The police found Kemmler in a nearby bar, and after a day of questioning, he confessed.[16]

Kemmler's trial was brief, and on May 14, 1889, Judge Henry Childs pronounced him guilty of first-degree murder, punishable by death. The judge said, "The sentence of the court is that for the crime of murder in the first degree, whereof you stand convicted, within the week commencing on Monday, June 24, 1889, and within the walls of Auburn State Prison, ... you suffer the penalty of death, to be inflicted by the application of electricity as provided by the Code of Criminal Procedure of the State of New York ...."

Westinghouse secretly hired a famous New York attorney, U.S. congressman, and friend of Winston Churchill, W. Bourke Cockran, to defend Kemmler on appeal. Kemmler's trial attorney, Charles Hatch, filed a brief contending that his client "was entitled to be released because the death penalty imposed under the statute constituted cruel and unusual punishment."

The appeal hearing started on July 8, 1889. Harold Brown was the first witness and testified that "the alternating or Westinghouse current [was] the deadliest current and the one which should be used for killing criminals."[17]

After several days of testimony by other witnesses, the state, probably at Brown's suggestion, called Thomas Edison as a witness. Edison not only agreed to appear, but he also invited the participants to his West Orange laboratories for relevant demonstrations.

Edison took the stand on July 23. Cockran recognized that Edison's reputation alone could sway the court, so he sought to undermine his credibility. He posed questions about the effects on the human body from the application of a strong alternating current. Edison answered that he did not know much "about that part of it."

Cockran asked Edison if he knew anything about anatomy. Edison said, "No, sir."

Cockran asked whether blood or muscular tissue was the better conductor of electricity. Edison replied that he thought blood was the better conductor but that he would have to experiment to be sure.

Finally, Cockran asked, "Do you know anything about the conductivity of the brain?" Again, Edison admitted, "No, sir."

Despite Edison's demonstrated lack of critical knowledge regarding the effects of electricity on the human body, Kemmler lost his appeal on October 9, 1889. Subsequent appeals also failed, and Kemmler's execution was set for August 6, 1890.[18]

"They Would Have Done Better Using an Axe"

On the designated date, Kemmler was awakened at 5:00 a.m. After dressing in a suit, white shirt, and tie, he ate breakfast and prayed. The top of his head was shaved, and at 6:38 a.m., he entered the execution room to face 17 witnesses.

He said to them, "Gentlemen, I wish you luck. I'm sure I'll get a good place, and I'm ready." Witnesses said the Kemmler was composed and did not scream, cry, or resist in any way.

Edison had suggested using a set of handcuffs as electrodes for applying the current. But Drs. MacDonald and Rockwell, the electric chair designers, had chosen head-and-spine electrodes, each ten-centimeter diameter brass plates covered with moistened sponges and enclosed in rubber cups. So, after cutting a hole in the rear of Kemmler's suit, attendants attached an electrode to his back. Kemmler sat on the chair, was strapped in, and the top electrode was lowered onto his head.

The warden, Charles Durston, ordered the switch thrown, sending 1,300 volts of AC into the murderer's body. After 17 seconds, during which the smell of burning flesh caused several spectators to flee the room, Kemmler was declared dead.

However, after the power was turned off, several witnesses noticed that Kemmler

Chapter 17. Dying for Electricity

Execution of William Kemmler, August 6, 1890, from the *New York Herald*. The switchboard was in the room behind the door.

was still breathing. The attending physicians examined Kemmler and, confirming he was still alive, called for the immediate application of 2,000 volts.

The current remained on for one to two minutes; in the confusion, no one kept track of the time. Blood vessels ruptured, blood rolled down his face, and Kemmler's body caught fire. This time Kemmler was dead.

Thus did William Kemmler become the first person to die in the electric chair. The *New York Times* headlined the story, "FAR WORSE THAN HANGING— Kemmler's Death Proves an Awful Spectacle." The *London Chronicle* characterized it as "worthy of the darkest chambers of the Inquisition in the 16th century."[19]

When told of the gruesome execution, George Westinghouse commented, "They would have done better using an axe."[20]

History has proven that both Edison and Westinghouse were wrong. Edison erred by insisting that AC would never be adopted, and DC would triumph because it was safer. Westinghouse was mistaken when he claimed that AC was no more dangerous than DC.

But William Kemmler's execution did little to resolve the AC vs. DC dispute. That resolution would have to await two far more significant events, the 1893 World's Columbian Exposition and the harnessing of Niagara Falls to produce electricity.

Chapter 18

The Worst of Times

> *"It was the best of times, it was the worst of times, it was the age of wisdom, it was the age of foolishness...."*
> —Charles Dickens in A Tale of Two Cities

Two tragic events, the Johnstown Flood and the Homestead Strike, bookended the period between Memorial Day, 1889, and Independence Day, 1892. Although they did not directly affect George Westinghouse, both severely impacted men who had been his friends, Andrew Carnegie and Henry Clay Frick. These events also highlighted the eternal struggle between rich and poor, haves and have-nots, owners and laborers. The first cost 2,209 lives; the second just 10, and both events caused significant repercussions and triggered lasting changes in western Pennsylvania and in the country as a whole.

The Johnstown Flood

The South Fork Fishing and Hunting Club

Lake Conemaugh, near the town of South Fork, Pennsylvania, was a remnant of the Pennsylvania Main Line Canal. South Fork Dam and Lake Conemaugh behind it had been built between 1840 and 1852 to supply water for the portion of the canal between Johnstown and Pittsburgh. Obsolete almost as soon as it was completed, the dam and lake stood virtually abandoned until, in 1879, Mr. Benjamin Ruff of Pittsburgh bought the dam, lake, and property for $2,000.

Ruff intended to create a summer resort for some wealthy Pittsburgh friends and sold memberships shares to 15 prominent Pittsburghers. On November 15, 1879, they obtained an Allegheny County charter for the South Fork Fishing and Hunting Club, even though the Club was in Cambria County.

The Club modified the earthen dam, including lowering it by about three feet to make the top wide enough to accommodate a road, building a wooden bridge across the spillway, and installing heavy bars and screens under the bridge to prevent the escape of fish. The dam's overflow pipes and valves had been removed and sold for scrap by a previous owner, so there was no way to lower the lake level to accomplish the repairs that are periodically required on any earthen dam. Lake Conemaugh, created and restrained by the dam, was about two miles long, one mile wide, covered 400 acres, and held back twenty million tons of water. What could go wrong?

By 1889, the Club had 61 members, mostly wealthy industrialists from the East End of Pittsburgh. Among the better-known names were:

- Andrew Carnegie, steel magnate and later an outstanding philanthropist;
- Henry Clay Frick, coke magnate and a poker-playing buddy of George Westinghouse;
- Philander Chase Knox, who would later become U.S. Secretary of State;
- John Caldwell, Jr., Treasurer of Westinghouse's Philadelphia Company;
- Andrew Mellon, banker and future U.S. Secretary of the Treasury; and
- Robert Pitcairn, superintendent of the Pittsburgh Division of the Pennsylvania Railroad and friend of George Westinghouse.[1]

George Westinghouse was not a member.

Rain and the Flood

At about 11 p.m. on May 30, 1889, pouring rain started. Very few members of the South Fork Fishing and Hunting Club were in residence, as the "season" would not begin for another three weeks. The rainfall in western Pennsylvania that day was the heaviest ever recorded in the area, with six to eight inches in 24 hours.

Numerous mountain streams, all overflowing their banks, fed Lake Conemaugh, which rose one inch every ten minutes. Logs, stumps, and other debris clogged the fish screens and prevented excess lake water from draining over the spillway. There was no place for the deluge of water to go but over the top of the dam.

Despite heroic efforts by workers and residents, that is precisely where it went. At 11:30 a.m. on May 31, water started to flow over the dam. At 3:10 p.m., the center of the dam disintegrated, and the resulting 200-foot-wide gap drained the lake in less than 45 minutes.

Telegraph messages warning Johnstown of the coming calamity were either ignored (similar warnings over the years had all proven to be false alarms) or failed to arrive because of downed lines. Carrying before it trees, small bridges, houses, dead animals, and assorted rubbish, the 60-foot-high wall of water rushed down the valley through the towns of South Fork and Mineral Point to Johnstown. At 4:07 p.m., after decimating the eastern suburbs of East Conemaugh and Woodvale, the torrent hit Johnstown.[2]

Water and a conflagration that burned in a thirty-acre debris field behind a stone bridge drowned or incinerated 2,209 victims. The 1889 Johnstown flood remained the worst civilian disaster in the United States until the 1900 Galveston hurricane.

Help arrived quickly. The South Fork Fishing and Hunting Club donated 1,000 blankets. About half of the club members contributed money, most notably Andrew Carnegie, who gave $5,000 and Henry Clay Frick, who donated $5,000. Carnegie later built the Johnstown Library, a building that is now the Johnstown Flood Museum.[3] Even though he had no direct connection to the South Fork Fishing and Hunting Club, George Westinghouse contributed $15,000 to the relief effort.[4]

Clara Barton and five Red Cross workers arrived within five days and led the first significant peacetime relief effort of the American Red Cross. Robert Pitcairn alerted Pittsburgh that Johnstown had been destroyed and asked for immediate help. The first relief train from Pittsburgh arrived on June 2. A total of over $3.7

The Schultz home with an extra tree. Six people were inside when the flood hit; all survived (Wikimedia Commons).

million was collected for the relief effort from across the U.S. and twelve foreign countries.[5]

The flood destroyed 1,600 homes and four-square miles of the downtown area and caused $17 million in property damage.

Survivors filed suit against the South Fork Fishing and Hunting Club for negligence in the modifications made to the dam and inadequate maintenance. Attorneys Philander Knox and James Hay Reed, both members of the Club, successfully argued that the flood was a natural disaster that was an Act of God. No one was held legally responsible for the disaster.[6]

The Homestead Strike

Homestead, Pennsylvania, the industrial town on the Monongahela River nine miles southeast of Pittsburgh, was the home of one of the world's most powerful corporations, Carnegie Steel Company, and the nation's strongest trade union, the Amalgamated Association of Iron and Steel Workers (AAISW).[7]

Unlike most modern unions, the AAISW represented only skilled tradesmen, who constituted just 800 of the 3,800 workers at Homestead, not the low-skill jobs typically held by recent immigrants.

Labor strikes in 1882 and 1889 had strengthened the union and caused Andrew Carnegie and his Director of Operations, Henry Clay Frick, to vow to break the union and regain control of the vast Homestead Works.

With the 1889 contract set to expire in mid–1892, Carnegie ordered the Homestead plant to increase production and build inventory so he could weather a strike. Negotiations for a new labor agreement started in February 1892 with the union asking for a wage increase. Frick countered with a demand for a 22 percent wage

Chapter 18. The Worst of Times

decrease. Carnegie, vacationing at his family home in Dunfermline, Scotland, gave Frick free rein to break the union.[8] After more posturing by both sides, Frick locked the workers out of the plant at the end of June. He had a high, barbed-wire-topped fence built to encircle the plant and exclude the workers.

On July 2, Frick fired all 3,800 workers.[9] The company placed newspaper ads for replacement workers as far away as St. Louis, but the strikers established 24-hour picket lines to keep out strikebreakers. On July 6, three hundred agents from the Pinkerton Detective Agency, hired by Frick to protect his mill and open access to strikebreakers, attempted to enter the Homestead Works by way of the river. But strikers had discovered the plan and met the Pinkertons as they tried to land. Gunfire erupted, and over the next six hours, seven AAISW members and three Pinkertons died.

Governor Robert E. Pattison finally sent the state militia to restore order, but on July 23, a New York anarchist with no connection to either side, Alexander Berkman, attempted to shoot Henry Clay Frick in Frick's office. Although seriously injured, Frick not only survived the attack but continued to work for the rest of the day.

The violence of the strike and the attempted assassination of Frick undermined public support for the strikers, and they voted to return to work on the terms dictated by Frick. The strike collapsed, and, as Frick and Carnegie desired, the AAISW was broken. Membership in the union declined from its high point of 24,000 in 1891 to 8,000 in 1895.[10]

The first troops in Homestead. The Eighteenth Regiment passing the office and works of the Carnegie Company (*Harper's Weekly*, July 23, 1892).

The Homestead Strike also destroyed the relationship between Carnegie and Frick. In a letter to Carnegie, George Westinghouse sought to mend the rift, calling it "a calamity by reason of the fact that the private affairs of your company will undoubtedly be made public.... I might add that Mr. Frick has recently spoken to me in such terms that I feel there must be a way to adjust matters between you and him."

But Carnegie refused the olive branch, and the two remained bitter enemies. Then, in the late spring of 1919, Carnegie, age 83 and seriously ill, wrote a brief letter to Frick. Both men were living in their New York City mansions, so it was a short walk for Carnegie's associate, James Howard Bridge,[a] to carry the letter to Frick.

When handed the letter, Frick tore it open and quickly read it. "So, Carnegie wants to meet me, does he? Yes, you can tell Carnegie I'll meet him. Tell him I'll see him in Hell, where we are both going."[11]

Comparing Westinghouse to Carnegie and Frick

The contrast in labor philosophy between Westinghouse and Carnegie-Frick was striking in both senses of the word.

Westinghouse cut the workweek for his Westinghouse Air Brake employees to 55 hours over 5½ days in 1881.

In 1886, Carnegie said, "At present every ton of pig iron made in the world ... is made by men working in double shifts of twelve hours each, having neither Sunday nor holiday the year-round. Every two weeks, the men change to the night shift by working twenty-four hours consecutively."[12] So the normal workweek, at least for workers involved in the production of steel ingots, was seven twelve-hour shifts, or 84 hours. Other parts of the steel mills worked shorter hours, ranging from 48 to 72 hours.[13]

There was also a vast difference in working conditions at Carnegie and Westinghouse facilities. Accidents among steelworkers were responsible for twenty percent of the deaths among men in Pittsburgh in the 1880s.[14]

Westinghouse "was a paternalist from the days of his work at his father's New England machine shop and his mother's religious training. He believed the treatment of his workers was directly linked to his and the company's success. Unlike his industrialist neighbor, Andrew Carnegie, Westinghouse's concern for the employees seemed more based on his love of his fellow man than any potential productivity gain."[15]

Samuel Gompers, president of the American Federation of Labor and the leading trade unionist in the late 1800s and early 1900s, once said, "If all business owners treated their workers as well as George Westinghouse, the American Federation of Labor would have to go out of business."[16]

Despite his progressive labor policy, Westinghouse did not support labor unions. In a 1903 letter to Gompers, he declined to unionize his workforce voluntarily. Westinghouse said that the fair and honorable treatment of employees, including the partial workday on Saturdays, the standard 54-hour week, and an openness to

---

(**a**) James Howard Bridge wrote a 1903 book titled, *The Inside History of the Carnegie Steel Company—A Romance of Millions.*

settling grievances, precluded the need for outside representation. He noted that he had refused to join anti-union organizations of employers and offered to work with Gompers to assure a comprehensive and beneficial system for retirement security.[17]

Carnegie and Frick were both called "Robber Barons"; Westinghouse never was. In fact, two decades after his death, over 50,000 current and former Westinghouse employees contributed money to build a memorial in his honor.[18]

Ernst H. Heinrichs, before he worked for Westinghouse as press agent and confidant, was a reporter for the *Chronicle-Telegraph*, a Pittsburgh newspaper. In that role, Heinrichs had often interviewed Andrew Carnegie, Henry Clay Frick, and George Westinghouse. In 1931, Heinrich reminisced on the differences among these industrial giants.[19]

Regarding Carnegie, he said:

> To see Mr. Carnegie was easy enough, but to interview him for news was another matter; in fact, it could not be done, because he invariably turned interviewer himself. While his Scotch shrewdness made him fully alive to the value of publicity as well as appreciative of the good will of the press, he never refused to receive newspapermen, but the news he gave them was not always what they wanted. If you asked him a question, and it referred to a subject about which he preferred to say nothing, he would look at you with a quizzical smile and remark: "Now supposing you ask me this question," and then he would form an interrogatory about something which might be as far removed from the matter in your mind as Tallahassee from Timbuktu; but as the reporter was after news from Mr. Carnegie, he readily compromised on the subject and Mr. Carnegie got his story into the paper.

Heinrichs said of Henry Clay Frick: "As for Mr. Frick, he was entirely different from Mr. Carnegie. Where the latter was quite voluble when it suited his purpose, Mr. Frick was always very reserved, very reticent. He carried this so far even that he appeared to be afraid to say anything, because he might be misunderstood or misconstrued. But withal, he was exceedingly polite and quite affable, and once you gained his confidence he would talk quite freely and without any reserve, but those who were in his confidence were very few in number."

Before he became his press secretary, Heinrichs had this to say about interviewing George Westinghouse:

> In the Westinghouse office, dealing with the press was very simple. To begin with, Mr. Westinghouse did not see newspapermen himself except on very rare occasions. They were received by his private secretary, Mr. W.D. Uptegraff, and he was a past master at his job. If there was any news, he would give it. If the reporters wanted enlightenment on any subject pertaining to Mr. Westinghouse or the Westinghouse interests, they would go to Mr. Uptegraff and he would handle the matter either by giving the information himself, from his own knowledge, or he would promise to find out what was wanted and then supply the information as soon as he could.

It's All About Money, Part 1

Sandwiched between the Johnstown Flood and the Homestead Strike were two other events that captured less public attention but transformed the empires of George Westinghouse and Thomas Edison.

By 1890, George Westinghouse controlled multiple businesses that he had

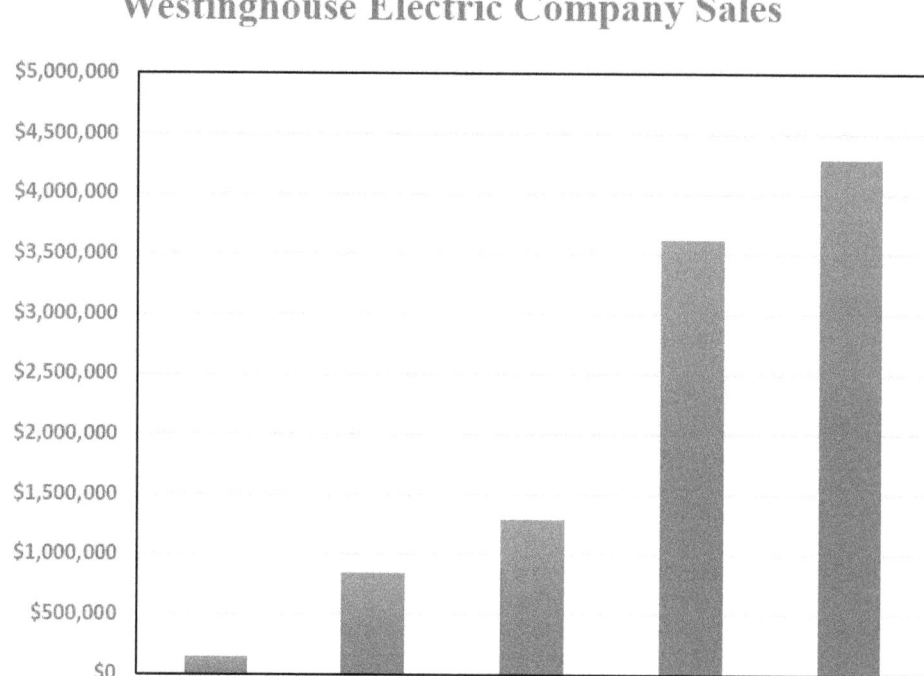

Westinghouse Electric Company sales (*1886 sales start on April 1).

founded and acquired. Some of the larger ones manufactured air brakes and signaling systems for railroads, furnished natural gas to Pittsburgh residents and industries, built equipment required for generating and distributing electricity, and operated power plants to supply electricity to people, businesses, and industries all over the world. Revenue from Westinghouse Electric alone reached over four million dollars.[20]

The amazing growth in sales by Westinghouse Electric obscured less favorable developments. Most manufacturing is capital-intensive, and the nascent electrical equipment manufacturing business was an extreme example. Production equipment did not exist and had to be designed from scratch and built, often in Westinghouse's own shops. Stock in the company was closely held by George Westinghouse and his associates, and banks were skeptical about this new electrical equipment business, so sources of funds were limited.

In October 1890, Westinghouse, Marguerite, and their seven-year-old son, George III, moved into their new home, Erskine Park, in the Berkshire Mountains of western Massachusetts. On November 16, the *Pittsburgh Press* brought disturbing news to George Westinghouse. Because of risky investments in Argentina, Barings Bank in London faced bankruptcy. Such a failure could have a ripple effect on financial markets worldwide, and the additional funds that Westinghouse Electric required to keep operating might be challenging to obtain.

Westinghouse immediately returned to Pittsburgh on his private railcar, Glen

Eyre, and called his Board of Directors together. The Board agreed to change the company name to Westinghouse Electric & Manufacturing and doubled the number of shares. They hoped that the sale of the new shares would raise the needed capital, but instead, the stock price dropped from $50 to $13. George Westinghouse then met with leading Pittsburgh bankers, who agreed to loan him the $500,000 he needed, but reserved "the privilege of naming the general manager."[21]

Westinghouse refused their offer and went to New York aboard Glen Eyre to negotiate with bankers there. Within a few months, August Belmont & Co. of New York; Lee, Higginson & Co. of Boston; and Brayton Ives, president of the Western National Bank of New York, formed a financial syndicate to underwrite the needs of Westinghouse Electric & Manufacturing. August Belmont, the company founder, had been an enemy of J.P. Morgan but had died in 1890. His son, August Jr.,[(b)] now ran the company and had close ties to the European Rothschilds. So, Belmont and Company were strong enough and willing to support Westinghouse in opposition to Morgan. More important, Belmont had none of Morgan's dreams of establishing an electrical trust and did not demand any role in controlling Westinghouse Electric.[22]

Additional support came from the Westinghouse employees. Upon learning of his financial distress, the employees of Westinghouse Electric proposed to work for half pay until the problems passed. Also, T.A. Gillespie, who had drilled the gas wells at Solitude, Westinghouse's home in Pittsburgh, offered to loan Westinghouse thousands of dollars if that would help. Westinghouse declined these generous offers but was encouraged to learn that his associates held him in such high regard.[23]

On April 7, 1891, the Westinghouse Electric & Manufacturing Board of Directors approved the plan for obtaining new capital. On April 23, George Westinghouse sent a circular to all stockholders, encouraging them to accept the provisions of the proposal, which asked them to surrender 40 percent of their holdings to the company in return for receiving a 7 percent dividend on their remaining shares. Primarily because of the personality of George Westinghouse, his hopefulness, and charisma, the plan received enthusiastic acceptance among the stockholders. Almost all of them surrendered the requested 40 percent of their holdings, looking to the future success of the company to recover their investment. Bankers were persuaded to accept an ownership position in the company in the form of preferred shares in return for working capital. From a debt of $3 million, carrying an interest charge of $180,000 before the reorganization, the debt after the reorganization was reduced to zero, and adequate funds were obtained to move the company forward.[24]

The problems of Westinghouse Electric & Manufacturing were caused by an optimistic business outlook, too-rapid expansion of fixed assets in the form of manufacturing capacity, and failure to maintain sufficient liquid assets. The reorganization solved the immediate problems but did not alter the underlying optimism of the man in charge, George Westinghouse.[25]

(b) August Belmont, Jr., financed the construction of the first New York City subway and built New York's Belmont Park racetrack, which became home of the third jewel in the triple crown of thoroughbred horse racing, the Belmont Stakes.

It's All About Money, Part 2

While George Westinghouse fought for control of Westinghouse Electric & Manufacturing, Thomas Edison was losing control of the Edison General Electric Company.

Unmatched as an innovator and manager of experimenters, Edison was a poor businessman. Time and again, J.P. Morgan had invested in the various Edison companies, and he was impatient when Edison continued to lose market share in the lucrative central power plant business. Morgan and many others urged Edison to admit that DC power distribution was doomed and that AC was the future. But Edison refused to acknowledge the inevitable. Edison General Electric president Henry Villard was negotiating a merger with competitor Thomson-Houston and thought that Edison General Electric's lamp patents placed it in a position to dictate terms of the merger.

But J.P. Morgan, in looking at the financial performance of the two companies, realized that the Thomson-Houston management team was more effective, and engineered a behind-the-scenes deal to make Charles Coffin, president of Thomson-Houston, head of the merged companies. After April 15, 1892, the merged Edison General Electric and Thomson-Houston Companies would be called the General Electric Company. Edison learned of the merger, and that his name would not be part of the new company, one day before the deal was finalized. Once again, the banker, J.P. Morgan, won.[26] Edison became a bitter man and never again rose to his previous level of creativity.

The Final Battles

The War of the Currents was almost over. George Westinghouse, with funding from August Belmont, was at full strength. Thomas Edison, wounded by J.P. Morgan, was off the battlefield, but the newly formed General Electric Company, under the leadership of Charles Coffin, was a worthy replacement. The war's final battles would be fought in Chicago at the 1893 World's Columbian Exposition, and at Niagara Falls, the massive waterfall that had fascinated young Nikola Tesla when he said, "I pictured in my imagination a big wheel run by the Falls."[27]

Chapter 19

The White City

Make no little plans; they have no magic to stir men's blood."
—Daniel Burnham, Chief Architect and Director
of Works of 1893 World's Columbian Exposition

World's Columbian Exposition—Opportunity Knocks!

The first mass gathering identified by the Bureau International des Expositions (BIE) as a World Exposition occurred in London between April and October 1851. That pioneering event was championed by Prince Albert and officially opened by his wife, Queen Victoria. At the center of the sprawling exposition grounds was the Crystal Palace, a soaring structure with the largest area of glass ever used in a single building up to that time.

Subsequent World Expositions took place roughly every five years and lasted as long as six months. Nine other World Expos were held from 1855 to 1889, prior to the World's Columbian Exposition in Chicago from May to October 1893.

Also known as the Chicago World's Fair, the event celebrated the 400th anniversary of Columbus' landing in the New World in 1492 (actually the 401st anniversary).

Chicago competed for the right to host the Exposition with other American cities, including New York City, Washington, D.C., and St. Louis. J.P. Morgan and Cornelius Vanderbilt backed the New York bid, while Marshall Field and Cyrus McCormick supported the Chicago effort.[1]

The final decision on which city would win rested with the U.S. Congress. The atmosphere on February 24, 1890, the day of voting in the House of Representatives, resembled a political convention. Partisans packed the House gallery, and people gathered in the streets of the competing cities.

Chicago received the most votes during the first round but did not achieve a majority. After six hours and eight rounds of voting, Chicago finally reached a majority. The Senate concurred with the decision in April, and Chicago was the official host city.[2]

Congress determined one other important detail of the Exposition—the opening date would be May 1, 1893, rather than a year earlier, the actual 400th anniversary. As 1892 was a presidential election year, they did not want the Exposition to distract from the vital political campaign.[3]

The extra year proved critical, as there were many unanticipated problems.

The Chicago architectural firm of Burnham and Root was selected to design and oversee construction of the World's Columbian Exposition. But on January 15, 1891, while the firm was heavily involved in planning the design, partner John Wellborn Root died of pneumonia. The firm's other partner, Daniel Hudson Burnham, assumed the role of Director of Works for the Fair. Burnham gathered a distinguished team of architects and landscape designers, including Louis Sullivan and Frederick Law Olmstead, and moved his personal residence to a wooden building on the fairgrounds so he could closely supervise construction.

With a too-short construction schedule, too many would-be decision-makers, and too-large committees, building and equipping the Exposition was challenging.[4] But the dedication ceremonies were held as planned on October 21, 1892, and the public opening was May 1, 1893.

George Westinghouse, Marguerite, and George III attended the opening celebration, with George staying with his family rather than seeking acclamation for his role in making the Exposition happen. They remained in Chicago for several weeks to enjoy the wonders of the White City.

Over 100,000 people gathered at the Grand Plaza of the Fair to see President Grover Cleveland press the button signaling the opening of the World's Columbian Exposition. A thousand flags unfurled, the orchestra played the Hallelujah Chorus, electrically operated fountains shot water into the sky, and cannons boomed.

The real thrill for the fairgoers was yet to come. At dusk, all of the exterior arc lights, the massive floodlights, and the thousands of incandescent lamps decorating the buildings burst on at once. No one had ever before seen such a sudden transformation from darkness to brilliant light.

Many features of the Fair were not completed on the opening day. For example, the iconic Ferris Wheel, designed and built by George Washington Gale Ferris, Jr., and standing 264 feet high, did not open until June 21, 1893. But once opened, it proved to be immensely popular and helped to save the Exposition from bankruptcy.[5]

George Ferris did his early work in Pittsburgh, evaluating metals for railroad bridges. He designed the Ferris Wheel as a signature attraction for the Columbian Exposition, much as Gustave Eiffel's Tower marked the 1889 Paris Exposition. The

The Statue of the Republic overlooking the brightly lighted Court of Honor and Grand Basin.

Chapter 19. The White City

Ferris Wheel at 1893 World's Columbian Exposition; 36 Gondolas Held Up to 2,160 People for the 20-Minute Ride (Wikimedia Commons).

finished wheel stood 264 feet high and held 36 passenger cars. Each car could hold up to 60 passengers, giving the wheel a capacity of 2,160 riders.

To this day, no other Ferris Wheel has eclipsed the passenger load of this first wheel. The Beijing Great Wheel, standing 693 feet tall, would have come the closest with a maximum of 1,920 passengers, but construction stalled in 2011.[6]

The 1893 Columbian Exposition covered 690 acres on the swampy south shore of Lake Michigan and included almost 200 buildings (nearly all temporary), with fourteen "great" buildings. All buildings around the "Court of Honor" were spray-painted with a mixture of oil and white lead whitewash, leading to the name "The White City."[7]

The event drew 27.5 million visitors, second only to the 1889 Exposition in Paris and more than any U.S. World's Fair until the 1933–34 Chicago World Expo. It cost $27 million, more than the previous five World's Fairs combined and 2.5 times as much as any previous exposition.[8]

The Palace of Fine Arts building (Number 3 on the lithograph) had the largest footprint of any building in the world, covering nearly five acres. It is the only building still surviving from the 1893 Exposition and is now the Museum of Science and Industry.[9]

Major Buildings at the 1893 Columbian Exposition

Key	Building	Key	Building
1	Forestry	7	Administration
2	Agriculture	8	Electricity
3	Palace of Fine Arts	9	Mines
4	United States	10	Horticulture
5	Driving Force	11	Women's Palace
6	Machine Hall	12	Illinois State

Lithograph providing a bird's eye view of the 1893 Columbian Exposition in Chicago.

Two references deserve mention for their vivid but disparate depictions of the Columbian Exposition. The video, *EXPO: Magic of the White City*, is a video tour with hundreds of vintage photographs and informative narration by Gene Wilder.[10] The excitement and danger surrounding the exposition are captured in the engrossing novel, *Devil in the White City*, by Erik Larson.[11]

Let There Be Lights

With Edison's invention of the incandescent light bulb in 1879 and the subsequent lighting of American cities by Edison, Westinghouse, and Thomson-Houston, electricity was sure to be a vital part of the 1893 World's Columbian Exposition. But only two bids for the 93,040-bulb incandescent lighting system for the Exposition were submitted in April 1892—one by General Electric for $1.68 million, and one by the obscure Southside Metal and Machine Works of Chicago for $558,000. Westinghouse Electric was focused on its rapidly expanding central power system opportunities and did not initially enter a bid for the lighting contract.

G.E.'s proposal was viewed as exorbitant, and Southside's as risky because they had no manufacturer to back them. The extreme difference in the two bids "attracted much attention," and fair officials conferred with G.E. and Southside Metal. G.E. reduced their bid, first to $930,000, and then to $554,000, suspiciously just under the Southside Metal bid.[12]

At that point, Charles F. Lockstadt, president of Southside Metal, approached George Westinghouse to ask for his support. After careful study, Westinghouse agreed to take over the Southside Metal bid and, if successful, pay them a suitable commission. Given this new input, all previous proposals were rejected, and the project was re-advertised for bids. Westinghouse had provided a $50,000 bond guaranteeing that his bid would not exceed the last G.E. bid of $554,000.[13]

In this round of bidding, G.E., based on its DC system, submitted the same proposal as previously, $554,000. Westinghouse, based on its AC system, bid $399,000 and received the contract on May 23, 1892.[14] Other manufacturers, such as General Electric, Siemens-Halske, and Western Electric, received contracts for specialized lighting and generators, including arc-lighting systems for outdoor illumination.

Working on a Vacuum

G.E. had lost the contract but still had an ace in the hole. They refused to sell their bulbs for use at the Exposition, and Edison's 1879 patent on incandescent lamps prevented Westinghouse Electric from manufacturing their own light bulbs.[15]

What was Westinghouse to do? He could not buy bulbs from G.E., and he could not make bulbs that infringed Edison's patent. Could he and his engineers design a non-infringing bulb and produce a quarter-million of them in less than one year?

Remember that Edison's patent specified, "The combination of carbon filaments with a receiver made entirely of glass and conductors passing through the glass, and from which receiver the air is exhausted, for the purposes set forth." Also remember that Sawyer and Man had developed and patented a two-piece glass structure, called the stopper lamp. Instead of being made "entirely of glass" as specified in the Edison patent, the stopper lamp consisted of two pieces of glass joined by a high-temperature glue. Westinghouse had the foresight to buy the company (Electro-Dynamic Light Company) that held the Sawyer-Man lamp patent rights. So the Sawyer-Man patent gave Westinghouse a starting point to solve the lamp problem.

The original drawing by George Westinghouse shows his concept for a stopper lamp. Beside it is a photograph of an actual lamp as manufactured by Westinghouse Electric Company.

But no machinery existed to perform the precision glass grinding required to manufacture the stopper lamps, so Westinghouse workers had to design and build it.

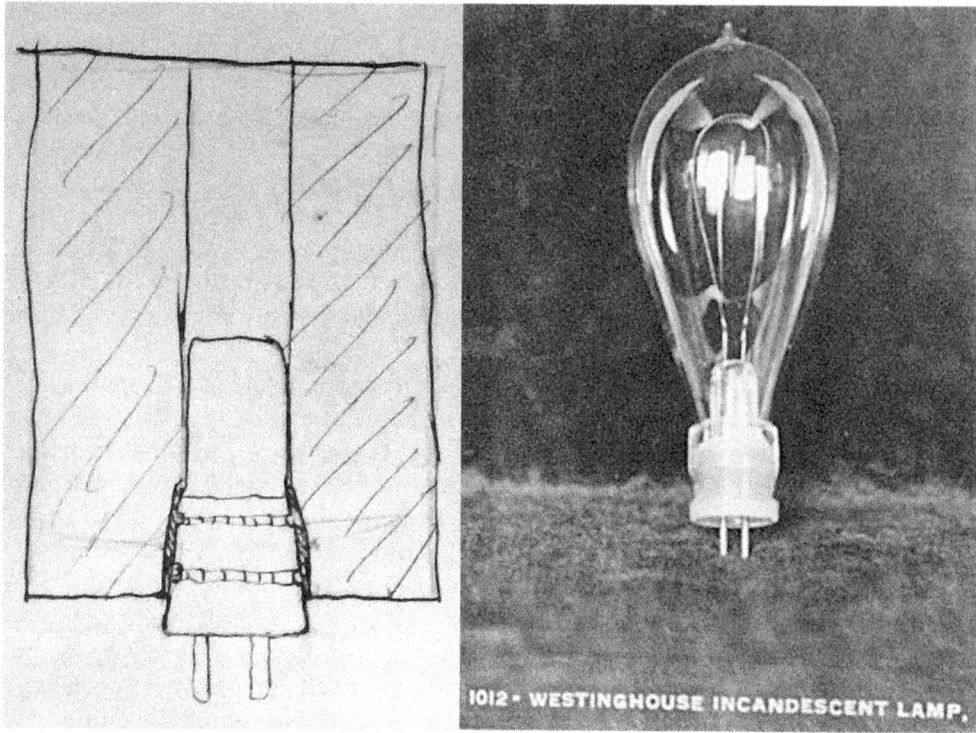

George Westinghouse drawing of a Stopper Lamp Base and an actual Westinghouse Stopper Lamp (Westinghouse Collection, Detre Library and Archives Division, Senator John Heinz History Center, Pittsburgh, PA).

Westinghouse converted his former Air Brake factory in Allegheny City into a glass factory. When he was in town, he often instructed the grinding machine operators how to make the stoppers fit as tightly as possible.

Because the stopper lamp could not achieve the perfect seal of the Edison one-piece bulb, air leaked in, and the stopper lamps had a shorter lifetime. Therefore, extra lamps were required so that burned-out bulbs at the Fair could be replaced immediately.[16] For the 90,000+ sockets at the Exposition, Westinghouse Electric manufactured 250,000 stopper lamps. All sockets were filled, and spares were available when the Fair opened.

Providing Power for the Fair

Electricity to power all of those lamps was generated by a massive AC power plant, the largest then in existence. Housed in Machinery Hall were fifteen steam engines totaling 13,000 horsepower. Twelve of these engines drove dual 500-horsepower AC generators with an aggregate output of 12,000 horsepower (almost 9,000 kilowatts), enough to simultaneously light 172,000 incandescent bulbs. Each dual generator stood ten feet high and weighed seventy-five tons. These twelve dual generators would be far more critical to the future of Westinghouse Electric & Manufacturing than the quarter-million stopper lamps.

Westinghouse AC Switch Board in Machinery Hall; Note the size of the operator on the platform (Westinghouse Collection, Detre Library and Archives Division, Senator John Heinz History Center, Pittsburgh, PA).

A one-thousand square foot marble-based switchboard to control and route power to forty primary circuits completed the installation.

Outside of each building to be served, AC transformers were installed in fireproof, water-proof pits. Virtually all wiring was in underground ducts.

Praise for a Job Well Done

A summary report of the Exposition commented, "...the Westinghouse Company fully complied with its contract, and performed its huge task in a manner that was entirely satisfactory."[17] Harlow Niles Higinbotham, third president of the Exposition, was more lavish in his praise, "I take pleasure in certifying to the fact that we were in every respect fully satisfied, not only with the plant, but with the work performed and the uniform gentlemanly and courteous treatment received at your hands as well as from your subordinate officers. For this work I beg to extend to your company the thanks of the Exposition Management."[18]

Financial Results

The 1894 Westinghouse Annual Report provided details of the financial results of the Exposition, as follows:

> All of the apparatus used in connection with the World's Fair lighting contract and for exhibition purposes has now been returned to the warehouses of the Company, and the final payments have been made by the World's Columbian Exposition Company. The amount of the contract for lighting was $399,000, and for extras under the contract $88,704.47, making the total received by the Company $487,704.47. All goods shipped to the Fair were charged at the cost of labor and material with 20 percent added. After charging the account with all costs of every nature in connection with the running of the plant and of the exhibition, and crediting it upon the return of the material with the net cost of labor and material, it shows a net cost to the Company of only $16,013.03 for a grand display of the Company's apparatus, the advertising effect of which has been invaluable. The price at which the lighting apparatus used at the World's Fair, and the apparatus used for Exhibit has been sold, will yield a handsome profit to the Company.
>
> It is gratifying to state in connection with this subject that the entire contract was carried out without friction of any character, and without a single deduction of any kind being made from the bills rendered by the Company.

Before credits for selling the surplus equipment, the net cost to the company was $16,013.03, a tiny expense for the favorable publicity received.[19]

Tesla's Polyphase System

In addition to the lighting power plant in Machinery Hall, Nikola Tesla designed and built, using Westinghouse equipment, a complete polyphase power system in Electricity Hall.

A large two-phase induction motor, powered by current from the Machinery Hall generators, drove a polyphase AC generator. The output of this generator fed step-up transformers to raise the voltage for transmission over a short transmission line, which connected to transformers for lowering the voltage for use by induction motors, a synchronous motor, and a rotating converter to supply DC to a railway motor.

In short, the exhibit demonstrated all components of a commercial power system for generating electricity and transmitting it over great distances to end-users.

А Bright Future

The overwhelming success of the World's Columbian Exposition and Westinghouse's AC electric system for powering the Fair dealt a near-fatal blow to Edison's hopes for DC as the standard for power distribution in the United States and indeed around the world.

The current war pitting AC against DC was almost over. Tesla's small but complete polyphase generating, transmission, and motor system brought together everything that would be required for the next giant step, harnessing the power of Niagara Falls.

Lord Kelvin was a distinguished visitor to the Exposition. There he witnessed the AC system in action and, though formerly a supporter of DC, would become a friend and vital advocate for Westinghouse on the Niagara Falls project.

Chapter 20

Over a Barrel at Niagara

"It seems that I have always been ahead of my time. I had to wait nineteen years before Niagara was harnessed by my system, fifteen years before the basic inventions for wireless which I gave to the world in 1893 were applied universally."

—Nikola Tesla

Water to Power

Niagara Falls ranks among the world's greatest waterfalls. While neither the tallest[a] nor the widest[b] waterfall, water flowing from the upper four Great Lakes gives Niagara the highest flow rate of any falls in North America at five million cubic feet per minute.

The first known effort to utilize the powerful flow of the Niagara River at the falls was by Daniel Joncaire in 1759. He diverted a small stream of water from the river to turn a waterwheel and thus power his sawmill.[1]

In 1877, Jacob Schoellkopf purchased an unfinished canal which had been constructed by a predecessor company that went bankrupt. He formed a company called the Niagara Falls Hydraulic Power & Manufacturing Company, which then completed the canal and built a powerhouse to generate DC electricity.

By 1882, Schoellkopf had seven power stations along the canal, each generating electricity that was used by factories within two miles of the generating stations. Schoellkopf remained president of the Niagara Falls Hydraulic Power & Manufacturing Company until his death on September 15, 1899.[2]

Independent of Schoellkopf's work, on July 1, 1886, a division engineer on the Erie Canal named Thomas Evershed issued plans for an ambitious generating system using water from the Niagara River. Evershed's idea was to dig a 2.5 mile-long tunnel parallel to the river starting above the Falls, running under the town of Niagara Falls, and ending below the Falls. A canal would convey water from the river to the beginning of the tunnel, where it would flow into thirty-eight 150-foot-deep shafts leading to pits containing turbine wheels. The wheels, driven by the falling water, would turn DC electric generators, which would send power to the surface for use by nearby factories. The plan called for generating 200,000

(**a**) Angel Falls in Venezuela, at 3,230 feet, is the world's tallest waterfall. But its flow rate is just 30,000 cubic feet per minute.
(**b**) Khone Falls in Laos near the Cambodian border, at 6.7 miles wide, is the world's widest waterfall. Its flow rate is at least twice as great as Niagara Falls.

Chapter 20. Over a Barrel at Niagara

Schoellkopf Mills along the Niagara Canal in 1900.

horsepower, which was about forty times larger than any other hydroelectric plant at the time.[3]

Wealthy residents attempted to raise the millions of dollars required to move the project forward, but such a large amount of money was not available in western New York. However, they did succeed in organizing the Niagara Falls Power Company and obtaining an 1886 New York state charter. Trying to keep the plan alive, a young New York attorney, William B. Rankine, discussed it with a friend and fellow attorney, Francis Lynde Stetson. Stetson happened to be the personal attorney of J.P. Morgan.[4]

Impressed by Evershed's plan to extract electrical power from Niagara Falls, Stetson presented it to Morgan as well as D.O. Mills, the prominent California banker; Edward Dean Adams, American businessman and banker; John Jacob Astor, American businessman and real estate developer; and several other men with large amounts of money to invest. Convinced that the plan had merit and could make money, they formed the Cataract Construction Company on June 12, 1889, to finance and execute the project.

Edward Dean Adams discussed the proposed project with foreign technologists during an 1890 trip to London. In June 1890, he organized the International Niagara Commission, which was to award $22,000 in prizes for the most useful ideas on:

- Building a central power station above the Falls to develop as much power as possible given the physical constraints of water volume and flow rate; and
- Transmitting that power overhead or underground using electricity, compressed air, water, or other means to manufacturing plants in a four-mile radius, and to the city of Buffalo, twenty-two miles away.

The chairman of the Commission was Sir William Thomson (the future Lord Kelvin), and it included prestigious engineers and academics from England, France, Switzerland, and the United States. The British Westinghouse Electric & Manufacturing Company and other major industrial concerns were invited to submit plans by January 1, 1891, to compete for the first prize of $3,000.

Twenty organizations and individuals submitted twenty-four proposals to the

International Niagara Commission.[5] The power transmission methods proposed included:

1. manila or wire ropes (to the Niagara Falls area);
2. hydraulic (water) transmission, again to the local area;
3. compressed air (the method favored by George Westinghouse at the time); and
4. direct current electricity.[6]

The Commission rejected all of the proposals but did express a preference for electrical transmission of power, with perhaps a partial use of compressed air as an auxiliary method.[7]

George Westinghouse refused to participate in the competition, explaining that the award was "an entirely inadequate sum to pay for one hundred thousand dollars' worth of advice." He added, "When the Niagara people are ready to do business, we shall make them a proposal."[8]

Meanwhile, Westinghouse was gaining practical experience with hydroelectric power in Colorado, 1,900 miles west of Niagara Falls.

The Ames Power Plant

The Gold King Mine near Telluride, Colorado, was discovered in 1887. Although abandoned in 1923, the mine made headlines in 2015. A contractor working for the Environmental Protection Agency accidentally released three million gallons of accumulated mine waste water heavily contaminated by toxic elements such as cadmium, lead, arsenic, and beryllium into the Animas River watershed. The spill, which was featured on network news broadcasts during the week of August 5, 2015, affected the water supply for cities and towns in Colorado, New Mexico, and Utah, as well as the Navajo Nation.[9]

George Westinghouse became involved in the Gold King Mine when "in 1890 a man from the west came east with a definite power transmission problem. His company was operating a stamp mill in the mountains of Colorado. The man from the west wanted to know whether electricity could transmit one hundred horse power a distance of three miles and replace the steam plant he was using."[10] The man from the west was Lucien Nunn, the manager of the Gold King Mine, and he was seeking help from the Westinghouse Electric & Manufacturing Company.

A stamp mill is essentially an automated set of sledge hammers that repeatedly beat on the potato-sized gold ore rocks to break them into small pieces that can then be chemically processed to retrieve the metallic gold. Before the advent of electricity, wood- or coal-fired steam boilers provided power for operating most stamp mills. However, at the high elevation of the Gold King Mine (above the timberline at 12,000 feet), timber was unavailable, and there was no railroad to deliver coal. Coal packed in by burros cost forty to fifty dollars per ton, contributing to the exorbitant cost for power of $2,500 per month. The fast-flowing water of the South Fork of the San Miguel River could provide the power required, but the river was almost three miles away and two thousand feet below the Gold King Mine.[11] How could that power be conveyed up the mountain?

Around 1890, power transmission methods included cable or rope drive, compressed air, high-pressure water, and electricity. But, as Lucien Nunn's brother, Paul Nunn, later wrote, "back in 1890, electricity was very much in doubt."[12] Adding to the uncertainty was the environment of the Gold King Mine. In the surrounding San Juan Mountains, temperatures of forty degrees below zero, avalanches, blizzards, and severe electrical storms were common. Given the rugged terrain, extreme environmental conditions, and distance involved, high voltage AC power transmission was the only viable alternative.

Lucien Nunn presented his problem and proposed solution to a group of Westinghouse engineers and managers. There was no precedent for such a project; no one had generated so much power and transmitted it over such a distance before, let alone in such a hostile environment. But Westinghouse management agreed to consider undertaking the challenge.

Nunn returned by train (the transcontinental railroad had been completed in 1869, so the trip took less than one week), and received bad news and good news when he reached Telluride. The bad news came in a telegram from Westinghouse Electric. After considering the multitude of issues, they concluded that the project would be too risky and expensive. The good news was that miners at the Gold King Mine had hit a rich vein of ore, and the mine was "in bonanza."

Wasting no time, Nunn gathered a satchel of gold ore and got back on the train to return to Pittsburgh. At his second meeting with Westinghouse personnel, he dumped the gold ore on the table and assured everyone that cost would not be a roadblock. Always eager for a technical challenge, and realizing that success would bolster the case for AC electricity, George Westinghouse agreed to tackle the project. Westinghouse was so impressed by Nunn's enthusiasm that he contributed $25,000 of his personal funds to augment the undertaking.[13]

Working with Louis B. Stillwell and other Westinghouse engineers, the Nunn brothers developed a plan for the project. Identical Westinghouse one-hundred horsepower alternators operating at 3,000 volts and 133 Hertz, the largest such units manufactured up to that time, arrived in the summer of 1890 and were installed that winter. The generator unit, situated near the San Miguel River, was connected by a belt to a six-foot Pelton water wheel.[(c)] The twin unit, at the stamp mill two thousand feet up the mountain, functioned as a motor. A power transmission line connecting the generator to the motor consisted of two No. 3 bare copper wires mounted on crossbars on poles. The cost of the wire for the AC transmission line was $700, while wire for an equivalent DC transmission system would have cost about $70,000.[14]

The initial operation of the system was in late 1890 or early 1891 and provided valuable experience for the Westinghouse engineers as they contemplated the challenge of harnessing Niagara Falls.

Even then, you could be sure of Westinghouse products. The power plant built to supply the Gold King Mine has undergone changes and upgrades over the years but still exists. It is now known as the Ames hydroelectric generating plant of the Public Service Company, a subsidiary of Xcel Energy.[15]

(c) A Pelton water wheel, developed by Lester Allan Pelton in the 1870s, extracts energy from the force of moving water (kinetic energy), rather than from its weight. Such wheels are ideal for low-volume, fast-moving streams of water.

Decision Time at Niagara

In January 1891, at the time the Ames power plant in Colorado came on line, the International Niagara Commission made its initial determination about the Niagara power plant. All except two members of the Commission rejected and condemned the use of alternating current for the new power plant.

Their decision must be viewed in light of then-present technology; the Commission members were not aware of the Ames hydroelectric plant, and the AC system at the World's Columbian Exposition was more than two years in the future.

Sir William Thomson, the chairman of the Commission, led the opposition to alternating current. The leading advocate for AC was George Forbes, an Edinburgh, Scotland-born, Cambridge-educated electrical engineer then living in London.

On December 14, 1891, the Cataract Construction Company changed Evershed's plan, substituting a central power station for his thirty-eight separate and dispersed turbine wheels. Based on the revised concept, they invited competitive plans and estimates for the power plant and transmission of power both locally and to Buffalo.[16]

Supersize It!

The generator at the Ames Power Station and the motor at the Gold King Mine were each rated at one-hundred horsepower.[d] Thomas Evershed's plans for harnessing Niagara Falls called for generating 200,000 horsepower (149 megawatts) of electricity!

As a tentative step toward the mammoth machines that would be required for the Niagara installation, in late 1892, the Westinghouse Electric & Manufacturing Company constructed and tested two 150 horsepower rotary converters.[e]

Westinghouse invited representatives from the Cataract Construction Company and the International Niagara Commission to visit Pittsburgh to inspect and test these machines. Dr. Coleman Sellers of Stevens Institute of Technology and Professor Henry Rowland of Johns Hopkins University accepted the invitation and tested the devices on January 10, 1893.[17] Professor George Forbes, who had been hired in 1892 as a consultant to the Cataract Construction Company, arrived somewhat later and performed his own tests. All of the visitors were favorably impressed, especially Forbes, who favored AC.

During these visits, Sellers, Rowland, and Forbes toured the Westinghouse facilities and met the lead engineers. They found the facilities to be ample for the proposed project. More important, the Westinghouse engineers, including mechanical engineer Albert Schmid and electrical engineers Lewis B. Stillwell, Oliver Shallenberger,

---

(**d**) At the dawn of the electric age, the size of electric machinery was measured in horsepower, because horses had been a major source of power up to then. Today we measure power in watts (or kilowatts). One horsepower is equivalent to 745.7 watts. The one-hundred horsepower machines at the Ames/Gold King site were equivalent to 74.57 kilowatts. Paul Nunn presented a 1956 retrospective look at the Ames/Gold King installation titled, "We Did Not Know What Watts Were." (General Electric Review, September 1956. Page 43.)

(**e**) A rotary converter was used to convert AC power to DC for those applications, such as streetcars, that required DC for their operation. Its structure is similar to but more complex than an AC motor of the same horsepower rating.

Chapter 20. Over a Barrel at Niagara

Benjamin Lamme, and Charles Scott, brought skill, experience, and creativity necessary to accomplish what had never before been attempted.

In March 1893, General Electric and Westinghouse Electric & Manufacturing both submitted proposals for three 5,000 HP AC generators and associated equipment. In the Westinghouse proposal, generators would be situated in a powerhouse at ground level with vertical shafts extending down 140 feet to water-driven turbines at the bottom of a wheel pit.

Westinghouse proposed to design the generators for 2,200-volt AC output and to use this voltage for distribution in the vicinity of Niagara Falls. For transmission

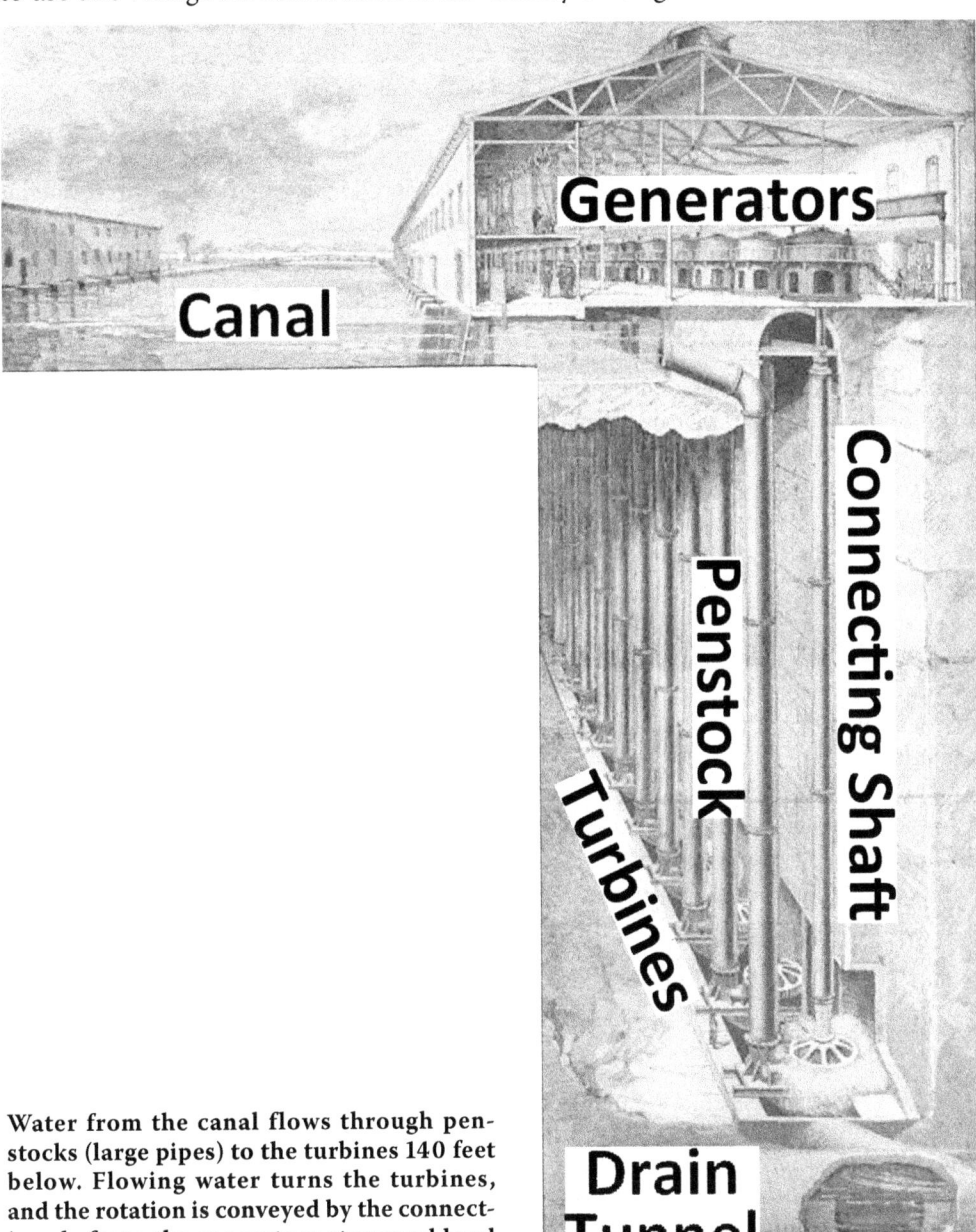

Water from the canal flows through penstocks (large pipes) to the turbines 140 feet below. Flowing water turns the turbines, and the rotation is conveyed by the connecting shafts to the generators at ground level (*Cassier's Magazine*, July 1895, p. 227).

to Buffalo, transformers would step-up this voltage to either 11,000 volts or 22,000 volts, depending on the capability of line insulators. At Buffalo, step-down transformers would reduce the voltage for industrial and residential use.

The frequency of the alternating current became the topic of much discussion. The Cataract Construction Company had already placed orders for turbines to run at 250 revolutions per minute (4.167 revolutions per second). To avoid gears, which would have wasted power and been a maintenance nightmare, the electrical frequency had to be an even number multiple of the revolutions per second of the turbines. So, the possible AC frequencies were 16⅔, 25, and 33⅓ cycles per second. Westinghouse chose 33⅓ cycles per second for their proposal.[f]

As was typical for virtually all motors and generators at the time, the Westinghouse design had an internal revolving armature (often called a rotor) and an external stationary field (often called a stator).

On May 6, 1893, the Cataract Construction Company, after extensive study, settled on alternating current as the means to deliver power both to the immediate vicinity and to distant customers in Buffalo and beyond. But even though GE and Westinghouse submitted proposals for AC systems, Cataract rejected both proposals.

Instead, they disclosed a revolutionary (pun intended) design from George Forbes, their new consulting engineer. Sometimes a consulting engineer believes that, to justify his fee, he has to advocate his own ideas in place of other solutions. We do not know whether that was the case with Forbes, but he proposed a generator with a stationary internal armature and a rotating external field, wound to produce 20,000 volts AC at a frequency of 25 cycles per second.

To understand the concept of a rotating external field, consider a modern ceiling fan. In most such fans, the center portion of the motor (the armature) is stationary, and the outer portion (the field) rotates, just as Forbes proposed. In the ceiling fan, the fan blades are attached to the rotating outer portion. Because the fan is overhead and out of reach, the rotating outer part of the motor, with its fan blades, is not hazardous. But Forbes was proposing a steel shell sixteen feet high, over eleven feet in diameter, weighing over 60 tons, and rotating at 250 RPM.[18] The physical structure was daunting, but Westinghouse engineers found the concept to have significant advantages.

However, Forbes's electrical design, with internal generator wires carrying 20,000 volts, was beyond impractical. Westinghouse engineers who evaluated the design wrote, "From our present knowledge of machine design it would have been a monumental failure."[19] The Westinghouse Electric & Manufacturing Company declined to accept any responsibility for the results if the machine were built.

Compromise

The Cataract Construction Company once again asked for bids. It specified the following parameters: two-phase AC at 2,200 volts and 25 cycles per second, the generator to rotate at 250 RPM and output 5,000 HP. Again, GE and

---

(f) Lower frequencies allow easier synchronization of generators so they can be connected in parallel. But if the frequency is too low, incandescent light bulbs produce an annoying flicker.

Westinghouse submitted proposals. In October 1893, Cataract executed a contract with Westinghouse for three 5,000 HP generators. The final design combined the electrical aspects of the Tesla polyphase system demonstrated at the 1893 Columbian Exposition with the mechanical structure of Professor Forbes's external rotating field design. GE received the contract for the step-up and step-down transformers.

By late 1893, AC generators as large as 1,000 HP had been built, but the leap to 5,000 HP was still a bold advance. George Westinghouse had full confidence in his engineers and machine shop. If they said they could complete the tasks, he believed them, and the results justified his faith. Westinghouse assigned Lewis B. Stillwell as the chief electrical engineer for the project, and Stillwell quickly became Westinghouse's man in charge of the entire Niagara project. He offered guidance to his co-workers, dealt directly with the Cataract Construction Company, and determined the contract conditions.

Westinghouse engineers Albert Schmid and Benjamin Lamme developed the modifications required to make the external rotating field design of Professor Forbes practical and manufacturable in the Westinghouse plant. Schmid was a Swiss mechanical engineer working at the French Westinghouse Air Brake Company when he was personally selected by George Westinghouse to come to the United States in 1885. He became chief designer and engineer when Union Switch and Signal started its work with electrical systems.

Benjamin Lamme received a degree in Electrical Engineering from The Ohio State University in 1888. He came to work for Westinghouse's Philadelphia (Natural Gas) Company in early 1889 but soon transferred to Westinghouse Electric, where he developed a practical version of Tesla's induction motor. He then designed much of the apparatus for the 1893 Columbian Exposition in Chicago. Lamme brought a systematic, analytical approach to electrical machine design, which previously had relied on cut-and-try methods. As Edward Dean Adams wrote of Lamme in 1927, "He did not imitate; he originated."[20]

By June 1895, Westinghouse Electric & Manufacturing had constructed the first of the massive generators. Designed to yield 5,000 HP, laboratory testing of the generator showed that it provided 5,315 HP.

The following photographs demonstrate the colossal size and unconventional design of the generators. The pictures were included in a 51-page paper titled "Electric Power Generation at Niagara,"[21] written by Westinghouse engineer Lewis Buckley Stillwell and published in *Cassier's Magazine* in July 1895. That magazine included 212 pages describing all aspects of the Niagara Falls project.

The next photograph shows two of the original 5,000 HP generators. The one at the right front is not operating; the bolt heads holding the field coils in place are visible. The generator at the left rear is running. The rotation of the field shell causes the bolt heads to be blurred. When rotating at the design rate of 250 RPM, the outer surface of the rotating field moved at 9,300 feet per minute, or 105 miles per hour!

A Chinese visitor to the power station, Mr. Li Hung Chang, fascinated by the massive rotating shell, probed it with the point of his umbrella. To his chagrin, a protruding bolt head caught the umbrella and propelled it with great force across the room. Fortunately, no one was injured.[22]

Chapter 20. Over a Barrel at Niagara   153

Rotating field ring ready to be lowered over the stationary armature; note the two men to the left (*Cassier's Magazine*, July 1895, p. 258).

Opposite, top: George and Marguerite Westinghouse at Niagara Falls, 1875 (Westinghouse Collection, Detre Library and Archives Division, Senator John Heinz History Center, Pittsburgh, Pennsylvania).
   Opposite, bottom: Rotating field ring with poles and bobbins in place (*Cassier's Magazine*, July 1895, p. 277).

Two 5,000 HP generators at the Niagara Falls power house (Westinghouse Collection, Detre Library and Archives Division, Senator John Heinz History Center, Pittsburgh, Pennsylvania).

Throw the Switch!

On August 26, 1895, the Niagara Falls power station delivered its first commercial power to the nearby Pittsburgh Reduction Company (now ALCOA) aluminum plant.

During the next year, bare copper wires were strung on wooden poles from Niagara Falls to the tiny Brace Street terminal house in Buffalo, New York.

At exactly midnight on November 15, 1896, Buffalo mayor Edgar B. Jewett threw the switch connecting the Niagara power station to the Brace Street station. An incandescent lamp glowed, and a small motor started turning. Alternating current at 11,000 volts flowed twenty-two miles, initiating a revolution that continues to this day.[23]

Official Dedication

The official opening ceremony for the Niagara Falls power plant occurred on January 12, 1897. As usual, partly because he did not enjoy speaking in public, George Westinghouse stood in the background and allowed others to take the credit. Nikola Tesla, upon whose inventions rested the successful operation of the

Chapter 20. Over a Barrel at Niagara

Niagara plant, often sought publicity and was more than pleased to appear in the spotlight. As part of his eloquent speech, Tesla said, "We have many a monument of past ages; we have the palaces and pyramids, the temples of the Greek and the cathedrals of Christendom. In them is exemplified the power of men, the greatness of nations, the love of art and religious devotion. But the monument at Niagara has something of its own, more in accord with our present thoughts and tendencies. It is a monument worthy of our scientific age, a true monument of enlightenment and of peace. It signifies the subjugation of natural forces to the service of man, the discontinuance of barbarous methods, the relieving of millions from want and suffering."[24]

Many years later, in 1938, Tesla again made clear Westinghouse's irreplaceable contribution: "George Westinghouse was, in my opinion, the only man on this globe who could take my alternating-current system under the circumstances then existing and win the battle against prejudice and money power. He was a pioneer of imposing stature, one of the world's true noblemen, of whom America may well be proud and to whom humanity owes an immense debt of gratitude."[25]

By 1904, ten of the 5,000 HP generators, all built by Westinghouse Electric & Manufacturing Company, had been installed in the Power House. The *Works of the Westinghouse Electric & Manufacturing Company, 1904*, reported,

> Power House No. 1 now contains ten generators of the outside revolving field, vertical shaft type. These ten generators are divided into banks of five each, and each bank has a

Edward Dean Adams Power Plant, circa 1960—Power House No. 1, site of the original generators, is on the right. Power House No. 2, constructed in 1903, is to the left rear. The Transformer Building is in the left foreground. All buildings except the Transformer Building were demolished in 1961 (Library of Congress).

separate switchboard. Current is generated at about 2,200 volts and a great deal of the service in and around Niagara Falls is served direct at this voltage. For a part of the more distant Niagara Falls service, voltage is stepped up to 11,000, and for the Buffalo, Lockport, and Tonawanda service, voltage is stepped up to about 22,000. The development has been so rapid that the demands on the long-distance service alone now take about two-thirds of the power developed by these first ten machines. It has been necessary, therefore, to build a new power house (Power House No. 2 in the Power Plant photograph) on the opposite bank of the canal for the large amount of local service which has to be rendered.[26]

There are no monuments to George Westinghouse at Niagara Falls except for the decaying Transformer Building. But there are two memorials to Nikola Tesla, one on the American side and one on the Canadian.

Based on the work of Westinghouse, Tesla, Stillwell, Schmid, Lamme, and countless others, electrical power is now available throughout the world, generated at power plants far removed from the consumers of that power and transmitted as alternating current at voltages up to 1,150,000 volts.

But in an ironic twist that would make Thomas Edison proud, some of the most

Two Statues Commemorating Nikola Tesla. At left: On Goat Island, Niagara Falls, NY; by Frano Krşiniç, 1976; Gift to the United States by the Yugoslavian government (Michael Gray, Flickr). At right: standing on his induction motor, Queen Victoria Park, Ontario, Canada; by Les Drysdale; sponsored by St. George Serbian Orthodox Church (Laslovarga, Wikimedia Commons).

modern and highest-capacity electric transmission systems use high-voltage direct current (HVDC).[g]

Tesla After Niagara Falls

Harnessing the energy of Niagara Falls would have been adventure enough for even a creative man, but Nikola Tesla was well beyond creative. Ideas that sprang from his mind continued to lead to patents, both esoteric and useful. One Tesla idea that impacts our daily life is remote control by radio waves. In 1898, he demonstrated the world's first wireless remote-controlled boat at Madison Square Garden. To say that his technology was ahead of its time would be a gross understatement.

Tesla was also well-known for his elegant tastes and poor money management skills.

In 1900, he visited George and Marguerite Westinghouse at Solitude, and the next day rode with George aboard Glen Eyre to New York City. As was often the case, Tesla was seeking money. This time he needed $6,000 to finance his experiments on transmitting power without wires. Westinghouse was loath to advance corporate money on ventures that could disrupt power transmission over wires, a technology essential to Westinghouse Electric. But he could not turn down an old friend, so he loaned the money to Tesla from his personal funds.

He also suggested that Tesla approach J.P. Morgan for additional backing. Tesla did so, and in February 1901, Morgan advanced $150,000 in return for 51 percent of the company formed around the venture as well as 51 percent ownership of Tesla's present and future patents in wireless transmission. However, when Tesla burned through that money, Morgan refused to lend him more to continue the venture.[27]

The irony here is that Anne Tracy Morgan, J.P. Morgan's youngest daughter, was once infatuated with Tesla. We can only speculate what would have happened if Tesla had encouraged Anne's romantic feelings and married her. Perhaps Morgan would have been more generous with his son-in-law, thus unleashing Tesla's genius.

In his later years, Tesla's ideas verged on the ridiculous, such as communicating with the inhabitants of Venus. But his tastes remained extravagant, and he lived in a series of expensive New York City hotels, moving on whenever the manager demanded payment.

Finally, on January 2, 1934, to avoid embarrassment to Tesla and possibly to the Westinghouse name, Tesla was placed on the payroll of the Westinghouse Company as a consulting engineer. The company paid for Tesla to stay (and eat) at the Hotel New Yorker until his death on January 7, 1943.[28]

(g) HVDC is often used to transfer power between two AC grids that are not synchronized. Also, losses can be less for a high-voltage DC transmission line as compared to an AC line operating at the same voltage. A well-known HVDC transmission line, called the Pacific DC Intertie, carries up to 3,100 megawatts of power from a generating station near The Dalles, Oregon, to Los Angeles.

Chapter 21

New Lands to Conquer

> *"Savages may indeed be a formidable enemy to your raw American militia; but upon the king's regular and disciplined troops, Sir, it is impossible they should make an impression."*
> —British General Edward Braddock
> spoken to Benjamin Franklin in 1755

George Westinghouse had recently celebrated two momentous victories for alternating current. The first came at the 1893 World's Columbian Exposition in Chicago, where his generators powered the entire Exposition, and his light bulbs turned night into day. The second occurred in 1895 at Niagara Falls, where he and Nikola Tesla successfully harnessed a small part of the hydraulic power of the Falls to generate and then supply electricity to users.

Based on those triumphs, the demand for equipment to create and utilize AC electricity was booming, and capacity at the Garrison Alley plant of Westinghouse Electric & Manufacturing quickly became overwhelmed.

When visiting his Air Brake facilities in Wilmerding, Westinghouse often passed the confluence where Turtle Creek emptied into the Monongahela River and thought it would be an ideal location for a manufacturing plant. In addition to the land's proximity to his Wilmerding works, railroads and water for transportation of raw materials and finished equipment were nearby, and the area was undeveloped.

The ground that caught Westinghouse's eye had a violent history.

Prelude to a War

In 1755, 15-year-old John Hendrick Wistinhausen, who would become George Westinghouse's great-grandfather, had settled in what is now Pownal, Vermont. On May 29, 1755, 450 miles southwest of Pownal, an arrogant General Edward Braddock, Commander-in-Chief of British forces for the Thirteen Colonies, led his troops west from Fort Cumberland, Province of Maryland.

Braddock's intermediate goal was to capture French-controlled Fort Duquesne at the junction of the Allegheny, Monongahela, and Ohio Rivers (the Point in present-day Pittsburgh). After that supposedly easy assignment, Braddock was to proceed north, capturing French positions all the way to Fort Niagara. These skirmishes were part of the fight for control of North America between France and Britain that would lead to the French and Indian War (1756–1763).

With Braddock were two regiments of British regular soldiers, local troops from the colonies, and a young Virginia Colonel named George Washington, who volunteered as an aide-de-camp to Braddock. Braddock and his forces marched over 125 miles northwest, over mountainous territory, carrying with them massive artillery and siege cannons. The primitive trail over the Alleghenies had to be widened to accommodate the artillery and supply wagons.

After 40 days of struggling through the wilderness, Braddock and his forward contingent encountered the Monongahela River about ten miles southeast of their target of Fort Duquesne. They crossed the river twice, first near present-day McKeesport and then again just west of its junction with Turtle Creek.

French troops, dispatched from Fort Duquesne, had wanted to ambush Braddock's forces while they were in the river and thus helpless. But their native American allies were not ready to attack, so the encounter was delayed until the next day, July 9, 1755.

Braddock's army was relieved to reach what is now the site of the Braddock Battlefield History Center in North Braddock without being attacked. Meanwhile, the French leader, Captain Daniel Lienard de Beaujeu, divided his forces into three columns. He led the center column in a frontal assault on Braddock's forces while the

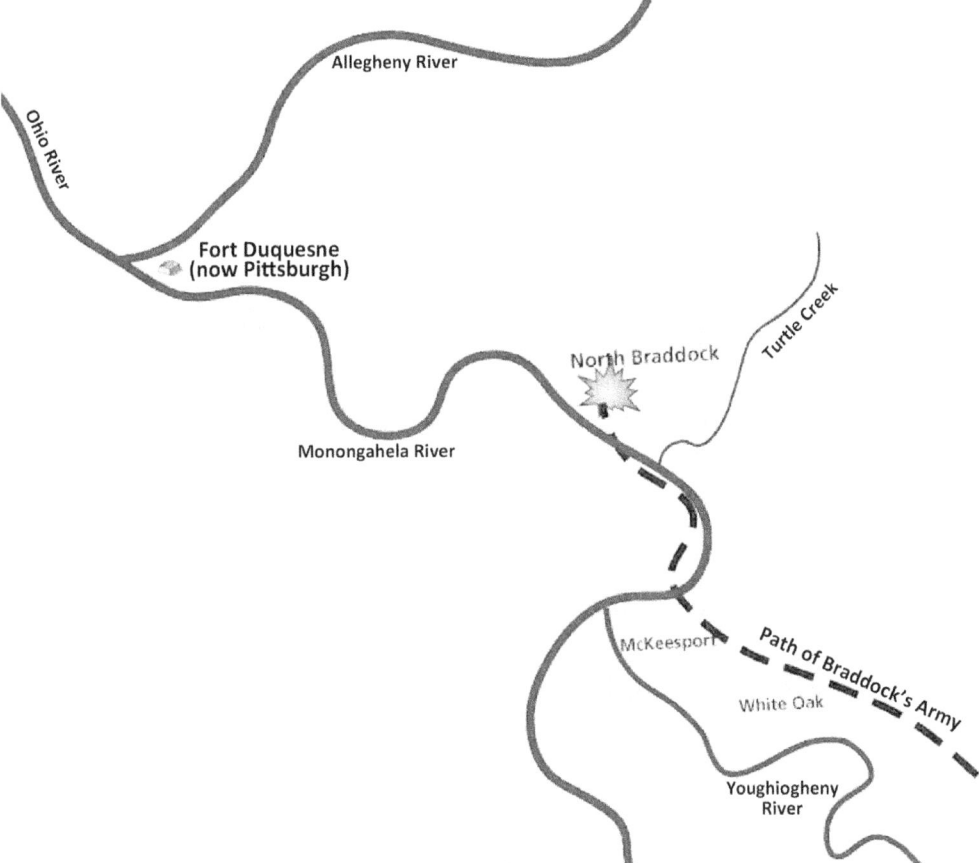

Route of Braddock's Army to the Battle of the Monongahela (drawn by Tom Markwardt, National Park Service).

two outer columns of native forces flanked the opposing army and pinched them in between. One of the French reported of the Indians, "They came like foxes, attacked like lions, and fled like birds." Beaujeu was killed at the beginning of the battle, but the rest of his plan worked to perfection.

Three hours into the ensuing battle, Braddock, who had four horses shot from under him, was shot in the lung while mounting his fifth horse.[1] Without an official order being issued, the British troops began to withdraw. When the retreating soldiers reached the Monongahela River, they were met by Indians who were allies of the French. Facing the threat of scalping, the British broke ranks and fled in every direction. The French and Indians decisively won what came to be called The Battle of the Monongahela.

Of the 1,469 British soldiers who crossed the Monongahela River, 976 (66 percent) were killed, wounded, or missing. Braddock died of his wounds four days later, on July 13, 1755.[a] The towns of Braddock and North Braddock and the Braddock Road (portions of which are now U.S. Route 40) commemorate the defeated general.[2]

A New Use for an Old Battleground

In 1894, George Westinghouse would start building his massive Westinghouse Electric & Manufacturing Company plant on forty acres of farmland just upriver from where Braddock had suffered his humiliating defeat. The site of General Braddock's failure on the banks of Turtle Creek, less than three miles from Wilmerding, would become Westinghouse's largest factory.

The new facility was enormous, with a total floor space of over two million square feet. The old Garrison Alley manufacturing space had narrow aisles and limited crane capacity that made producing the massive Niagara Falls generators a significant challenge.

The East Pittsburgh buildings would eliminate those problems. The new main shop building measured 373 feet by 1,184 feet and was divided into four sections, each with aisles seventy feet wide to allow the manufacture of the largest conceivable machines. Traveling cranes with capacities from ten to fifty tons traversed each section. The sections were organized and equipped to produce a specific type of device: Section A for AC and DC street railway motors; Section B for motors and generators larger than 100 KW; Section C for AC and DC motors and generators up to 100 KW; and Section D for gigantic equipment.

Raw materials and sub-assemblies entered at one end of the section. Additional sub-assemblies and parts were added, and machining was done as the assembly moved along the aisle. Finished equipment was loaded onto railroad cars resting on standard-gauge tracks at the far end of the aisle. Rigorous testing was done after every stage of the assembly process. Correspondence, design sketches, and small drawings

---

(a) The Battle of the Monongahela marked the beginning of George Washington's reputation for courage, bravery, and resourcefulness under duress. As Peter Stark relates in *Young Washington: How Wilderness and War Forged America's Founding Father*, "... the hero of the battle—if there could be one—was one George Washington. Word spread among Virginians and other American colonists of his bravery in the face of battle, his recovery of the wounded Braddock, and his remarkable escape with bullet holes through hat and coat and horses shot out from under him." "Even Lord Halifax, who shared the British low regard for the colonial forces, was impressed."

Chapter 21. New Lands to Conquer

Westinghouse Electric & Manufacturing plant along Turtle Creek in East Pittsburgh, 1895 (Westinghouse Collection, Detre Library and Archives Division, Senator John Heinz History Center, Pittsburgh, PA).

were distributed throughout the facility via a pneumatic tube mail system with the longest tube measuring 1,550 feet.[3]

Pittsburgh and the rest of the nation were in the middle of a recessionary period that started with the Panic of 1893 and did not end until 1897. But Westinghouse had recently cemented deals with August Belmont, Jr., and other bankers, so he was on relatively firm ground. He relied heavily on that support for the massive amount of capital necessary to build and equip the East Pittsburgh plant.

Because the new factory was to manufacture electrical equipment, including motors and generators, it was natural for Westinghouse to construct an on-site power plant to provide electricity to run the machinery. Point-of-use electric motors replaced the previous maze of long and dangerous leather belts connecting each machine to an overhead, steam-engine powered drive shaft. The Westinghouse Electric factory was far different from other factories of the day, such as the one in Chapter 7 showing the Westinghouse Air Brake Wilmerding facility. The East Pittsburgh factory marked the transition from steam power to electricity for manufacturing plants.

Manufacturing and engineering employment at Westinghouse Electric & Manufacturing, mostly at the East Pittsburgh plant, stood at 5,000 in 1895; 8,400 in 1900; and grew to 15,300 in 1905.

As had been the case since 1881 at the Allegheny City Westinghouse Air Brake factory, worker safety and comfort were vital to George Westinghouse. Skylights provided natural lighting, which was augmented by incandescent, arc, and mercury vapor lamps. Fans feeding large-diameter pipes distributed heating and ventilation. Westinghouse even subsidized meals in employee dining rooms.

Main aisle of shop, Westinghouse Electric & Manufacturing Company in East Pittsburgh; Note one of the Niagara Falls rotating field generators at left foreground. (Westinghouse Collection, Detre Library and Archives Division, Senator John Heinz History Center, Pittsburgh, PA).

Education and social development of Westinghouse employees had always been priorities. For the professional staff, the Electric Club was the primary vehicle to accomplish these objectives. Membership consisted of about 500 office, engineering, and apprentice personnel at all Westinghouse facilities in the vicinity. A lending library featured both popular and technical books and magazines. Prominent men, from both inside and outside the Westinghouse organization, occasionally gave lectures on non-technical topics. Nearly every evening, a 300-person assembly hall hosted engineers from the company staff as they presented lectures and discussions, both elementary and advanced, on subjects related to general and electrical engineering.

Benjamin G. Lamme, who later became chief engineer at Westinghouse Electric & Manufacturing, said this about the education of Westinghouse engineers: "In my thirty-five years of work with the Westinghouse Company I have seen many young men grow from pupils to assistants and associates. This has been one of my greatest pleasures. I have aimed to instill in them fundamental ideas of engineering honesty and honor, square dealing and fair fighting—that there should be pride in accomplishment and that true engineering means more than merely making a living—that it means advancement of the art for the benefit of mankind."[4]

The technical presentations from the Electric Club were published in a monthly newsletter, which eventually grew into a respected technical journal. Volume 1 of The

Electric Club Journal, covering the period from February to December 1904, is currently available as a reprint online.[b]

The Casino Building provided recreational and educational programs aimed at the shop workers. Recreational facilities included bowling alleys, pool rooms, and auditoriums for movies, lectures, and live shows. Night school courses included English, math, science, and shop practices, each with two terms of about five months each, with twelve hours of instruction per week. The company supported 70 percent of the cost of these courses, but students paid $2 per month, reflecting Westinghouse's view that students should pay something for their education.[5]

Westinghouse wanted all products that bore his name to maintain the highest standard of quality and reliability. To that end, he introduced rigorous testing and quality control practices at every step of the manufacturing process. Inspectors checked all incoming material to ensure that it met specifications and performance objectives. Random samples of finished products were evaluated to make sure they met performance and appearance standards. The entire lot could be rejected if the sample failed. Customer site inspectors were welcomed to do their own observations and evaluations.

Westinghouse Electric & Manufacturing was the largest of the Westinghouse companies, and the East Pittsburgh plant was reported by the *Pittsburgh Press* to be the largest and most modern workshop in the world.[6]

The American Mutoscope Biograph Company filmed short (~6 minutes) movies inside and outside of the East Pittsburgh plant as well as at the Wilmerding and Trafford plants. Twenty-two of these fascinating and revealing movies are available for viewing online at https://www.loc.gov/collections/films-of-westinghouse-works-1904/.

When Westinghouse died in 1914, reminiscences of his employees were solicited, and hundreds were collected. One anecdote had to do with the East Pittsburgh plant prior to its opening and reflects the generosity that characterized George Westinghouse. The story relates to the Grand Army of the Republic (GAR), which was a fraternal organization composed of veterans who had fought on the Union side during the American Civil War. The GAR was founded in 1866 in Springfield, Illinois, and grew to include hundreds of "posts" (local community units) across the nation. It was dissolved in 1956 at the death of its last member, Albert Woolson (1850–1956).[7]

Former Westinghouse employee G.S. Davidson wrote:

> In September 1894, Pittsburgh entertained the Grand Army of the Republic. We were getting along very comfortably with the thought in our minds that we would have nothing to do in our jobs when Mr. Westinghouse suddenly sent for us. In effect he said, "The new main buildings of the Westinghouse Electric & Manufacturing Company at East Pittsburgh (twelve miles from the Pennsylvania Railroad terminal in Pittsburgh) will be ready in ample time for the plans I am proposing to you gentlemen. I shall turn these buildings, and with them the services of their architects, over to you when the buildings are completed. You will plan a Grand Ball for an evening during Grand Army Week to the extent of five thousand guests. You will make such changes in the buildings as may be necessary to adapt them to this affair. You will arrange and pay for the necessary transportation of the guests by way of the railroads. Among the alterations to be made you will build a grand stairway to the Ball Room on the second floor. Send all bills for work to me. In addition, I will give

---

(b) https://www.amazon.com/Electric-Club-Journal-Vol-February/dp/1333935846.

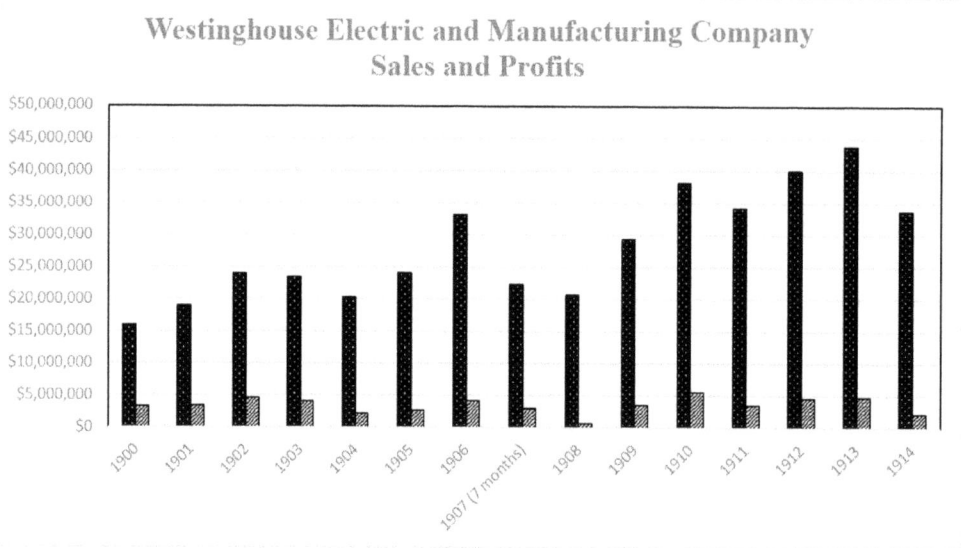

you … (naming a large sum) to cover the cost of furniture and equipment necessary. Our veteran visitors who are coming from all over the world must see how Pittsburgh plays the part of host."

A week before the event, Mr. Westinghouse came to Pittsburgh unexpectedly and with us went to view the progress that was being made on his order. He approved everything that had been done except the color scheme in the carpeting and rugs in connection with the grand stairway. He bluntly told us what we must do to give that part of our work the proper tone.

I might add that the general opinion was that this was the outstanding incident of Grand Army Week in Pittsburgh.[8]

Business was booming for the Westinghouse Electric & Manufacturing Company. Sales in 1890 were $4.3 million and grew to $16 million in 1900. Sales and profits for subsequent years are shown in the graph.

The year of 1907 shows sales and profits for just seven months. The events of that financially tumultuous year will be covered in detail in Chapter 24.

Chapter 22

Rotary Redux

> *"The time will come when people will travel in stages moved by steam engines, from city to city, almost as fast as birds fly,—fifteen or twenty miles an hour. Passing through the air with such velocity, changing the scene in such rapid succession, will be the most exhilarating exercise."*
> —Oliver Evans, American inventor, 1840

Rotary Engines, Turbines, and Turbo-Generators

From his early years in his father's machine shop in the 1860s, George Westinghouse had been fascinated by rotary engines. His first patent, issued on October 31, 1865, was for a rotary steam engine,[a] and Westinghouse tinkered with rotary engines for the rest of his life.

Conventional engines include a piston which is driven back and forth (reciprocating) by the expansion of a gas, whether it is steam or produced by burning natural gas or gasoline. The reciprocating motion of the piston must be transformed into rotary motion to drive loads such as locomotive wheels or electrical generators. The advantage of a rotary engine is that it skips the reciprocating motion and thus can be far more efficient.

The first rotary engine was described in the 1st century AD by Hero of Alexandria in Roman Egypt and was more like a toy. As shown by the diagram, the spherical chamber rotated on its axis because of the force of steam escaping through angled jets.

Over the next eighteen centuries, numerous inventors made practical but sporadic improvements on Hero's toy.

Finally, in 1883, Sir Charles Algernon Parsons invented the modern steam turbine and used it to generate 7.5 KW of electricity. Parsons patented his turbine in 1883, and, by 1889, he had built 300 steam turbines with capacities up to 75 KW. Parsons continued his development of turbines, and in the next five years, achieved power levels of 350 to 500 KW.[1]

Foreseeing the application of steam turbines to propel ships, Parsons and five associates established the Parsons Marine Steam Turbine Company in 1893. To demonstrate the potential of steam turbine propulsion, the company built the *Turbinia*, the world's first steam turbine-powered ship. The *Turbinia* was far faster than any vessel at that time (34 knots, or 39 mph) and revolutionized marine power plants.[2]

(**a**) U.S. Patent No. 50,759.

Although the steam turbine was virtually unknown in the United States, George Westinghouse had been following the work of Parsons and, in 1895, purchased a license to the Parsons patents for manufacturing in the U.S. for other than marine applications.

The original Parsons turbine operated at a high rotational speed and had low efficiency. Westinghouse and his engineers at the Westinghouse Machine Company modified the design to increase the power, lower the speed, and improve the efficiency.

The revised designs compared favorably in efficiency to reciprocating steam engines, but they weighed much less, were much smaller, and required less care and maintenance.

The improved turbine design was coupled to an electrical generator designed and built by the Westinghouse Electric Company to create the first commercial turbine generators built in the U.S. Three of these 400 KW machines, running at 3,600 RPM and producing 440 volts at 60 cycles, were installed in 1898 to provide electricity for the Westinghouse Air Brake plant at Wilmerding.[3]

The Westinghouse Machine Company had been founded in 1881 to manufacture gas and steam engines. Their primary facility was on a fifty-acre site in East Pittsburgh, adjacent to the Westinghouse Electric & Manufacturing East Pittsburgh works. Westinghouse Machine

Hero's Aeolipile steam turbine (Wikimedia Commons).

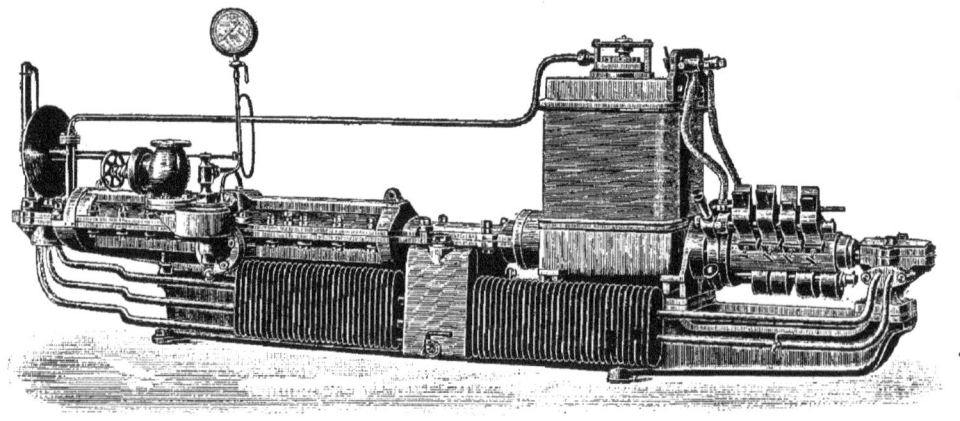

Parson's 1887 steam turbine-generator (Wikimedia Commons).

Chapter 22. Rotary Redux

Turbinia at speed, 1897 (Wikimedia Commons).

and Westinghouse Electric together continued to build larger and larger machines, culminating in 1910 in a massive 1,000 KW steam turbine.

Turbine structures are complex, and the terminology describing the technology is dense with terms having special meaning to those in the field and no meaning to the rest of us. Appendix V—How Does a Turbine Operate provides an overview of turbine structure and operation and will help in understanding Westinghouse's contributions. He was not just a manager of engineers creating steam turbines; he was an active, significant contributor to the development of the technology.

George Westinghouse obtained at least thirteen patents related to steam engines and steam turbines, starting with his very first patent on October 11, 1865, and ending on June 20, 1911.

Two groups of his patents were of significant importance to turbine development.[4] The first[b] focused on the close clearances between the rotor blades and the surrounding structure, and the impact of blade deformation and water droplet-enhanced pitting on that clearance. The second group[c] described new turbine architectures, including the single-double flow and reaction-impulse structures, substantially extending the size and speed limits of turbine construction.

(b) U.S. Patents 807,003 and 807,145 and 807,146, all issued on December 12, 1905; and 866,171 issued on September 17, 1907.

(c) U.S. Patents 787,485 issued on April 18, 1905; 816,516 issued on March 27, 1906; 935,569 issued on September 28, 1909; and 995,508 issued on June 20, 1911.

As a manager, Westinghouse stimulated many others to invent and pushed forward research and development.

Herbert T. Herr, vice-president of Westinghouse Electric & Manufacturing, wrote, "Whether he was in Pittsburgh or New York or Lenox, he would invariably call me on the telephone several times a day to inquire how things were going. His usual questions would be: 'How are you now? Did you get that turbine running again?' Then would follow a great many terse and direct questions."[5]

Turbo-generators (steam turbines sharing a common rotating shaft with electrical generators) multiplied in size and scope of application. The following table summarizes some early turbo-generators installations[6]:

_____			
Turbo-Generator Evolution			
Year	*Power Output*	*Location*	*Manufacturer*
1898	400 KW	Westinghouse Air Brake, Wilmerding, PA	Westinghouse
1901	1500 KW	Hartford Electric Light Company, Hartford, CT	Westinghouse
1903	5000 KW	Commonwealth Edison, Chicago, IL	GE
1905	3750 KW	New York, New Haven & Hartford RR	Westinghouse
1923	10,000 KW	Hartford Electric Light Company, Hartford, CT	GE
1928	80,000 KW	United Electric Light & Power, Hell Gate Station, New York, NY	Brown-Boveri

General Electric currently provides steam turbines capable of generating up to 1,200 MW (1,200,000 KW) of electricity in fossil fuel plants and 1,900 MW (1,900,000 KW) in nuclear plants.[7] In 2018, 61 percent of utility-scale electricity generation was by steam turbo-generators.[8]

All the Ships at Sea

As Charles Parsons had demonstrated in 1894 with his ship *Turbinia*, turbine engines with their rotary motion are a natural complement to the propulsion of naval ships by rotating propellers. The primary problem with turbines propelling ships is that they are most efficient when operated at high rotational speeds. Conversely, ship's propellers must run at relatively low speeds to prevent cavitation, the formation of air bubbles in the water around the propeller, which severely limits efficiency.

Despite the speed incompatibility issue, turbine engines were installed on some of the largest and fastest ships being built, specifically the *Lusitania* and *Mauretania*, both launched in 1906. Because of the turbine engines, these massive ships were able to cruise at speeds up to 25 knots (29 mph), and they successively captured the record for the fastest transatlantic crossing. Fuel consumption at top speed was reasonable, but operation at lower speeds consumed excessive amounts of fuel because of the low turbine efficiency.

The obvious answer to the speed mismatch was to add reduction gears to allow the turbine and propellers to both rotate at their preferred rates. But it was no simple task to transmit such large amounts of power from a high-speed turbine.

George Wallace Melville, George Westinghouse, and John Macalpine, circa 1904 (Library of Congress).

To solve the problem, Westinghouse enlisted the aid of two marine experts, Rear-Admiral George W. Melville and marine engineer John H. Macalpine, to report on the state-of-the-art and suggest improvements.

In May 1904, Melville and Macalpine reported back to Westinghouse, and he was so impressed by their work that he authorized the construction of a full-size model designed to transmit 3,000 horsepower.

At the time, Westinghouse Machine Company had been placed under the control of receivers acting for its creditors, and they strongly resisted the expenditure of funds to complete the model. As was his nature when he saw an opportunity for a breakthrough, Westinghouse persisted. The model ended up costing $75,000 but performed beyond the most optimistic hopes of the developers. It proved capable of transmitting 5,000 HP and demonstrated its capabilities when installed in the collier, USS *Neptune*, which launched on January 21, 1911.[9]

The work of Melville and Macalpine earned the following headlines on the front page of the October 3, 1909, edition of the *New York Times*:

NEW INVENTION TO TRANSFORM SHIPBUILDING;
Device of Melville, Westinghouse, and Macalpine May Cause the Rebuilding of All Navies.

USES TURBINES' FULL FORCE
By Solving the Problem of Hitching a Fast Turbine to a Slow Propeller.

SAVES SPACE, FUEL, POWER
Will Give a *Mauretania* 2 Knots More Speed,
2,150 Tons Less Weight, and Save $2,000,000 in Construction.[10]

The timing of Melville and Macalpine's work and Westinghouse's demonstration of it were fortunate, as turbine propulsion would be available to support the U.S.

USS *Neptune*, circa 1912 (George Grantham Bain collection at the Library of Congress via Wikimedia Commons).

effort in World War I. By the spring of 1920, turbine engines were powering twelve destroyers, three battleships, and two auxiliaries for the U.S. Navy. In total, 211 turbine-equipped ships had been launched, and 101 more were under construction.[11]

Thanks primarily to Westinghouse's insight and perseverance, steam turbines have become the preferred solution for generating massive amounts of power, both on land and in the water.

Chapter 23

Trolleys and Trains

"I went to lose a jolly hour on the trolley
And lost my heart instead."
—From *"The Trolley Song,"* Ralph Blane, Lyricist

Horsecars and Trolleys

In the mid–1880s, about 18,000 "horsecars" traveled over 6,000 miles of track in the United States.[1] These horsecars rode on iron or steel tracks and were pulled by horses or mules. By the mid–1890s, most of those animal-drawn streetcars transitioned to electric motors for their motive power, and by 1895, almost 11,000 miles of track were available for electrified streetcars.[2]

George Westinghouse did not initiate the transition from horse power to electric power but benefited from the pioneering work of two other men.

The first significant contributor to electric streetcar technology was Frank Julian Sprague, an 1878 graduate of the U.S. Naval Academy. Sprague left the Navy to work for Thomas Edison in 1883. He brought mathematical methods to Edison's trial-and-error experiments but grew unhappy with his salary and assignments at Menlo Park. After just eleven months, he left Edison in 1884 and founded the Sprague Electric Railway & Motor Company (SERM) later that year.[a]

Two of Sprague's key inventions were a non-sparking motor with fixed brushes[b] and a system for regenerative braking.[c] After trying unsuccessfully to interest New York City's Manhattan Elevated Railroad in his new motor, Sprague and SERM, in the spring of 1887, contracted to build a twelve-mile-long street railway in Richmond, Virginia. Despite grades of up to 10 percent, muddy terrain, and tight turns, the Union Passenger Railway opened on February 2, 1888.[3]

The magnitude of Frank Sprague's contributions to advancing electric trolleys is

---

(a) The son of Frank and Harriet (Jones) Sprague, Robert C. Sprague, founded the Sprague Electric Company in 1926. Sprague Electric became known for manufacturing high-quality electronic components, primarily capacitors.
  In 1962, near the end of my Bachelor of Science program at the University of Pittsburgh, I visited North Adams, Massachusetts, to interview for an entry-level engineering position at Sprague Electric Company. I received an offer from Sprague but elected to join Bell Telephone Laboratories instead. Sprague Electric struggled through the 1970s and closed the North Adams facility in 1985.
(b) U.S. Patent 428,732. "Electric Motor and Generator." Filed July 19, 1884. Issued May 27, 1890.
(c) U.S. Patent 318,668. "Method of Operating Electric Railway Trains." Filed December 22, 1884. Issued May 26, 1885.

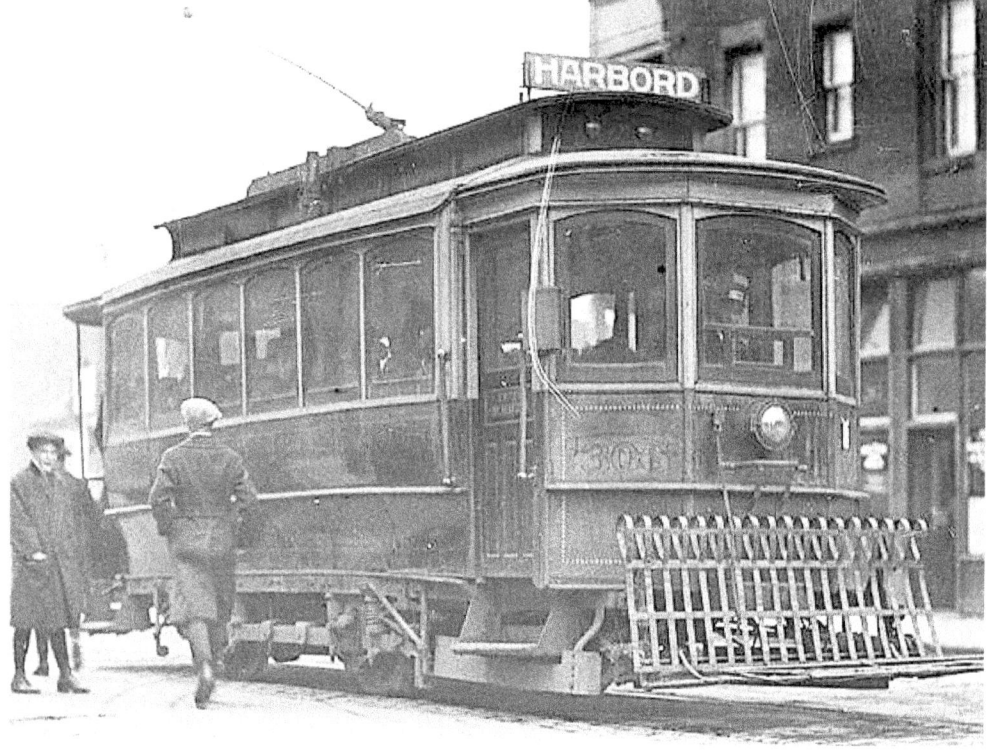

Top: Horse-drawn streetcar in Toronto, circa 1890 (Toronto Public Library via Wikimedia Commons). Bottom: Electric streetcar in Toronto, circa 1892 (Wikimedia Commons).

captured in a letter written by Elihu Thomson, the co-founder of Thomson-Houston Electric Company, on the occasion of Sprague's 75th birthday:

> My Dear Sprague,
>
> Now I have been more or less familiar with your subsequent work and career and take this occasion to emphasize the fact of its great importance, especially to the art of electric railway control and propulsion. I think it can be truly said that your trials of the trolley system in Richmond, Virginia, were a definitive starting point in the development of trolley systems in the United States. Needless to say, the subsequent electrification of the whole of the street car lines itself followed and created a profound extension of the electric railway systems throughout the country. Better than all this, I have appreciated our many years of acquaintance and friendship.
>
> Very truly yours,
Elihu Thomson[4]

The second major inventor of electric streetcar technology was Charles Joseph Van Depoele. Born in Belgium in 1846, Van Depoele immigrated to the United States in 1869 and started to experiment with electric motors. He soon realized that such motors could be applied to power streetcars, and demonstrated the concept at fairs and expositions from 1882 to 1884. Through his Van Depoele Electric Manufacturing Company, he created the earliest commercial streetcar lines in the U.S. in South Bend, Indiana, in 1885; and Appleton, Wisconsin, and Scranton, Pennsylvania, in 1886. Van Depoele's work on commercial systems forced him to solve real-world problems, and he was diligent in patenting the equipment that he used to solve those problems. Van Depoele attained 243 patents, over half of which focused on electric traction equipment.

The success of Van Depoele's systems created a huge demand for new installations, but his small company lacked the resources to meet the need. At the end of 1887, he considered selling out to Frank Sprague's company, but Sprague was also resource-limited and in no position to buy. So, Van Depoele sold his patents to Thomson-Houston Electric, and Van Depoele himself moved to Lynn, Massachusetts, to work as an expert and inventor for Thomson-Houston. In 1889, Sprague sold his company to Edison General Electric.[5]

As we saw earlier, J.P. Morgan forced the merger of Edison General Electric and Thomson-Houston to form General Electric on April 15, 1892. Thus, the patents of Van Depoele and Sprague gave General Electric a dominant position in the electric traction business.

George Westinghouse was encouraged to enter the competition for the burgeoning electric traction business by the successes of Sprague and Van Depoele, and he did so in late 1889. Based on their extensive experience with electric motors, in 1891, Westinghouse engineers developed a DC motor combined with a single reduction gear. The hinged cover contained field windings and enclosed the armature to provide protection from the elements. The single gear drive, with its pinion and gear, were enclosed in their own oil-filled case. The combination, simpler and more reliable than previous designs, drove all other motor structures out of the electric streetcar market almost overnight.[6]

The most significant problem for electric streetcars was how to connect the moving car to stationary sources of electricity such as wires or rails. Energizing the two rails that carried the wheels of the streetcars with positive and negative voltage was a simple answer, but the lethal danger that presented for humans and steel-shoed horses made the approach untenable. Indeed, any solution placing both positive and negative electric terminals at street-level, such as a third rail or an electrified trench, suffered the same limitation.

Westinghouse personnel performed early experiments to gauge the hazard. In his 1921 Westinghouse biography, Prout reported one eye-witness account, "I recall that they used to lead one of the old horses across the track to see whether he would jump if he chanced to get a front foot on one rail and a hind foot on the other while the rails were charged."[7] Even at relatively low voltages, the horses would indeed jump.

Moving at least one of the electrical terminals to above the streetcars would reduce the danger but would add significant and unsightly infrastructure. But despite the cost and appearance, that solution turned out to be the best alternative, and virtually all streetcar installations adopted it.

The spring-loaded trolley pole required to contact the overhead wire was covered

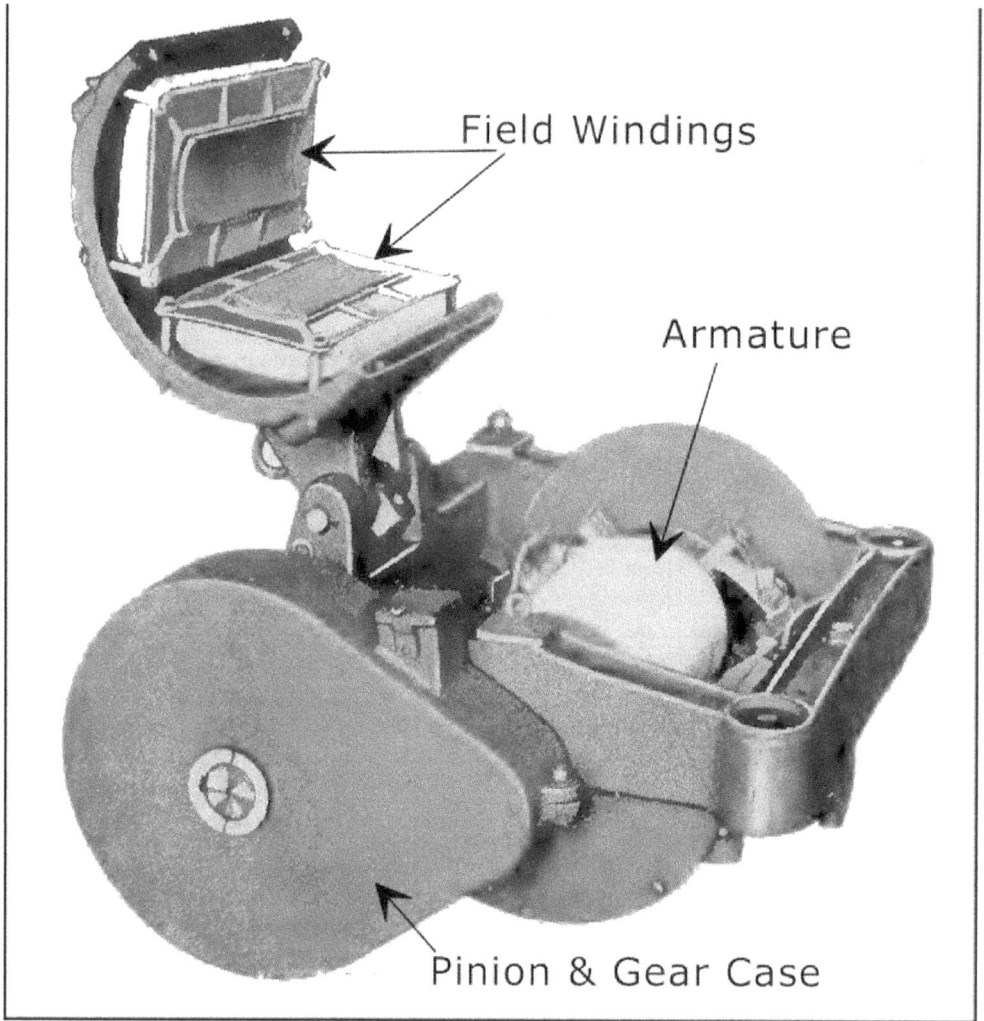

Westinghouse Number 3 traction motor of 1891 (*Cassier's Magazine*, August 1899, p. 344).

Advertisement for a trolley pole (*Street Railway Journal*, 1896).

by two of the patents issued to Charles Van Depoele. The advertisement shows such a trolley pole.[8]

On March 12, 1887, Van Depoele filed a patent application describing overhead wiring and a method of connecting a moving trolley with a spring-loaded trolley pole. At the top end of the pole was a rotating metal wheel that rode on the overhead wire.

As sometimes happens (see Appendix VI—Patent Law Primer for a brief explanation of patent law), the single application resulted in two separate but quite similar patents, the first issued April 1, 1890,[(d)] and the second issued April 11, 1893.[(e)]

In 1895, the Thomson-Houston Electric Company, by then part of General Electric and owner of Van Depoele's patents, filed suit for infringement of the 1893 patent against the Winchester Avenue Railroad Company (the user of the allegedly infringing equipment) and the Westinghouse Electric & Manufacturing Company (the manufacturer of the equipment).

The suit was heard by Judge William Kneeland Townsend of the United States District Court for the District of Connecticut, sitting in New Haven. Attorneys for Westinghouse represented the defense. Their primary argument was that the 1890 patent covered the same material, and therefore the 1893 patent was invalid under the prohibited practice of double patenting.

On December 7, 1895, Judge Townsend ruled for Thomson-Houston that all disputed claims (claims 6, 7, 8, 12, and 16) of the 1893 patent were valid. The judge's ruling led Thomson-Houston to file for sixteen injunctions in various courts to prohibit companies (both manufacturers and users) from using the equipment covered by the 1893 patent.[9] We will discuss the results of the filing for injunctions a bit later.

Stop the Fighting

The ruling put Westinghouse Electric & Manufacturing in a bind. It needed access to the Van Depoele patents to move forward with what appeared to be a lucrative electric railway business. Conversely, General Electric required the Tesla motor patents, held by Westinghouse, to manufacture superior AC induction motors.

George Westinghouse always respected intellectual property rights, whether those rights were held by an individual or a company. He recognized the competing

---

(d) U.S. Patent 424,695. "Suspended Switch and Traveling Contact for Electric Railways."
(e) U.S. Patent 495,443. "Traveling Contact for Electric Railways."

interests of GE and Westinghouse Electric as an opportunity to end the costly patent litigation that had been sapping the resources of both companies. Executives of GE also desired to end the needless expense, and discussions between the companies started in mid–1895. The initial results were disappointing.[10]

But on March 12, 1896, Westinghouse attorney Charles A. Terry announced

> Negotiations between the General Electric Company and the Westinghouse Electric & Manufacturing Company have resulted in an arrangement with respect to a joint use of the patents of the two companies, subject to existing licenses on terms which are considered mutually advantageous.
>
> It has been agreed that after certain exclusions the General Electric Company has contributed 62 percent and the Westinghouse Electric & Manufacturing Company 37 percent in value of the combined patents, and each company is licensed to use the patents of the other company, except as to the matters excluded, each paying a royalty for any use of the combined patents in excess of the value of its contribution to the patents.
>
> The patents are to be managed by a board of control, consisting of five members, two appointed by each company and a fifth selected by the four so appointed. Both companies have acquired during their existence a large number of valuable patents, and numerous suits have been instituted in consequence of the infringement of these patents by one party or the other or by their customers. In the prosecution of these suits large sums of money have been expended and the general expenses of the companies have in this manner been greatly increased. It is expected that the economies to be effected will be very considerable and that the two companies and their customers will be mutually protected.

The excluded patents were those covering cable and underground railway materials, which were already covered by previous licenses, and also foreign patents.[11]

The Board of Patent Control, headed by the presidents of the two companies, George Westinghouse for Westinghouse Electric & Manufacturing and Charles Coffin for GE, met periodically to verify compliance with the agreement. Just two disputes requiring the participation of the fifth board member to break a tie vote arose in the first thirteen years of the agreement.

At that time, the U.S. Department of Justice started an investigation to determine if the provisions of the Sherman Antitrust Act, which had been passed in 1890, were being violated. After an exhaustive examination, the DoJ found no violations and no reason to suspend the agreement, which expired at the end of its fifteen-year life in 1911.[12]

If Only Foresight Matched Hindsight

Now let us go back to the filing for injunctions resulting from Judge Townsend upholding the 1893 Van Depoele patent claims.

Sixteen injunctions were requested, and all were granted based on the written opinion of Judge Townsend. The sixteenth injunction request was filed in the U.S. Northern District Court of New York against The Hoosac Railway Company and the Walker Company, users and manufacturers of the infringing equipment, respectively.

The New York court granted the injunction as requested. But the defendants (Hoosac and Walker) appealed that decision to the U.S. Court of Appeals for the Second Circuit, the same Circuit where Judge Townsend was a District Court judge. In layman's terms, the appeal went to Judge Townsend's boss. Judge William James Wallace heard the appeal and disagreed with the ruling made by Judge Townsend.

In late 1897, Judge Wallace ruled that the 1893 Van Depoele patent was null and void because of double patenting based on the 1890 Van Depoele patent. Double patenting was the exact argument advanced by the Westinghouse attorneys when they unsuccessfully presented their case to Judge Townsend! The effect of Judge Wallace's overturn of Judge Townsend's ruling was to allow free use of the claims no longer covered by the 1893 Van Depoele patent.[13]

Had George Westinghouse been a bit more patient, he could have been able to manufacture streetcar equipment unfettered by Van Depoele's patents and without having to surrender the use of Tesla's motor patents to GE.

What might the futures have been for GE and Westinghouse Electric had GE been blocked from using the Tesla patents?

Patent law at the time set the term of a patent at 17 years from the date of issue. Tesla's earliest critical patent was issued on May 1, 1888, and therefore expired on April 30, 1905. So, without the GE-Westinghouse patent agreement, Westinghouse could have blocked GE from using Tesla's disclosures for another nine years.

Growth of Street Railways

After about 1900, every growing town or aspiring city had to have a streetcar line. In larger cities, streetcar systems extended their lines farther and farther out into the suburbs and to other nearby towns. The advent of inexpensive and reliable transportation meant that workers could live further away from their jobs.

Consider George Westinghouse's adopted hometown, Pittsburgh, as an example. In 1902, the Pittsburgh Railway Company operated 1,100 trolleys on 400 miles of track, carried 179 million passengers, and had revenue of $6.7 million. The year of 1918 was the peak year for the company, with 99 routes over 606 miles of track. But automobiles started to replace streetcars in the 1930s, and most, but not all, cities eliminated their streetcar systems or converted them to bus lines by 1960.[14]

(From 1945 to 1955, my family lived within walking distance of a trolley line that stretched from Pittsburgh to Washington, Pennsylvania [called "Little Washington" to avoid confusion with the nation's capital], a distance of 29 miles. My father often rode the trolley to work in downtown Pittsburgh. As a youngster of 12, I well remember taking swimming lessons at the Pittsburgh YMCA on Wood Street, searching for LP "hi-fi" records at Gimbels department store or the National Record Mart, attending a movie at the Warner or Stanley Theatre,(f) and then riding the trolley home.

In August 1953, service to Little Washington was eliminated, and the streetcar line then terminated at Drake stop on the Allegheny County boundary, which was where I lived.)

Trolleys played a significant role in the lives of many older Americans, and now their successor, light-rail systems, have brought back the convenience of mass

---

(f) The YMCA moved from Wood Street in 1986; Gimbels closed their iconic Pittsburgh store in 1987; National Record Mart, founded in Pittsburgh in 1937, closed its last store in 2002; the Warner Theatre was demolished in 1983; and the old Stanley Theatre is now the Benedum Center for the Performing Arts. Everything changes!

transportation. Frank Sprague, Charles Van Depoele, Nikola Tesla, and George Westinghouse all contributed to developing the required technology, much of which is still in use today.

Electrifying Trains

George Westinghouse's early experience with railroads via his air brake inventions, combined with his later successes with applying electric motors to streetcars, naturally led him to consider using electricity as the motive force for trains. But it took a unique problem to make electric trains attractive enough to get his attention.

Under the River

We have to go back in history to understand how the unique problem came to be.

In the 1840s, Montreal had become a significant site for trade between Canada and Great Britain. But the St. Lawrence River became impassable because of ice for four to five months per year. An all-weather route to a port on the Atlantic Ocean was needed, and Boston and Portland, Maine, both wanted to be that port.

The Montreal Board of Trade selected Portland in 1845. Both financing and terrain were obstacles, but by 1853 the Grand Trunk Railway of Canada completed the link from Portland to Montreal. To reach markets in the Midwest United States, by 1859, the Grand Trunk had extended its tracks west to Sarnia, Ontario, on the east bank of the St. Clair River, which connects Lakes Huron and Erie.

In March 1858, Grand Trunk officials had incorporated the Chicago, Detroit & Canada Grand Trunk Junction Rail Road Company to connect Port Huron and Detroit, Michigan. Port Huron is just across the St. Clair River from Sarnia, Ontario. The 63-mile line from Port Huron to Detroit was completed on November 21, 1859,

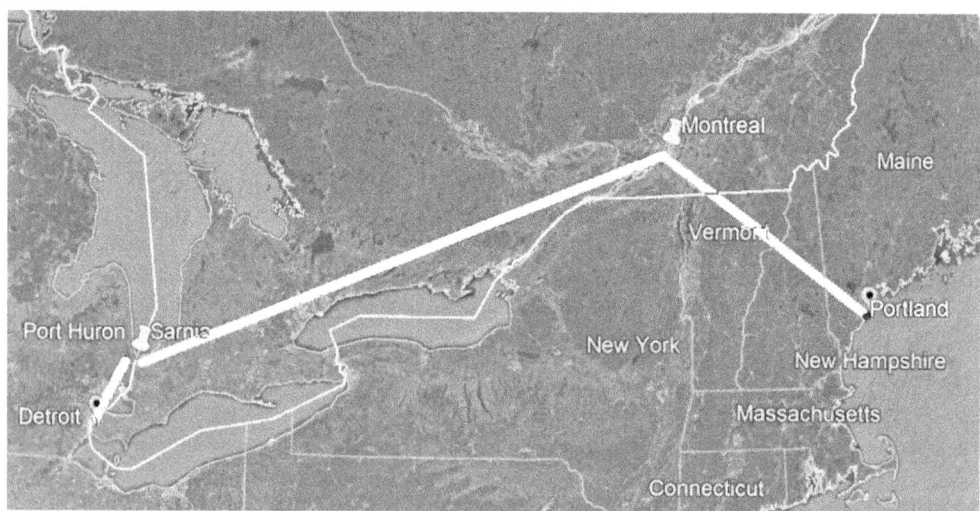

Map showing the idealized route of the Grand Trunk Railway of Canada from Portland, Maine to Detroit, Michigan (Google, Image Landsat/Copernicus, Image: NOAA).

and, as discussed in Chapter 15, twelve-year-old Thomas Alva Edison jumped at the chance to sell newspapers and magazines on trains along the route.

The lack of a rail link to cross the St. Clair River from Sarnia to Port Huron was a glaring gap in the route of the Grand Trunk. Everything on the train, freight and passengers, had to be unloaded in Sarnia, transported across the river by ferry and then reloaded onto the cars of another train for the trip to Detroit and points west. As time- and labor-consuming as that process was in good weather, it became impossible when ice blocked the river in winter. Even in good weather, heavy river traffic slowed the ferries.

The answer was a tunnel, but no one had ever before built a railroad tunnel underwater. Once again, financial and physical obstacles challenged the Grand Trunk Railway, and once again, they persevered. With the Canadian government providing 15 percent of the cost, the St. Clair Tunnel Company, a subsidiary of the Grand Trunk Railway, started construction of the tunnel in January 1888.

The tunnel was an engineering marvel and is today a National Historic Landmark (even though it is closed and flooded). Excavation proceeded in a compressed air atmosphere within twenty-one-foot diameter digging shields. Hydraulic jacks advanced the shields as digging progressed. A one-inch thick, segmented cast-iron lining was installed behind the shields to support the tunnel. Work started from both sides of the river, and crews, surrounded by the cast-iron pipes, met under the center of the river on August 30, 1890.

The 6,025-foot tunnel, built at the cost of $2.7 million,[g] opened for freight service on September 19, 1891, and for passenger service in 1892.[15]

Running a steam-driven train through such a long tunnel was problematic. Smoke from the burning coal used to heat water to make the steam could endanger or even suffocate the crew and passengers in any train stalled in the tunnel. Ten men died in the tunnel from asphyxiation between January 1892 and October 1904.[16] The need for a safe means of propulsion through the St. Clair Tunnel was the unique problem that encouraged the development and use of electric trains.

The Grand Trunk Railway solicited bids for locomotives to electrify the railway through the tunnel. Westinghouse joined with the Baldwin Locomotive Works for the proposal, with Baldwin building the bodywork and running gear and Westinghouse providing the electrical components.[17]

The Westinghouse-Baldwin team won with a proposal for a double-unit 3,300 Volt AC locomotive, with each unit weighing 132 tons and rated at 1,500 HP.[18] The winning bid included a power house, electrification of the tunnel water pumps, and lighting for the tunnel. The first electric locomotives passed through the tunnel on February 20, 1908.[19]

The success of the installation at the St. Clair Tunnel stimulated the use of electric locomotives in other tunnels and in urban areas where smoke pollution was becoming a problem.[20] New York City passed laws that led to the electrification of all trains serving the city effective in 1908.[21] The New York, New Haven, and Hartford Railroad was at the forefront of electrification, starting the process on a small scale in 1895 and accelerating in 1907. Westinghouse-Baldwin supplied most of the locomotives for the system.

(g) $76 million in 2019.

Railroad electrification in the United States reached its peak in the late 1930s when diesel locomotives entered the picture and eventually displaced most electrified lines.[22] Westinghouse Electric & Manufacturing Company continued to build electric locomotives until 1952.[23]

George Westinghouse's foresight provided his companies with electric streetcar and then locomotive business for almost 60 years.

Chapter 24

Panic!

"Bankers—pillars of society who are going to hell if there is a God and He has been accurately quoted."
—John Ralston Saul, Canadian Writer

Bank Crises in Antiquity

The earliest recorded bank crisis occurred in 33 AD, as reported in *The Annals* by Tacitus.[1] In ancient Rome, the practice of usury created and amplified class inequity and fomented anger among those who were forced to borrow at exorbitant rates. To combat the practice of usury, the maximum interest rate on loans was set at 10 percent. That rate was subsequently reduced to 5 percent, and later, compound interest was banned altogether. The unintended consequence was a severe scarcity of money, and merchants and landowners demanded full payment for all transactions.

Commerce ground to a halt until Emperor Tiberius himself contributed one-hundred-million sesterces and permitted borrowing with no interest for three years, provided the borrower provided land amounting to double the amount borrowed as collateral.

Thus started the banking practice of loaning money only to people who did not need it.

Modern Banking History

The law of unintended consequences has never been repealed. Man invented banks as a safe haven for savings, and other people looked to those banks as a source of funds for various ventures. But if the money had been given out for loans, it would not be available for depositors who wanted to withdraw it. So, if too many people wanted to withdraw their deposits at the same time, a liquidity crisis could occur.

The least severe banking crisis is a bank run, where depositors lose confidence in a single bank and collectively withdraw their funds within a short time (one or two days).

The next crisis level is a panic, where the run extends to multiple banks and can lead to the demise of financial institutions, the suspension of stock markets, and a general lack of financial liquidity.

The most extreme level is a systemic banking crisis in which an entire country experiences multiple financial institution defaults, and liquidity issues severely impede trade, both domestic and foreign. The U.S. sub-prime mortgage problem in 2007–2008 rose to the level of a systemic banking crisis.

Between the end of the Civil War in 1865 and the creation of the Federal Reserve System in 1913, the U.S. banking system suffered five waves of panic in 1873, 1884, 1893, 1896, and 1907.[2]

The Panic of 1907

The immediate cause of the 1907 panic appears to be the failed attempt by F. Augustus Heinze to corner the market for the stock in his United Copper Company.

When his scheme collapsed, the price of United Copper stock plunged from $30 to $10 per share in one day, from October 15 to October 16, 1907. The State Savings Bank of Butte, Montana (owned by Heinze), held shares of United Copper as collateral for loans and was forced to declare insolvency. The Mercantile National Bank in New York City (of which Heinze was president) was the correspondent bank[a] for the Butte, Montana, bank. When people heard of the failure of the Butte bank, they began to withdraw their funds from the Mercantile National Bank. The Mercantile had sufficient funds to withstand a run for a few days. But the concern spread like a virus to other banks, especially the Knickerbocker Trust Company, whose president was Charles Tracy Barney.

J.P. Morgan, through his National Bank of Commerce, had been loaning money to the Knickerbocker to keep it afloat. But on October 21, the National Bank of Commerce announced that it would no longer clear checks for the Knickerbocker. In less than three hours the next morning, depositors withdrew $8 million from Knickerbocker, and the trust was forced to suspend operations shortly after noon.

Fueled by rumors and inflammatory newspaper stories, the worries grew to a full-fledged panic, and depositors withdrew $47.5 million from Trust Company of America and more from other financial institutions.[3] Ultimately, thirteen New York City banks closed their doors.[4] The panic soon spread beyond New York City. For example, the Pittsburgh Stock Exchange closed for three months, starting on October 23, 1907.[5]

The panic spread to the New York Stock Exchange, which required money for short-term loans, and the City of New York, which faced bankruptcy.

To stem the tide, J.P. Morgan took charge, either loaning money himself or twisting arms at late-night meetings in his library to obtain financial commitments from other bankers. John D. Rockefeller deposited $10 million to shore up the Union Trust Company. Secretary of the Treasury George Cortelyou deposited over $37.6 million from the U.S. Treasury[b] in New York national banks.[6]

As in the Roman crisis of 33 AD, the emperor (government) provided funds to increase liquidity.

(**a**) A correspondent bank provides services such as conducting business transactions, accepting deposits, or gathering documents on behalf of another financial institution.
(**b**) Theodore Roosevelt, the "Trust-Buster," objected to government involvement, but had no tools available to resolve the crisis and was forced to acquiesce to government help. Another result of the bank crisis was the takeover by U.S. Steel of its competitor, Tennessee Coal, Iron, and Railroad. In early November 1907, Roosevelt agreed not to oppose the takeover on antitrust grounds.

By mid–November, the crisis subsided, but money for loans remained unavailable. For capital-intensive businesses such as manufacturing, such loans were essential. The Panic of 1907 eventually led to the formation of the Federal Reserve System on December 23, 1913.

Westinghouse Electric & Manufacturing Company before 1907

Westinghouse Electric & Manufacturing Company emerged from its financial difficulties of the early 1890s as a financially strong and growing organization. By 1894, the company had about $9 million in common stock outstanding and an operating surplus of $4 million. Profits for the previous year were $1.6 million; floating debt was just $0.5 million; the company had no outstanding bonds; and business prospects were favorable.[7] But as large as the electrical equipment business was and the rate at which it was growing, there were risks.

In particular, expensive factories were required to manufacture the end products; the manufacturing equipment itself had to be invented and built; and the electric utilities that would buy the products were speculative and under-capitalized.

Both Westinghouse Electric and General Electric resolved the last of these problems by loaning money to the electric utilities in return for their stocks and bonds as collateral. The loaned money was used as part payment for the generating equipment they wanted to buy.

To obtain the funds for these loans, Westinghouse Electric increased its common stock to $15 million in 1896 and to $25 million in 1901. In 1895, Westinghouse Electric started issuing bonds to raise additional capital. Despite the cash influx, notes payable rose to $5 million in 1901 and $14 million in 1905. Meanwhile, Westinghouse Electric established subsidiaries in Canada and Europe. Although these ventures brought in a large amount of additional business, the costs exceeded revenues.

Growth does not necessarily indicate prosperity. Most manufacturing businesses experience lower profits as a percentage of sales as the sales grow. Westinghouse Electric was no exception. Profits as a percentage of sales were about 20 percent in 1901. In 1906, sales had increased by 50 percent, but profits as a percentage of sales dropped to 13 percent.

Finally, there was the cost of capital. Dividends on Westinghouse Electric common and preferred stock were 7 percent in 1902, rose to 9 percent in 1903, and reached 10 percent from 1904 to 1907. Interest on bonds was at a lower rate but still amounted to millions of dollars.

Westinghouse Electric Company and the 1907 Panic—Financial Impact

The fiscal year ending March 31, 1907, was a year of "distinct business prosperity" for Westinghouse Electric. But because of heavy borrowing, the amount available after paying interest on the debt was $2.77 million. Of that amount, stock dividends consumed $2.5 million, and other non-operating expenses left nothing to add to the permanent surplus account.

In a "normal" financial environment, Westinghouse Electric could have borrowed short-term funds to alleviate its cash-flow problem. But late 1907 was anything but normal. Because of the bank panic, funds were unavailable, and on October 23, 1907, the Westinghouse directors had no option but to apply to the United States Court for Receivers.[8]

Late that day, George Westinghouse instructed E.H. Heinrichs, his press agent,[9] "In speaking of the affairs of the Electric Company to your newspaper friends, do not forget to make it very emphatic to them that this receivership is not the end of the company. Tell them the Westinghouse Electric & Manufacturing Company is fundamentally as sound and solid as ever, and it will emerge out of this unfortunate situation a greater and more prosperous concern than ever.

In fact, I believe that in ten years from now it will be more than twice as large as it is now, and will give employment to twice as many men as we have at present."[(c)]

Liabilities of Westinghouse Electric at the time of the receivership were as follows:

- $5 million to merchandise creditors, those vendors who had supplied raw materials, sub-assemblies, and other products to Westinghouse Electric.
- $8 million to banks in the form of short-term notes and floating debt. The banks involved were primarily small banks extending from Maine to California.
- $29 million to the bond holders with bond maturities beyond one year.
- $28 million to stockholders.

In addition to Westinghouse Electric & Manufacturing, two other Westinghouse holdings, Westinghouse Machine Company and the Nernst Electric Lamp Company, also entered receivership. Significantly, Westinghouse Air Brake and Union Switch and Signal were financially sound, and George Westinghouse retained control of them.[10]

On November 13, 1907, the officers advanced the first plan of adjustment, which was quickly rejected by the banks. In December, the creditors agreed to form a Readjustment Committee composed mostly of bankers and headed by James N. Jarvie of the New York National Bank of Commerce. On January 20, 1908, the Readjustment Committee presented its "Plan for the Adjustment of Debt." Despite initial enthusiasm by all parties, the severely depressed value of Westinghouse Electric stock made it untenable, and on April 2, 1908, the plan was withdrawn.

A series of proposals were advanced and rejected by one or more of the creditor groups. Finally, at the end of March 1908, the "Merchandise Creditors Modified Plan" appeared to satisfy all parties. Its significant provisions were:

- Conversion of $4 million of Merchandise Creditor debt into common stock;
- Subscription to $6 million of new stock at par value by old and new stockholders;
- Conversion of the bank floating debt to new stock, long-term convertible bonds, and notes with maturity averaging five years; and
- Old bonds and short-term debt were undisturbed.

(c) For once, Westinghouse proved to be pessimistic. Westinghouse Electric & Manufacturing Company sales in 1906 were $33 million (because of the panic and reorganization, 1907 sales were much lower). Sales in 1917, ten years later, were $95.7 million!

After difficult negotiations, especially with the banks and stockholders, in December 1908, Westinghouse Electric & Manufacturing Company emerged from receivership.

The most significant financial results of the reorganization were:

- Net debt of the company was reduced from $44 million to $31 million;
- Interest charges were reduced by $1 million;
- Capital stock was increased from $29 million to $41 million[d]; and
- The large, rapidly maturing debt, carrying heavy interest charges, was converted into a stock liability with no fixed costs.

Westinghouse Electric Company and the 1907 Panic— Personal Impact

The blame for the financial disaster fell on George Westinghouse and his over-optimistic manner of managing the Westinghouse Electric & Manufacturing Company. Of course, the management team had to be replaced. A new Board of Directors was formed in 1909, with sixteen members selected from those who had been instrumental during the receivership. Six of the board members, three of them bankers, controlled financial policy. Robert Mather, formerly president of the Chicago, Rock Island and Pacific Railway, was elected Chairman of the Board. George Westinghouse remained as President of the Company but had no financial authority.

In August 1910, the bankers took over, electing Edwin F. Atkins to replace George Westinghouse as President of the Company. Westinghouse, the largest stockholder in the Company, remained on the Board, with his term set to expire in July 1912.[11] The bean-counters were entirely in charge.

By mid–July 1911, George Westinghouse had seen enough. Among other conservative measures, the Board had suspended stockholder dividends, thus depressing share prices. Westinghouse solicited proxies asking stockholders to support six director candidates, enough to cause a change in company management. Westinghouse Electric & Manufacturing Company President Robert Mather issued a strong rebuttal of George Westinghouse's position.[12]

At the July 26–29, 1911, annual stockholders meeting, George Westinghouse lost his bid to regain control of the company he founded. Over 650,000 proxies were submitted. Mather's candidates won the contested Board of Directors positions by about 490,000 to 187,000.

Edwin F. Atkins declined to accept another term as company president. Edwin M. Herr,[e] vice-president of Westinghouse Electric since June 1, 1905, and a longtime associate of George Westinghouse, was elected to that post.[13]

Had George Westinghouse been a vindictive man, he might have gained a

---

(d) Five thousand Westinghouse Electric & Manufacturing Company employees demonstrated their loyalty to the company and its founder by subscribing to $611,000 of stock.

(e) Edwin Musser Herr (1860–1932) and Herbert Thacker Herr (1876–1933) were half-brothers. Edwin was the son of Theodore Witmer Herr and Annie Musser. Herbert was the son of Theodore Witmer Herr and Emma Musser, who was the sister of Annie. Both men became executives at Westinghouse companies, Edwin at Westinghouse Air Brake and Westinghouse Electric, and Herbert at Westinghouse Machine and Westinghouse Electric.

glimmer of satisfaction upon hearing that Robert Mather, his successor as Chairman of the Board of Westinghouse Electric & Manufacturing Company, died after a short illness on October 24, 1911.[14]

But Westinghouse was far more concerned with the loss of his beloved company. His press secretary, E.H. Heinrichs, commented, "The loss of the Electric Company was to Mr. Westinghouse a disappointment from which he never recovered. There is no doubt that it broke his spirit."[15]

As she had for the previous 44 years, Marguerite encouraged and supported George. Without her constant presence, her normally taciturn husband would have surrendered to depression over the loss of his "baby."

The Crisis in Retrospect

In the financial crisis of 1891, George Westinghouse had refused to cede control of Westinghouse Electric & Manufacturing to Pittsburgh bankers, many of whom were allied with J.P. Morgan. Instead, he obtained financing from August Belmont and Company, New York bankers who were competitors of Morgan.

But Morgan remembered and got his revenge. He used the 1907 panic (which some say he caused) to take down a profitable Westinghouse Electric, depose its legendary founder, and place bankers in control. While he never achieved the complete Morganization of the electrical equipment industry by merging Westinghouse Electric and General Electric, he and his fellow bankers now controlled the dominant players in the industry.

Arthur Dewing, in his 1914 analysis of the reorganization of Westinghouse Electric, summarized the conflict between the natural optimism of George Westinghouse and the pragmatic views of his creditors as follows:

> Negotiations looking toward a successful reorganization were much more complex than was indicated by the somewhat prosaic printed circulars, which, from time to time, reached the eyes of the public. The failure of the Company had been brought about by the policy of the Westinghouse management in securing new business and borrowing money to finance it. Mr. Westinghouse himself, through his untiring energy and great optimism, was the actual power in the business. In the emergency he had turned to all quarters for succor, borrowing every available cent. When the crash actually came he maintained the same hopefulness and the same lack of conservatism. Opposing him were the various creditor interests, which, although affected by the contagious optimism of Mr. Westinghouse, were concerned primarily in the relief of the present emergency, and the immediate payment of their obligations. Thus in the background of the long and tedious negotiations, of which only rumors reached the public ear, lay the conflict of interests between the generous hope of the founder in the wonderful future of his business, a future not to be embarrassed by over-conservatism or delay, and the stern practical necessity that confronted the creditors of forcing the payment of debts past due.[16]

Niles Carpenter, in his 1916 study, provided a more positive view of George Westinghouse:

> It used to be a financial commonplace that "Westinghouse was a great inventor but no financier, and so he went broke." This seems to have been far from accurate. Mr. Westinghouse was a great inventor, but he was more. He was a consummate organizer, not only industrially but commercially. The first reorganization of 1891 bears ample witness to his

financiering ability. He must have been a wonderful administrator, for he kept under his personal supervision over a score of enterprises, employing approximately three hundred thousand persons. In this earlier part of his career he marketed his own products' and built up his own companies in the face of indifference and opposition such as must have crushed any but a gifted salesman and a skillful promoter. The remarkable thing about the personality of Mr. Westinghouse is that it combined several personalities. Displaying extraordinary ability as an inventor, as an administrator, as a salesman, as a promoter, and, let it not be forgotten, as a financier, he established a group of industrial organizations which bear eloquent witness to his genius.[17]

It is a testimony to the corporate legacy left by George Westinghouse that Westinghouse Electric survived the panic of 1907 and the resulting strict financial controls to remain a vital player in the electrical equipment business. The concluding chapter (Chapter 28) will examine the significant contributions of Westinghouse Electric without George Westinghouse at the helm.

Chapter 25

Homes and Family

"A very rich person should leave his kids enough to do anything, but not enough to do nothing."
—Warren Buffett

Return to Solitude

We last visited the Westinghouse home, Solitude, in Chapter 9. It was soon after the May 20, 1883, birth of George Westinghouse III.

George and Marguerite Westinghouse moved to Solitude in 1871. Separated by several miles from downtown Pittsburgh, Wilmerding, and East Pittsburgh, but connected to all points by the adjacent main line of the Pennsylvania Railroad, Solitude afforded a sanctuary from the problems of running multiple industrial organizations.

On most evenings, the young family shared their home with dinner guests—friends, neighbors, business associates and their wives, and distinguished people from the world over. If the dinner topic turned to technical issues, George would sketch diagrams that the men would later discuss over the table in the Billiard Room.

If George was the Captain of Industry, Marguerite was the General of Domestic Life. She was the consummate hostess. An 1888 publication heaped praise on her, saying: "(She) lives in greater style, entertains more splendidly and wears more gorgeous, varied, elegant toilets, has more and finer diamonds than any woman in Pittsburg. Her table appointments are simply superb, the entire service being of solid silver and gold and the cut glass, Sevres, Dresden, and other fine porcelains are worth a small fortune. Their whole style of living is after the plan of the household of an English lord. The house is a perfect palace, and the grounds worthy of it."[1]

Among the staid society of wealthy Pittsburghers, Marguerite was alternately viewed as a style-setter or scandalous. She wore her hair in an elaborate style with curls, bangs, puffs, and knots. Some described its color as canary yellow, but she also had wigs in colors to suit her various outfits. Her dresses, from New York and Paris, reflected the utmost in fashion. She wore plumed tiaras, diamond-trimmed slippers, ermine capes, and imported silks. An auction sale of personal items after her death listed, among other items, a cigarette case, a revolver, and a poker set.[2]

Despite her flamboyance, Marguerite loved, supported, and encouraged her husband throughout their marriage. They were an unusually devoted couple whose affection, trust, and faith in each other were constant. When George was away in the states, they communicated by phone every day. If she could not accompany him on an overseas trip, they substituted daily telegrams for phone calls.

George's parents often visited Solitude, sometimes staying for as long as two months. Their concern that George Jr. would never succeed at anything was replaced by amazement and pride. As adults, George, the father, and George, the son, developed a warm friendship, far different than their parent-child relationship had been. George Sr. often questioned his son's lavish homes, but he enjoyed the luxury. George Sr. died at Solitude on December 28, 1884.[3] George's mother, Emeline Vedder Westinghouse, lived at Solitude in her last years and died there on December 10, 1895.[4]

In his 1918 biography of Westinghouse, Leupp tells of visiting Solitude soon after the deaths of George and Marguerite: "When I was at Solitude early in 1915, the house stood just as he and his wife had left it, except that it had been stripped of most of the finer furniture, and the bric-a-brac and curios with which they had filled it as souvenirs of their repeated trips to the old world. The walls sent back echoes of every footstep, and there was a ghostly suggestion as one walked through it and came suddenly upon a huge photograph of Mr. and Mrs. Westinghouse as they appeared during their first sojourn in London."[5]

From the mid–1890s, the Westinghouse family spent less time at Solitude and more time at their home in Massachusetts, Erskine Park. After the crisis of 1907, their choice of residence became far more pronounced, as being in Pittsburgh reminded George of his beloved but lost Westinghouse Electric & Manufacturing Company.

When George and Marguerite Westinghouse died, both in 1914, Solitude passed to their only child, George III. In 1918, George III sold the house with its 10.2 acres

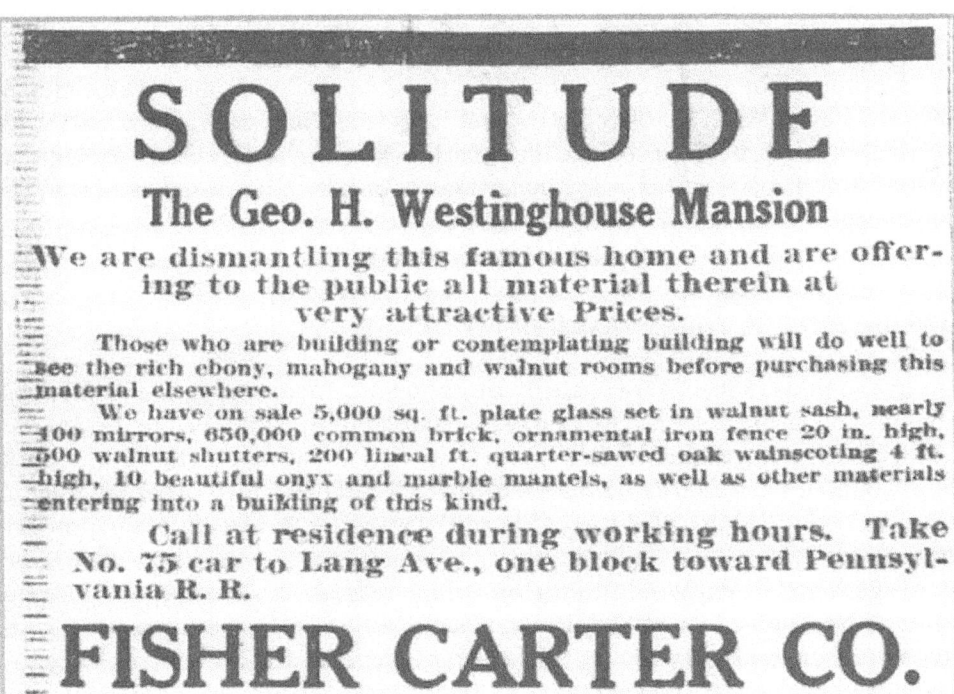

Advertisement for building materials to be salvaged from Solitude, 1919.

of property to the Engineers' Society of Western Pennsylvania for $100,000.[6] On November 30, 1918, in exchange for $1, the society deeded the property to the city of Pittsburgh for use as a park to be called Westinghouse Park. The Society retained the right to erect a memorial to George Westinghouse in the park but never did so. A condition of the transfer was that the house had to be razed within six months. The demolition was carried out, reducing the splendor of Solitude to a pile of bricks collapsed into the basement.

It is not clear why Solitude had to be destroyed. Maybe continued maintenance would have been too costly, or perhaps the then-raging Spanish flu pandemic of 1918–1920 warped people's thinking.

Westinghouse Park served as a vibrant community gathering place through the 1950s. But as the wealthy residents died off and the grand homes were torn down, the immediate neighborhood changed, and use of the park declined. In 1965, the old stable building was demolished, and a utilitarian cement block structure was erected in its place. With little maintenance by the city of Pittsburgh, the park became decrepit, with no hint of its former residents or glory.

In 2018, David Bear, a longtime local resident and former *Pittsburgh Post-Gazette* editor, realized that the park was approaching its 100th anniversary. Bear and other interested individuals organized a Centennial Celebration for the park on December 1, 2018. The celebration included a Mayoral Proclamation and led to the formation of the Westinghouse Park 2nd Century Coalition (WP2CC).

The WP2CC has identified four main objectives: (1) Restore and preserve the park, (2) Erect a new park building, (3) Perform professional archaeological research of the site, and (4) Create a physical and virtual memorial to focus on the history of Solitude and the world-changing contributions of George Westinghouse. More information on WP2CC is available on their website, *www.westinghousepark.org*.

A proposal for a more comprehensive city-wide program called Westinghouse 175 grew out of WP2CC. October 6, 2021, was the 175th anniversary of George Westinghouse's birth. To recognize the occasion, Westinghouse hosted 175 regional exhibits and displays of his many inventions, along with items unearthed by the archaeological exploration at Westinghouse Park.

Erskine Park in the Berkshires

In the mid–1880s, Marguerite Westinghouse's doctors suggested that clean mountain air, as opposed to the smoky blackness of Pittsburgh, would benefit her health. George Westinghouse had planned to visit Great Barrington in the Berkshire Mountains of western Massachusetts to inspect the AC distribution system that William Stanley had installed there.[7] He and Marguerite traveled by train, probably aboard Glen Eyre.

Together they explored the region, which was about 70 miles southeast of George's boyhood home at Schoharie, New York. After inspecting several sites, they settled on the Henry de Bois Schneck farm between the settlements of Lenox, Lee, and Stockbridge. In November 1887, they bought the property, which included a solidly-built hilltop house on 100 acres. The next year they added 41 acres of the adjoining Clark farm and added 12 more acres from the Smith estate in 1889.

Chapter 25. Homes and Family

Westinghouse home, Erskine Park, in Western Massachusetts (courtesy Selena Lawrie).

Hall at Erskine Park—Note the tufted cloth ceiling and walls and the light bulbs near the ceiling (courtesy Selena Lawrie).

George, Marguerite, and George III moved into the enlarged Schneck house in October 1890.

The Smith property sat on the shore of Laurel Lake, and for the next ten years, they waited for an opportunity to add more lakefront property. Then a series of purchases, mostly along the lake, nearly doubled their land area. Other additions brought their holdings to almost 600 acres by 1911.[8]

They called their new home Erskine Park, in honor of Marguerite's maternal grandmother, Margaret Erskine (1791–1849).

Marguerite had definite ideas about decorating and applied them inside and outside the house. She transformed the Schneck farmhouse into an impressive Queen Anne–style mansion, complete with dining rooms, a billiard room, and a bowling alley.

The 1893 Columbian Exposition, with its thousands of brilliant Westinghouse stopper light bulbs and its Court of Honor, nicknamed the White City, inspired architects and decorators to beautify their buildings and to emphasize white for walls. Marguerite Westinghouse needed little encouragement, as she hated the black smoke and soot of Pittsburgh.

She chose white as the predominant color (more precisely, lack of color) for both interior and exterior decoration. Inside, the walls and ceiling were covered by white and gold tufted satin, which appeared either luxurious or like a padded cell, depending on your mood. White mahogany furniture and silverware, including a seven-piece Tiffany tea and coffee set, adorned the rooms.[9]

Marguerite encouraged George to electrify the house. First, he had to supply electricity to the estate, so in 1892, he constructed a stone powerhouse at the north end of the grounds. Inside were a steam plant and generators for producing alternating current. Westinghouse sent the excess power to the town of Lenox via an underground conduit for electric street lighting and, in emergencies, to power the water system. Power from Erskine Park also lighted another nearby mansion, Elm Court, owned by furniture executive William D. Sloane and his wife, Emily. Emily was the granddaughter of Commodore Cornelius Vanderbilt.[10]

Then Westinghouse designed and had his workers build and install 1,500 lamps, mostly placed in decorative molding, where the walls met the ceiling. The resulting lighting gave the appearance of daylight without shadows. Nothing like it had ever been done in a private home.

The grounds at Erskine Park stretched over 120 acres of open lawns, a natural lake, and an artificial pond (converted at Marguerite's direction from a swampy area), with nearly 400 white pine trees, white birches, and carriage roads made of crushed white marble. Two marble bridges carried pathways over the pond, which featured white swans and five fountains. Marguerite did not just provide a general concept to be executed by others; she designed and, in some instances, participated in building fences and roadways. Extensive flower beds and ornamental trees provided color everywhere.

In addition to horticulture, Marguerite's hobbies included raising livestock. Her farm horses regularly won prizes at the annual Lenox horse show. Her herd of 20 to 25 dairy cows had the benefit of a modern dairy barn equipped by George with the latest electrical equipment and surfaced throughout with the ever-present crushed white marble.

With nothing to do on one rainy day, Marguerite explored a large, unused barn. She decided that the barn could be converted into a recreation hall for George III and the family. She had the lower level gutted and installed a kitchen, pantry, and dressing rooms. The upper level became one large room, equipped with bookshelves, card tables, lounging chairs, gymnastic apparatus, a bowling alley, and a billiard table. Most of the area could be cleared for receptions, parties, or dancing.[11]

Westinghouse also added a machine shop, laundry, and boarding house for his employees. A lighted tennis court and golf course provided space for outdoor recreation.[12]

The completed estate cost Westinghouse $1 million for the land and $200,000 for improvements and additions to the original farmhouse. George and Marguerite spent their summers, and often more extended periods, at Erskine Park for 24 years.

In 1917, three years after his parents died, George Westinghouse III sold Erskine Park to Margaret Emerson, the widow of Alfred Gwynne Vanderbilt. Vanderbilt had been killed in the sinking of the *Lusitania* by a German U-boat on May 7, 1915. Recall that the *Lusitania* was one of the first ships outfitted with turbine engines. As with Solitude, the deed for Erskine Park required the demolition of the house that Marguerite had so carefully decorated. But the recreation hall where George III played remains to this day. The Playhouse was listed in 2017 as a six-bedroom, 6,500 square-foot property for sale for $1.4 million.[13]

After demolishing the main house at Erskine Park, Margaret Emerson built a large Colonial Revival house called Holmwood on the grounds. From 1939 to 1976, the property was the Foxhollow School for Girls. In 1976, the property was converted to a resort, then a health spa, and is now a timeshare called The Ponds at Foxhollow.[14]

Blaine House, Washington, D.C.

Because George Westinghouse was in Washington for business so often, he and Marguerite decided to buy a home there. Although it was never their permanent residence, the couple visited and entertained in what was called the Blaine House.

Located at 2000 Massachusetts Avenue, N.W. on Dupont Circle, the mansion was completed in 1882. The original owners were James Gillespie Blaine and his wife, Harriet Stanwood.[15] Blaine, a journalist turned politician, was one of the founders of the Republican Party. He was Secretary of State under Presidents James Garfield and Benjamin Harrison. In 1884, Blaine was the Republican nominee for president, but he lost the election to Grover Cleveland.[16]

The home turned out to be too large for the Blaines, so they leased it to George and Marguerite Westinghouse in 1898. Marguerite Westinghouse bought the house in 1901.

As they did at Solitude and Erskine Park, George and Marguerite entertained frequently and extravagantly at Blaine House. The most notable occasion was an 1899 reception for the American Society of Mechanical Engineers (ASME),[a] which was holding a convention in the city. Soon after invitations had been sent, George was summoned to Europe to confront a crisis. With no preparations made, Marguerite took over. She realized that even her spacious ballroom would be inadequate for the expected crowd. So she arranged to open up the back of the room and had a

---

(a) In 1910, George Westinghouse would be named President of the American Society of Mechanical Engineers.

Blaine House, circa 1900 (Library of Congress via Wikimedia Commons).

temporary but substantial addition constructed over the rear garden area. Elaborate decorations and arches concealed the junction with the house, and no one unfamiliar with the residence suspected that the extension was temporary. Marguerite supervised every detail of the reception and, with President George Wallace Melville[b] of the ASME, proudly stood at the head of the receiving line. Over 3,000 guests, mechanical engineers, government officials, and diplomats, mingled, ate, drank, and had a marvelous evening.[17] A 1903 book on the city of Washington calls it "the finest reception ever given in a private residence in Washington."[18]

George Westinghouse III inherited Blaine House upon the deaths of his parents, but never lived there and sold it in 1919.[19]

Glen Eyre

A far less conventional home, but one where George Westinghouse spent many hours, was Glen Eyre, his custom-made private railcar. It was built for Westinghouse

---

(**b**) This was the same George Melville, who, with John H. Macalpine, would later develop the reduction gear that allowed turbine engines to work efficiently on ocean-going ships, as discussed in Chapter 22.

by the Pullman Company, and he used it to journey in style all over the eastern United States.

In 1890, the Ways and Means Committee on the House of Representatives was considering legislation on the protection of railroad workers.[c] William McKinley, chairman of the committee, sought expert advice from George Westinghouse. The two men met at Solitude, and, as Westinghouse was headed for Chicago, he invited McKinley to join him aboard Glen Eyre. Thomas Kerr, Westinghouse's longtime attorney, joined the party and reported on the trip: "Mr. Westinghouse began, and for an hour held us spellbound with a wonderfully comprehensive and convincing exposition of the dangers attendant on railway service, their cause, their remedies, and the importance of remedying them, from the point of view not only of human safety, but of economy and efficiency of railroad operation, citing statistics and facts which were startling, and expressing views and making recommendations that showed his wide knowledge and the maturity of his conclusions."[20]

McKinley, much impressed, left the train for his home in Canton, Ohio. He and Westinghouse were both veterans of the Grand Army of the Republic, both ardent supporters of the YMCA, both Republicans and both Presbyterians. They became close friends and often met at the White House and all of Westinghouse's homes. On July 4, 1897, Westinghouse augmented his traditional fireworks extravaganza with a light show, including a profile of then-President William McKinley.[21]

Tragically, McKinley was shot by anarchist Leon Czolgosz on September 6, 1901, at the Temple of Music in Buffalo, New York. McKinley, surrounded by friends, relatives, and his wife, Ida, died of gangrene on September 14.

The crowd at the Temple of Music attacked Czolgosz, but the police rescued and jailed him. His two-day trial ended on September 24. Because Czolgosz refused to cooperate with his court-appointed attorneys, no defense witnesses were called. The jury deliberated less than one-half hour before declaring him guilty. On October 29, 1901, the unrepentant Czolgosz was electrocuted at Auburn State Prison in New York, the same place where William Kemmler became the first person to be legally electrocuted 11 years earlier.[22]

Although we have no record of Westinghouse's response, he must have been devastated at the loss of his friend and president. McKinley's funeral train from Washington, D.C., to his burial place at Canton, Ohio, would have passed through Wilmerding, which McKinley saw as the positive work of Westinghouse and a model for progressive capitalism.[23] McKinley's last ride would also have been on the same tracks traveled eleven years earlier with Westinghouse aboard Glen Eyre.

Glen Eyre's most-traveled route was the ten-hour trip from Westinghouse's home at Solitude via the Pennsylvania Railroad mainline to New York City for pleasure or business. That trip was usually uneventful but turned tragic soon after midnight on June 25, 1892.

Two sections of the Western Express No. 9 left New York at 6:30 p.m. and approached Harrisburg, Pennsylvania, about 12:30 a.m. The first section, composed of a baggage car, an express car, three day coaches, and Glen Eyre in the rear, stopped for a few minutes to permit some shifting in the train yards.

(c) These hearings led to the passage of the Railroad Safety Appliance Act in 1893, as discussed in Chapter 6.

Glen Eyre Railcar—The car was 70 feet long and had two large private compartments separated by a dining room. Up to ten guests could be accommodated (Westinghouse Collection, Detre Library and Archives Division, Senator John Heinz History Center, Pittsburgh, PA).

On board Glen Eyre were George Westinghouse, his wife and son; Westinghouse's close friend, Robert Pitcairn, Superintendent of the Pittsburgh Division of the PRR; a porter named W.H. Woodyard; and a cook. As the first section started forward, the second section of the train, with an engine, two baggage cars, and five Pullman sleeper cars, failed to stop as it should have and plowed into Glen Eyre at 40 miles per hour, telescoping it into the rest of the train. Westinghouse and his family, Pitcairn, the porter, and the cook all escaped with only minor injuries, but 12 passengers in the day coaches were killed and 23 injured. The accident destroyed Glen Eyre. The coroner's jury investigating the accident determined that negligence on the part of the 22-year-old Steelton block operator, the flagman on the first train section, and the engineer of the second section all contributed to the disaster.[24]

A new Glen Eyre replaced the one destroyed in the Harrisburg accident and was used by George Westinghouse for many more years. The second Glen Eyre now resides at the California State Railroad Museum in Sacramento, California.[25]

George Westinghouse III

George and Marguerite Westinghouse's only child, George Westinghouse III, was born on May 20, 1883. Marguerite had departed for New York City on one of her frequent visits around August 1882, and when she returned to Solitude in the spring of 1884, George III was with her. Because George III did not resemble either of

George Westinghouse
New York, New York City Births, 1846-1909

Name:	George Westinghouse
Event Type:	Birth
Event Date:	20 May 1883
Event Place:	Manhattan, New York, New York, United States
Gender:	Male
Race:	White
Father's Name:	George Westinghouse, Jr.
Father's Birthplace:	Central MDFC, U.U.
Father's Age:	36
Mother's Name:	Margaret Erskine Walker
Mother's Birthplace:	Roebury, Delaware Co. N.Y.
Mother's Age:	36

Birth Record of George Westinghouse III, from New York, New York City Births, 1846–1909. Note that Marguerite's age should have been recorded as 41.

his parents, the household staff and some Westinghouse acquaintances believed that George III had been adopted.[26] But there is evidence that George III was the natural son of George and Marguerite, as the Birth Record shows.[27]

As might be expected of an only child with wealthy parents, George III was undeniably spoiled. He often dressed in Lord Fauntleroy costumes and wore long, golden curls. He once expressed a desire for a donkey and a donkey he received. The donkey and a pet raccoon resided in the stable on the grounds of Solitude.

George III received his early education at Shady Side Academy, just ten blocks from Solitude. Shady Side Academy, an all-male day school, was established in 1883, the same year that George III was born. The school was intended to educate the sons of the wealthy industrialists living in Pittsburgh's Point Breeze neighborhood. Among the founders were Henry Clay Frick, Robert Pitcairn, and George Westinghouse.[28]

For his college education, George III attended the Sheffield Scientific School ("Sheff") at Yale University in New Haven, Connecticut, from 1902 to 1905.[29] The undergraduate program at Sheff was three years long, and combined classical learning with technical courses such as Chemistry, Mechanical Engineering, Electrical Engineering, Zoology and Botany, and Biology preparatory to Medical studies.[30]

The *Yale Daily News* reported on-campus activities, and the Westinghouse name appeared just four times during the time George III was at Yale. George III joined the Yale Corinthian Yacht Club in February 1903 and the Yale University Tennis Club in

September 1903. His father arrived aboard Glen Eyre to visit on November 14, 1903. And George III joined a Yale Military Drill Company in January 1904. There is no mention of any academic honors earned by George III.

George III attended Yale for at least three years, but he never received a degree. Instead, perhaps emulating his father, who had lasted just three months in college, George III left Yale and went to work at his father's Westinghouse Air Brake in Wilmerding.

George III met his future wife, Evelyn Violet Brocklebank, when both were children. Their parents, George and Marguerite Westinghouse and Sir Thomas and Lady Brocklebank, had become acquainted in the mid–1870s when George was displaying his automatic air-brake to the English railroad industry. The two couples became close friends and exchanged visits to Solitude and Erskine Park in the states, and to Irton Hall and the Liverpool townhouse of the Brocklebanks in England. On one of those visits, the children, George III and Evelyn, met and played together. With more trans-Atlantic visits, the childish affection grew to love and, eventually, marriage.[31]

The wedding was held on March 4, 1909, at St. Paul's Church in Irton, Cumbria, England, where Sir Thomas was a patron. It was a simple but elegant ceremony, presided over by the bride's uncle, the Rev. Charles H. Brocklebank. Light snow covered the trees, lawns, and roads.

The reception was held at historic Irton Hall, the country home of the Brocklebanks, which had previously hosted Oliver Cromwell and King Henry VI.

The young couple traveled from the reception by coach-and-four driven by the bride's twin sister, Sylvia Brocklebank, to the train station at Drigg to embark on their Italian honeymoon.[32]

Marguerite Westinghouse and Lady Brocklebank arriving at St. Paul's for the wedding, trailed by George Westinghouse. The ladies do not appear to be happy. Mr. and Mrs. George Westinghouse III after the ceremony.

Mr. and Mrs. George Westinghouse III traveling by coach-and-four to Drigg Station after their wedding, with Sylvia Brocklebank, the bride's twin sister, driving (Westinghouse Collection, Detre Library and Archives Division, Senator John Heinz History Center, Pittsburgh, PA).

On their return from Europe, George and Evelyn moved to a rented home at 201 Maitland Avenue (street name changed to Kinsman Road in 1911[33]) in Pittsburgh, just one mile from Solitude. The couple had four servants.[34]

Although his father had built a house for him at 201 North Murtland Street, within one block of Solitude, there is no evidence that George III ever lived there. By 1907, Emil E. Keller occupied the house.[35]

George III went to work as a clerk at Westinghouse Electric & Manufacturing Company in East Pittsburgh.

Family of George Westinghouse III

Between 1911 and 1925, George and Evelyn Westinghouse had six children, three boys and three girls. The birthplaces of the children trace the moves of the family.

George Thomas Westinghouse, their first child, was born while they lived in Pittsburgh. The two births at Lee, Berkshire Co., Massachusetts, occurred when the family resided on Stockbridge Road,[36] near Erskine Park, the Westinghouse estate

Children of George and Evelyn Violet (Brocklebank) Westinghouse

Name	Birth Date	Birth Location	Death Date (Age)
George Thomas	April 27, 1911	Pittsburgh, Allegheny Co., PA	September 22, 1983 (72)
Aubrey Harold	March 4, 1913	Lee, Berkshire Co., MA	March 11, 2003 (90)
Agnes Sylvia	October 28, 1914	Lee, Berkshire Co., MA	June 3, 2006 (91)
Violet Louisa	September 9, 1917	Altadena, Los Angeles Co., CA	May 21, 2001 (83)
Richard Lawrence	March 4, 1919	Victoria, British Columbia, Canada	June 25, 1996 (77)
Margaret Virginia	December 25, 1925	Santa Barbara, Santa Barbara Co., CA	January 30, 2020 (94)

discussed above. After the death of George III's parents, the family moved to Altadena, California, just north of Pasadena, in 1916, where Violet Louisa was born.

Around 1919, the Westinghouse family moved to an estate at 912 Mt. Newton Cross Road on Vancouver Island, British Columbia. Richard Lawrence Westinghouse was born there.

George III commissioned a massive gymnasium on adjacent land, which earned the nickname Valhalla from the children. The gymnasium had over 6,700 square feet (plus 3,700 square feet in the unfinished basement).

The outstanding feature of the gymnasium was a central great room measuring 35 feet by 53 feet and open to the roof 30 feet above.[37] Its design was reminiscent of the recreation hall that his mother, Marguerite, had built for George III at Erskine Park.

The last child of George and Evelyn Violet Westinghouse, Margaret Virginia, was born on Christmas Day, 1925, in Santa Barbara, California.

Additional residences of the George Westinghouse III family were in the Bahamas and at Bainbridge Island, Washington.

Starting in 1927, George III, Evelyn, and the youngest four children wintered in Tucson, Arizona.[38] In 1930, they bought property there (now 10500 East Tanque Verde Road) and built a three-story house, the "Flying W" Ranch, in 1931. The name, Flying W, was a tribute to the aviation exploits of eldest son George Thomas Westinghouse, who often landed at nearby Davis-Monthan field in Tucson.

In 1936, the Westinghouse family leased the property to Julia Bennett, who operated a dude ranch there, the "Diamond W," until 1950.[39] The property is now a drug and alcohol rehabilitation center called the Circle Tree Ranch.

Evelyn Violet Westinghouse died on June 9, 1943, at age 61 in Santa Barbara, California.[40] George Westinghouse III survived until November 5, 1964, when he died in New Westminster, British Columbia, Canada, at age 81. His later years were tumultuous, even scandalous.

By 1959, George III was confined as mentally incompetent in the Hollywood Sanitarium in New Westminster, British Columbia. The sanitarium was well-known in British Columbia as a facility that treated prominent and wealthy members of society, including several celebrities.[41] The fact of George's confinement was revealed in a spectacular newspaper article on July 16, 1959. Datelined London, the story told of a British court ruling that Mrs. Grace Margaret Smith of Northwood, Middlesex, England, had to return over $200,000 that George Westinghouse III had given her between November 1954 and July 1955. Mrs. Smith did not oppose the ruling but explained that she and George were engaged to be married until she broke off the engagement in November 1954. She further said that they had been close and intimate friends and had known each other for seventeen years. Westinghouse's attorneys added that he had also given $151,000 to a Fort Worth, Texas, housewife, and $182,000 to a female ex–taxi driver from Denver, Colorado, before "relatives had him put away."[42]

Chapter 26

Retirement, Honors, and Death

"I hate biographies. They always end badly."
—William R. Huber

Accomplishments

During his lifetime, George Westinghouse received 361 patents, the first at age 19 and the last in 1918, four years after his death. He founded more than 60 companies, including two that were Fortune 500 companies—Westinghouse Air Brake, ranked between 185 and 267 up to 1968; and Westinghouse Electric & Manufacturing, ranked between 13 and 37 up to 1993.[1] His companies employed well over 100,000 people and were valued at $200 million.

He was an outstanding inventor, entrepreneur, industrialist, and person. Despite losing control of Westinghouse Electric & Manufacturing in 1907, he left an estate valued at $50 million when he died in 1914.[2] That amount would be equivalent to $1.3 billion today.[3]

How did a man with virtually no formal education accomplish so much? Clearly, George Westinghouse possessed a rare combination of talents.

In Prout's 1921 biography, he identifies imagination as the foremost of those talents.[4] Given a problem, Westinghouse could not only imagine one or more solutions but also recognize the long-term consequences and financial impact of those solutions.

Sometimes the solution proposed by Westinghouse went against conventional thinking and established experts. A prime example is his advocacy of alternating current electricity when not only Thomas Edison but also Sir William Thomson (Lord Kelvin) were convinced that direct current was superior.

Westinghouse demonstrated two more of his gifts, audacity and perseverance, in proposing and defending his idea despite the opposition of these great men. Through years of patience and persistence, he was able to achieve concrete results that others argued were impossible. Because Westinghouse kept the debate focused on technical and financial issues rather than personal attacks, he won the battle of the currents and gained the respect and friendship of Lord Kelvin.

Not everything went according to plan, as external financial challenges, such as the crisis of 1891, sometimes disrupted progress. But Westinghouse also had the gift of fortitude. His courage in adversity encouraged his colleagues and workers to forge ahead despite the obstacles.

Chapter 26. Retirement, Honors, and Death

George Westinghouse and Lord Kelvin (Sir William Thomson) at London Offices of British Westinghouse Company, 1903 (Westinghouse Collection, Detre Library and Archives Division, Senator John Heinz History Center, Pittsburgh, PA).

George Westinghouse had two more attributes that contributed to his success. He could concentrate on a given problem for hours at a time. Other issues receded into the background as he studied, calculated, and drew. And he had a phenomenal memory. Without notebooks or other aids, he remembered with equal facility details of the lives of his employees and the plans he had drawn years before to solve some problem.

Graham Moore's terrific novel, *The Last Days of Night*,[5] is told from the viewpoint of Paul Cravath, Westinghouse's friend, partner, and patent attorney. Cravath was the founding partner of the law firm Cravath, Swaine, and Moore and worked closely with Westinghouse for several decades, through good times and bad. In a 1936 remembrance of Westinghouse, Cravath said:

> I saw him thus intimately under almost every conceivable condition—in his home, at his office, in his factory, in his private car which was almost another home, abroad, as well as in this country. I saw him when he was elated by successful achievement, amid disappointments and discouragement, and more than once in the face of threatening disaster. I saw him when he was carrying a load of responsibility under which any other man whom I have ever known would have fallen. He was always the same; simple, unassuming, direct, frank, courageous, unfaltering in his faith, and supremely confident in the ultimate triumph of his plans. I have seen him wearied almost beyond endurance; disappointed beyond expression over some miscarriage of his plans; wounded in his feeling because he had discovered stupidity where he expected intelligence; discouragement where he expected encouragement; disloyalty where he had a right to expect loyalty. I had seen him more than once when every man about him despaired of his being able to attain the ends for which he was striving and advised surrender or compromise, but I have never known him to acknowledge defeat nor to yield to discouragement, nor to falter in his efforts to accomplish his main objectives.[6]

Andrew Carnegie said it in fewer words: "George Westinghouse is a man that can't be downed."⁷ Another Carnegie statement captures Westinghouse's business philosophy (and how it contrasted with Carnegie's own): "George Westinghouse could have made a lot more money if he hadn't treated his workers so well."⁸

Could Westinghouse Ever Retire?

George Westinghouse's expulsion from Westinghouse Electric & Manufacturing in mid–1911 signaled the end of his illustrious career as an industrialist. Although he was still president of Westinghouse Machine Company and Westinghouse Air Brake, as well as overseas subsidiaries, he was more of a figurehead, as other people managed those companies. When his younger brother, Henry Herman Westinghouse, suggested that George and Marguerite take a pleasure trip around the world, George replied, "I have not had a long vacation in more than forty years, and I don't need or want one now. Work is my pleasure."⁹

As with many people who have been active and productive during their working careers, retirement did not suit George Westinghouse. While he and Marguerite "retired" to Erskine Park, he continued to search for problems to solve.

Westinghouse had been opposed to the automobile since its introduction in the United States by Charles and Frank Duryea in 1893. But in 1904, when his French Westinghouse Works presented him with a specially-built limousine, he became a car owner if not an enthusiast.

His first trip in the new car had been with Marguerite and a chauffeur from Erskine Park to Kingston, New York. The driver failed to see a deep pothole, and the car dropped into it. George's head collided with the roof of the vehicle, gouging a wound that left a permanent scar.

With less work to fill his time, Westinghouse's thoughts turned to how to smooth the ride of automobiles, especially given the rough roads that still predominated. He made drawings of two telescoping cylinders with oil-impregnated leather packing between them to create an air-tight joint. He included an integral pump to return any oil that leaked out during operation.

George had expert machinists at Westinghouse Machine Company construct models from his drawings and had them installed on the French limo and another car. The initial models were far from perfect, but he and his team systematically worked out the flaws and succeeded in creating springs that used compressed air as cushions against shocks from uneven roads.

In doing so, Westinghouse had gone back to his work of 40 years earlier and applied compressed air to solve a new problem.¹⁰ He filed patent applications[a] to protect his inventions and established the Westinghouse Air Spring Company to manufacture the springs.¹¹

His son, George Westinghouse III, became vice president and advanced to president in 1914. The new air spring soon became standard equipment in all cars.¹²

(a) U.S. Patent 1,031,759—"Vehicle-Supporting Device," issued July 9, 1912; U.S. Patent 1,036,043—"Fluid-Pressure Device," issued August 20, 1912; and U.S. Patent 1,185,608—"Automobile Air Spring," issued May 30, 1916.

Honors

The genius that was George Westinghouse, and his contributions to society, were recognized and honored at many times and places during his lifetime.[13]

- In 1874, he was awarded the John Scott Legacy Medal by the City Council of Philadelphia. The Scott Medal was presented to men and women whose inventions improved the comfort, welfare, and happiness of humankind in a significant way.
- Leopold II, King of the Belgians, awarded him the Order of Leopold in 1884. The honor, named for its founder, King Leopold I, originated in 1832 and is the oldest and highest order of Belgium.
- In 1889, Umberto I, King of Italy, recognized Westinghouse with the Order of the Crown of Italy for outstanding civilian merit.
- Union College, which Westinghouse attended for just three months in 1865, awarded him an honorary Doctor of Philosophy degree in 1890.
- Westinghouse was made a member of the French Legion of Honor in 1895. The Legion was established to recognize high military and civil merit by Napoleon Bonaparte in 1802.
- The American Association of Engineering Societies awarded Westinghouse the John Fritz Medal in 1905. The Medal was established in 1902 to honor outstanding scientific or industrial achievements. Some describe it as the Nobel Prize for engineering.
- The Berlin, Germany Konigliche Technische Hochschule awarded Westinghouse the honorary degree Doctor of Engineering in 1906.[14]
- In 1910, Westinghouse was elected President of the American Society of Mechanical Engineers. Despite his lack of formal education, he always considered himself to be a mechanical engineer and was delighted to be so honored by his peers.
- In 1911, the American Institute of Electrical Engineers[b] chose George Westinghouse to receive the Edison Medal, named for his long-time antagonist, Thomas Edison. The Medal is for a career of meritorious achievement in electrical science, electrical engineering, or the electrical arts. It is the oldest and most coveted Medal in electrical engineering in the United States.[15]
- Westinghouse became the first American recipient of the Grashof Medal from Germany during a ceremony for a group of 300 touring American engineers in Leipzig on June 23, 1913. The Medal is awarded annually to a person who has done pre-eminent work in the engineering field.[16]
- The Engineering and Science Hall of Fame enshrined George Westinghouse on November 6, 1986, at a ceremony in Dayton, Ohio.
- Westinghouse was inducted into the National Inventors Hall of Fame on February 12, 1989, in Washington, D.C.[17]

(b) AIEE merged in 1963 with the Institute of Radio Engineers, IRE, to form the Institute of Electrical and Electronics Engineers, IEEE. I am a Life Senior Member of IEEE.

Decline

George Westinghouse had neglected his health for years. While he ate and drank in moderation and did not smoke, lack of exercise had added pounds to his already large frame.

His first admission of mortality came in early 1911 when he was ascending the steps to the Air Brake plant in Wilmerding. Then 64 years old, he had to pause midway and remarked, "I must be getting old; it tires me to walk up these steps."[18]

Later that year, while residing at Erskine Park, he was awakened by a fit of uncontrollable coughing that lasted for two hours.

In the summer of 1913, while at Solitude, a similar coughing episode occurred at the dinner table. Marguerite was at Erskine Park at the time, but the servants were so alarmed that they summoned Dr. William A. Stewart, his Pittsburgh physician. Despite his objections, Dr. Stewart conducted a thorough physical examination; the first Westinghouse had undergone since his Army induction over 50 years earlier. The doctor discovered an enlarged heart and other ailments that caused him to order Westinghouse to withdraw for one month to Erskine Park to relax and recuperate.

A business associate urged Westinghouse to make a will, and he reluctantly did so on January 13, 1914.[19] Unlike his contemporaries, such as Carnegie and Frick, Westinghouse felt no compulsion to memorialize his name by creating libraries or art collections. He was a simple man at heart and was ambivalent about giving wealth to others. He said, "I am convinced from observation and experience that the greater part of the money which is given for benevolence is a detriment rather than a help, for it tends to pauperize the recipient by destroying his honest pride of independence, and adds to the burden of society by the development of a class of people who are willing to accept charity rather than to exercise their own ability. I think, as a rule, a dollar given to a man does him ten dollars' worth of harm, while a dollar honestly earned by his own efforts does him ten dollars' worth of good; so my ambition is to give as many persons as possible an opportunity to earn money by their own efforts, and this has been the reason why I have tried to build up corporations which are large employers of labor, and to pay living wages, larger even than other manufacturers pay, or than the open labor market necessitates."[20]

Despite this overall philosophy, both George and Marguerite gave generously to charities and individuals, but they usually did so anonymously. Leupp, in his 1919 biography, reported: "While he disclaimed belief in the efficacy of benevolent giving, and shrank from acknowledgment of his kindness, those of us who were closely connected with him knew of many instances where he was supporting whole families and doing other deeds of helpfulness in an unostentatious way. Mrs. Westinghouse was very sympathetic and loved to relieve distress, and Mr. Westinghouse made her a regular allowance for the gratification of her desires in this respect. The amount was stated to me, and it was large."[21]

An unusual gift in December 1909 marked one time that Marguerite revealed herself as the donor. When riding in her carriage, she saw two horses slip and fall on the icy streets of Pittsburgh. She contacted Superintendent James Bell of the Western Pennsylvania Humane Society and instructed him to acquire, at her expense, "a great number of horse overshoes" to be supplied to "any horse found in distress on the streets."[22]

Further Decline of George Westinghouse

The one-month hiatus at Erskine Park, although extended to three months, resulted not in rejuvenation but in further decline. A near-disaster, which George concealed until later, contributed to his malaise.

In what indicates the degree of his boredom, Westinghouse went fishing at the Erskine Park pond. He often boarded his rowboat either for a place to sit or for a leisurely ride on the water. On this day, he stepped onto the boat, not realizing that his regular boat had been removed for repairs and another boat substituted. The replacement boat had no keel to stabilize it. As soon as he placed his weight on the boat, it capsized, dumping Westinghouse into the chin-deep water.

He managed to remain upright and climb up the steep bank, but the extreme exertion strained his heart. He headed to the house in his drenched clothes but by bedtime, had a severe cough. The subsequent cold lingered for weeks. Because Marguerite was suffering from her own health issues, including strokes in September 1912 and the summer of 1913, Westinghouse did not reveal his problems to her.

At the end of three months, Westinghouse returned to Pittsburgh, where his business associates were shocked at his change in appearance and stamina. His face lacked color; his brisk manner was now slow and thoughtful; he often dozed off during conversations; and he walked haltingly. Dr. Stewart, after another examination and upon hearing of George's unplanned swim in the pond while fishing, insisted that Westinghouse cease work immediately and return to Erskine Park. George agreed as long as the doctor accompanied him, and they left Pittsburgh in November 1913.

Back at Erskine Park, George recovered some of his former enthusiasm. But he fluctuated between his old energetic self and a listless, sleepless replica. He lost his appetite, and the only food that appealed to him was a concoction composed mainly of raw eggs.

After Christmas, he and Marguerite planned to move to Blaine House in Washington. They got as far as New York City, where they rented a suite at the Hotel Langham, overlooking Central Park, for the rest of the winter. George's condition remained stable until early March when he declined severely.

On March 12, 1914, at the age of 67, George Westinghouse dozed off while resting in a reclining chair and never woke up. His wife, his son and daughter-in-law, and his brother Herman and his wife were with him when he died.

The funeral was held two days later at the Fifth Avenue Presbyterian Church at 7 West 55th Street at 2 p.m. Honorary pallbearers included distinguished businessmen and public figures, including Senator George T. Oliver and Rear Admiral Robert E. Peary. The active pallbearers were eight old and loyal employees, including Christopher Horrocks, the first employee Westinghouse hired in 1869 when he started Westinghouse Air Brake Company.[23] Many other Westinghouse employees and members of leading scientific and engineering societies filled the church. Interment at Woodlawn Cemetery was private, attended only by the immediate family.[24]

By 1914, Westinghouse's parents and siblings, except for his youngest brother, Henry Herman Westinghouse, had all died. George's will, probated on March 18, 1914, specified the following bequests[25]:

- Two-thirds of the capital stock of the Westinghouse Air Spring Company to Marguerite; the remaining one-third to their son, George Westinghouse III;
- One-year's salary to nine long-term Westinghouse employees and to the many servants in his several homes;
- 40 percent of his residual estate to Marguerite; 40 percent to George Westinghouse III; and 20 percent to Herman Westinghouse;
- Cancel all personal debts owed to him by other individuals.

The value of the estate was estimated at between $35 and $50 million.[c]

After Westinghouse's funeral, Marguerite returned to Erskine Park but continued to grieve. They had been a devoted couple throughout over 46 years of marriage. When George was away, he would telegraph or telephone daily.

Her family, including son George III, his wife Evelyn and her two sisters, and Herman Westinghouse and his wife, helped to distract and encourage her. But on June 15, 1914, just after her guests departed, Marguerite suffered a third and massive stroke, leaving her unconscious.

She died on the morning of June 23, 1914, a little over three months after her husband had passed. She was buried beside him at Woodlawn Cemetery.[26]

The Westinghouses did not stay long at Woodlawn. Because of his military service, they were entitled to burial at Arlington National Cemetery. On December 15, 1915, their remains were moved from Woodlawn to Arlington.[27]

With the death of his father and then his mother, George Westinghouse III inherited the portion of the estate that went directly to him as well as the part that initially went to his mother.

Westinghouse Marker at Arlington National Cemetery.

(c) $35 to $50 million in 1919 would correspond to $900 million to $1.3 billion in 2020.

Chapter 26. Retirement, Honors, and Death

The March 21, 1914, issue of *Electrical World* published heartfelt tributes to George Westinghouse from eight of his friends and associates. These reminiscences show the respect that these men felt for Westinghouse as a friend, an inventor, an industrialist, and as a devoted family man. Among the contributors were several of his leading engineers, along with inventors Nikola Tesla and Peter Cooper Hewitt.

Tesla said,

> The first impressions are those to which we cling most in later life. I like to think of George Westinghouse as he appeared to me in 1888, when I saw him for the first time. The tremendous potential energy of the man had only in part taken kinetic form, but even to a superficial observer the latent force was manifest. A powerful frame, well proportioned, with every joint in working order, an eye as clear as a crystal, a quick and springy step—he presented a rare example of health and strength. Like a lion in a forest, he breathed deep and with delight the smoky air of his factories. Though past forty then, he still had the enthusiasm of youth. Always smiling, affable and polite, he stood in marked contrast to the rough and ready men I met. Not one word which would have been objectionable, not a gesture which might have offended—one could imagine him as moving in the atmosphere of a court, so perfect was his bearing in manner and speech. And yet no fiercer adversary than Westinghouse could have been found when he was aroused. An athlete in ordinary life, he was transformed into a giant when confronted with difficulties which seemed insurmountable. He enjoyed the struggle and never lost confidence. When others would give up in despair he triumphed. Had he been transferred to another planet with everything against him he would have worked out his salvation. His equipment was such as to make him win easily a position of captain among captains, leader among leaders. His was a wonderful career filled with remarkable achievements. He gave to the world a number of valuable inventions and improvements, created new industries, advanced the mechanical and electrical arts and improved in many ways the conditions of modern life. He was a great pioneer and builder whose work was of far-reaching effect on his time and whose name will live long in the memory of men.
>
> New York
Nikola Tesla[28]

Peter Cooper Hewitt was the son of New York City Mayor Abram Hewitt and the grandson of Peter Cooper, the famed industrialist who founded Cooper Union.[d] Peter Hewitt invented the highly-efficient mercury vapor lamp, the precursor to the fluorescent light bulb. He and Westinghouse together formed the Cooper Hewitt Electric Company to commercialize Hewitt's invention. Hewitt had the following observations upon the death of George Westinghouse,

> In George Westinghouse the world has suffered the loss of a great and valued citizen and much should be said in honor of his accomplishments and advancement of modern industry. Essentially an American, he had universal reputation and respect. The vast and diverse works which he pioneered give an idea of his force and energy and illustrate dominant characteristics. Every effort of his genius was expended for humanity in advancing the arts of civilization. As the problems involved in nature's secrets unfolded themselves before him he strained every effort and means at his disposal to turn their use to the benefit of

---

[d] Cooper Union, a private college founded in 1859, was based on Peter Cooper's fundamental belief that an education equal to the best technology schools should be accessible to those who qualify, and should be open and free to all. It retained that tuition-free model until 2014.

mankind. He appreciated the enormous danger to humanity that accompanies the harnessing of vast forces of nature for public use, and by his foresight and skill prevented disasters which might conceivably have been of enormous extent. The memory of George Westinghouse will live through his appreciation of nature's forces and his love of mankind.

New York
Peter Cooper Hewitt[29]

Just as George Westinghouse avoided public speaking and having his picture taken, he left us with few quotable lines. He was a doer, not a talker; he sought achievement, not fame.

Chapter 27

Memorials

"Since it is not granted to us to live long, let us transmit to posterity some memorial that we have at least lived."
—E. Joseph Cossman, Pittsburgh-born businessman and author

Westinghouse Memorial

On November 30, 1918, when the Engineers Society of Western Pennsylvania deeded the ten acres, including Solitude, the Westinghouse home, to the city of Pittsburgh, one condition of the transfer was that a memorial to George Westinghouse was to be erected on the site. But World War I had ended a few weeks earlier, and then the Spanish Flu pandemic disrupted everyone's plans. By 1926, no such memorial had been completed, let alone started. At that time, Pittsburgh City Council accepted a proposal to move the memorial site to Schenley Park, near the Phipps Conservatory and the Carnegie Institute of Technology (now Carnegie-Mellon University). The Westinghouse Memorial Society was formed to oversee the project, which was funded by donations from over 50,000 current and former Westinghouse employees. Contributions totaled $200,000 (equivalent to about $3 million today).[1]

Two Pittsburgh–based architects, Henry Hornbostel and Eric Fisher Wood, designed the memorial.

Hornbostel had designed over 225 buildings, bridges, and monuments across the United States, 22 of which are listed on the National Registry of Historic Places.[2] Eric Wood was a civil engineer and part-time architect who also co-founded, along with Theodore Roosevelt, Jr., The American Legion.[3]

Sculptors for the project were Daniel Chester French and Paul Fjelde. One of the most acclaimed sculptors of the late nineteenth and early 20th century, French is best known for his iconic statue of Abraham Lincoln at the Lincoln Memorial in Washington, D.C.[4] Fjelde headed the Sculpture Department at Carnegie Tech and also created a memorial to Abraham Lincoln, his being in Oslo, Norway.[5]

Photographs, even ones in color, struggle to capture the size and scope of the massive Westinghouse Memorial. At the center front stands the statue of a young man on a pedestal. Titled "The Spirit of American Youth," he is facing and drawing inspiration from Westinghouse, who is portrayed in the center of three massive sculptured bronze panels. The memorial sits at the end of a tranquil lily pond.

The Westinghouse Memorial in Schenley Park, with "The Spirit of American Youth" gazing on George Westinghouse and his accomplishments (courtesy Harvey Butts).

George Westinghouse over his drawing board, flanked by a mechanic and an engineer; note the train at the bottom, symbolizing Westinghouse's massive contributions to the railroad industry (author's photograph).

Chapter 27. Memorials

The depiction of Westinghouse on the center panel is from a photograph, taken surreptitiously on October 6, 1910.

Westinghouse was famously averse to being photographed, but a picture of him "at work" was needed to illustrate a *New York Times* article. He had a small office at Westinghouse Machine Company, where he often went to sketch drawings for his latest piece of equipment. Herbert T. Herr, vice-president of the Machine Company, arranged for Ed Kilchenstein, a company photographer, to hide in the restroom of that office with his camera. Herr then took Westinghouse to the office to discuss a project. When Westinghouse sat down at the drawing board, Kilchenstein cracked open the restroom door just wide enough to capture the image (without flash, of course). The photograph, one of very few of George Westinghouse, was banned by him during his lifetime, but it became famous after his death.[6]

The balance of the Westinghouse Memorial consists of triple panels on each side of the center sculpture depicting six of the most significant achievements of Westinghouse: hydroelectric plant at Niagara Falls; illumination at the Columbian Exposition; steam turbine; electric locomotive; air brake, and railway signaling and control.

Work on the memorial extended over five years. The back surfaces of all sections carry the same level of detail as the front, thus creating three-dimensional panels. The statues are cast in bronze, from wax patterns carved by renowned Italian sculpture artist Masaniello Piccirilli. Gold leaf once covered the bronze material of the statues to protect them from the notorious smoke of Pittsburgh, but vandalism forced removal of the gold in 1941.

Unveiling of the memorial took place on October 6, 1930, in celebration of the 84th anniversary of the birth of George Westinghouse. Almost 15,000 people attended the dedication ceremony, which included the combined Westinghouse bands, and choruses of Westinghouse Electric and Union Switch and Signal employees singing patriotic songs.

George Westinghouse at work, October 6, 1910 (Westinghouse Collection, Detre Library and Archives Division, Senator John Heinz History Center, Pittsburgh, PA).

The dedication plaque, on the back of the center panel, reads: "This memorial unveiled October 6, 1930, in honor of George Westinghouse is an enduring testimonial to the esteem, affection and loyalty of 60,000 employees of the great industrial organizations of which he was the founder. In his later years rightly called 'The Greatest Living Engineer,' George Westinghouse accomplished much of first importance to mankind through his ingenuity, persistence, courage, integrity and leadership. By the invention of the air brake and of automatic signaling devices, he led the world in the development of appliances for the promotion of speed, safety and economy of transportation. By his early vision of the value the alternating current electric system, he brought about a revolution in the transmission of electric power. His achievements were great, his energy and enthusiasm boundless, and his character beyond reproach; a shining mark for the guidance and encouragement of American youth."[7]

Secretary of the Treasury Andrew Mellon sent a message to be read at the ceremony, "George Westinghouse earned an important and permanent place in history by his many contributions to the advancement of civilization."[8]

Former President Calvin Coolidge wrote, "George Westinghouse had that combination which is so rare of both inventive and business genius. Because he lived, industrial life is more humane, more safe and more productive. He ranks as one of the great benefactors of mankind."[9]

Over the decades, environmental damage, vandals, and neglect combined to disfigure the memorial. Around 2012, an effort was organized, under the direction of the Pittsburgh Parks Conservancy, to refurbish the memorial. After several years of planning, fundraising, and restoration costing over $2 million, the Westinghouse Memorial was rededicated in front of a large crowd on October 6, 2016.[10] Sixteen direct descendants of George and Marguerite Westinghouse attended the ceremony.

Today, the George Westinghouse Memorial in Schenley Park stands again as an inspiration to all who would strive, as Westinghouse said, to "contribute something to the welfare and happiness of my fellow man."

George Westinghouse Bridge

Pittsburgh is known as "The City of Bridges." A 2006 study found 446 bridges within the city limits.[11] But until 1932, one critical bridge was missing. The Lincoln Highway, U.S. Route 30, included what the *Pittsburgh Post-Gazette* called "the worst section of any road in the state." A trip from the east of the city, for example, from Greensburg to downtown Pittsburgh required navigating steep hills (up to 9 percent grades), crowded streets, countless stoplights, and multiple railroad crossings. It took 40 minutes or more to travel just two miles.[12] There was no reasonable way to avoid Turtle Creek valley, the deep gash in the earth where George Westinghouse had built his massive Westinghouse Air Brake and Westinghouse Electric & Manufacturing plants.

To relieve this bottleneck, in 1929, Vernon R. Covell, chief engineer for Allegheny County, designed a five-span arch bridge with what was then the longest concrete arch span in the world. That central arch is 460 feet long, and the five spans total 1524 feet. The bridge deck sits 215 feet above Turtle Creek. The bridge cost $1.75 million to build (about $33 million in 2020 dollars).

Massive pylons at the ends of the bridge carry 10-foot by 18-foot art deco images depicting important aspects of history and industry in the Turtle Creek Valley: Braddock's defeat by the French and Indians in 1755; the steel industry introduced by Andrew Carnegie; electricity promoted by George Westinghouse; and Westinghouse himself. The plaque on the Northwest pylon reads:

Chapter 27. Memorials

Bridge under construction, circa 1931 (Westinghouse Collection, Detre Library and Archives Division, Senator John Heinz History Center, Pittsburgh, PA).

September 10, 1932, dedication of the George Westinghouse Memorial Bridge. Note Andrew Carnegie's Edgar Thomson Works at top left (Westinghouse Collection, Detre Library and Archives Division, Senator John Heinz History Center, Pittsburgh, PA).

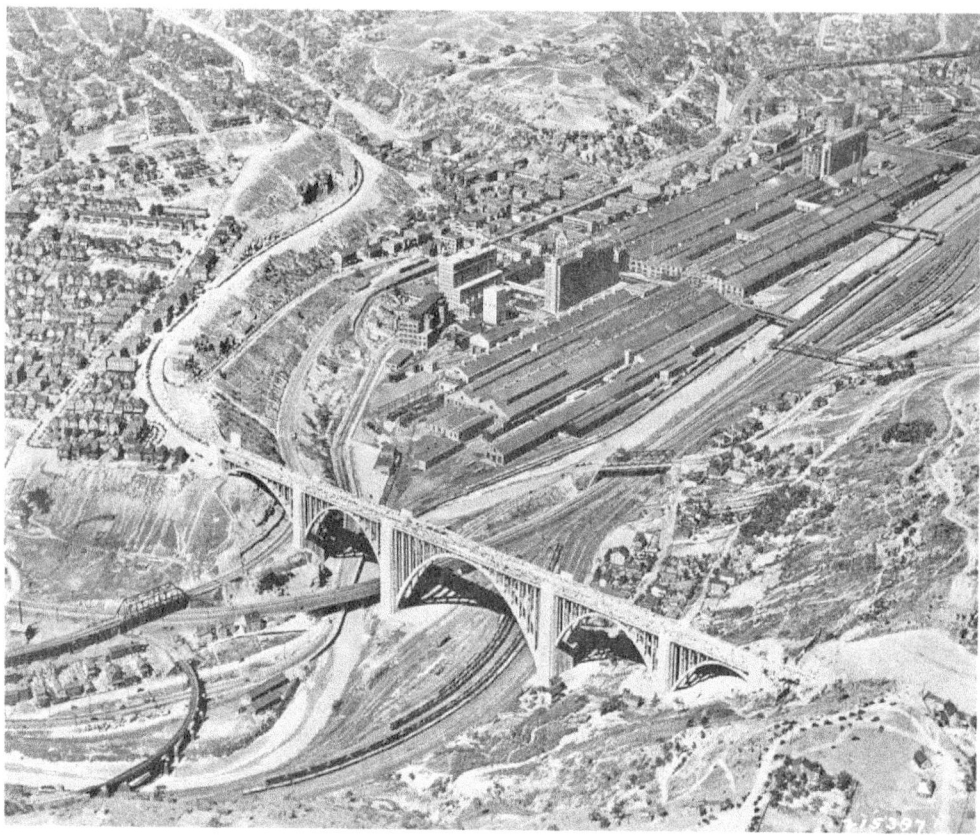

George Westinghouse Memorial Bridge and the Turtle Creek Valley; Westinghouse Electric & Manufacturing plant sits to the right of the bridge (Westinghouse Collection, Detre Library and Archives Division, Senator John Heinz History Center, Pittsburgh, PA).

IN BOLDNESS OF CONCEPTION, IN GREATNESS AND IN USEFULNESS TO MANKIND THIS BRIDGE TYPIFIES THE CHARACTER AND CAREER OF GEORGE WESTINGHOUSE 1846–1914 IN WHOSE HONOR IT WAS DEDICATED ON SEPTEMBER 10, 1932[13]

On the night before the bridge dedication, famous radio commentator Lowell Thomas featured the story on his nationwide radio broadcast. A crowd estimated at 30,000 attended the opening ceremony on September 10, 1932. Bands played, politicians talked, and Boy Scouts roamed the crowd to assist anyone overcome by the heat. At 4 p.m., Henry Herman Westinghouse, George's youngest and only surviving sibling, cut the ribbon as factory whistles in the valley below shrieked, and drivers laid on their car horns to celebrate the event.[14]

The Westinghouse Bridge is as challenging to photograph as the Westinghouse Memorial in Schenley Park. The bridge is so large that no earth-bound vantage point allows a full view. Driving across the bridge once revealed the bustling industrial valley below. But now that industry is mostly gone, and trees at the ends of the bridge obscure what remains. So, an historical aerial photograph provides the best perspective of this magnificent structure and the valley below.

Chapter 28

The Next Century

"You can't go home again."
—Thomas Wolfe

Westinghouse Companies after 1914

During his lifetime, George Westinghouse founded at least 60 companies, many of which endured long after he died. Let us examine what happened to his two largest companies, as well as two Westinghouse companies formed after his death.

Westinghouse Air Brake

When George Westinghouse died in 1914, his brother, Herman, took over as President of Westinghouse Air Brake. Herman later became Chairman of the Board and remained in that position until he died in 1933.[1]

In 1917, Westinghouse Air Brake bought another Westinghouse company, Union Switch and Signal. General prosperity with up and down periods continued for another decade. But the maturity of the railroad industry, coupled with the Great Depression of 1929, started a decline in Air Brake's employment.

In 1968, American Standard acquired the company. A management buyout in 1988 took the company private, but it went public again in 1995.

Westinghouse Air Brake spawned two currently operating successors. In 1999, Westinghouse Air Brake and MotivePower merged to form Wabtec Corporation (Wabtec is the acronym for Westinghouse Air Brake Technologies Corporation). Wabtec continues to design and manufacture railway air brakes in Wilmerding, and in 2006, employed about 1,000 people there.[2] On May 21, 2018, GE and Wabtec confirmed the merger of GE Transportation with Wabtec in an $11 billion deal, completed on February 25, 2019, that saw Wabtec shareholders take a 50.8 percent share in the merged company, with GE shareholders owning 24.3 percent and GE itself 24.9 percent.[3] In fiscal 2019, Wabtec had revenue of over $6 billion.[4]

WABCO Holdings, Inc., the other successor company to Westinghouse Air Brake, designs and builds control systems for commercial road vehicles, including air brakes, and is headquartered in Bern, Switzerland. WABCO was recently purchased for $7 billion by ZF Friedrichshafen AG.[5]

Westinghouse Electric & Manufacturing Company

After George Westinghouse was pushed out of Westinghouse Electric & Manufacturing in 1911, the company continued to operate and prosper.

In 1917, Westinghouse Electric bought the Copeman Electric Stove Company and started to manufacture and market home appliances.[6] The company became famous for its reliable appliances such as toasters, fans, and washing machines. Westinghouse Electric remained in the appliance business until 1974, when it sold that business to White Consolidated Industries (WCI). WCI had originated in 1858 as a sewing machine company, based on patents by Thomas H. White. Appliances manufactured by WCI were sold under the name White-Westinghouse.[7]

In 1954, Gwilym A. Price was President of Westinghouse Electric, and he projected a bright future. An article titled "Westinghouse Expansion to Continue Into 1955," he wrote: "The expansion scheduled for completion in 1955, (total investment to be $296 million over three years) has produced increased manufacturing capacity for both established and new products." "Major new laboratory and development facilities include:

1. A new main Research Center now under construction near Pittsburgh;
2. A steam and gas-turbine research laboratory at the South Philadelphia Works;
3. Experimental testing and model facilities planned for the Aviation Gas Turbine Division; and
4. Announcement of a new metals development plant at Blairsville."[8]

In 1955, the first year that the Fortune 500 list was published, Westinghouse Electric ranked 13. The company continued to be ranked at 42 or above until 1994, but its core businesses were in trouble.

A story in the February 16, 1985, *New York Times* reported, "The Westinghouse Electric Company said yesterday that it would gradually shut down its steam turbine plant in Lester, Pa., a move that would result in the layoffs of 1,200 management, salaried, and hourly workers."[9]

The lead story in the May 2, 1987, *Pittsburgh Post-Gazette* was headlined, "WE closes East Pittsburgh plant." It starts, "Westinghouse Electric Corp yesterday said it would permanently close its historic East Pittsburgh turbine plant, putting nearly 800 workers out of work."[10]

In 1995, Westinghouse Electric bought CBS, the radio and TV broadcasting company, and took CBS Corporation as the corporate identity in 1997.

Westinghouse, as a corporate brand name, has followed a tortuous path since then.

The new CBS Corporation, which included the remaining industrial portions of Westinghouse, sold off all of its non-broadcasting assets to focus on media and broadcasting. Siemens purchased the non-nuclear power generation assets in 1998; other firms bought the defense electronics, office furniture, and mobile refrigeration business units.

In December 1998, CBS Corporation sold its commercial nuclear power business to British Nuclear Fuels Limited (BNFL).

BNFL formed Westinghouse Electric Company LLC, to operate the nuclear power business, but then sold it to Toshiba and two partners for $5.4 billion in 2006. Based on worldwide growth in nuclear power plants, the outlook was positive, but by 2015, red flags went up. An accounting scandal led to the CEO of Toshiba resigning, and by the end of 2016, Toshiba took write-downs in its Westinghouse investment of "several billion."[11] In 2017, parent company Toshiba filed for Chapter 11 bankruptcy because of $9 billion in losses from nuclear reactor construction projects.[12]

Part of the 1998 sale to BNFL gave BNFL the rights to the Westinghouse trademarks. BNFL formed a new licensing subsidiary, confusingly called the Westinghouse Electric Corporation, to handle trademark licensing transactions.

The Westinghouse name has since been licensed worldwide for use by a variety of companies and their products. But these licensees have, at best, a tenuous connection to the original companies founded by George Westinghouse. Licensees include:

- Southern Telecom of New York City became the Westinghouse brand licensee in 2017 for a series of audio products, such as home theater, boom boxes, headphones, soundbars, and accessories such as chargers and cables.[13]
- Amerex, as Westinghouse Small Appliances India, sells small appliances such as sandwich makers, electric kettles, mixer-grinders, and irons in India.[14]
- Westinghouse Lighting Corporation, formerly Angelo Brothers Lighting, sells light bulbs, ceiling fans, lighting fixtures, and lighting accessories.[15]
- "The NCC" markets "smart" electrical accessories and power strips, surge protectors, and timers under the Stanley and Westinghouse names.[16]
- Westinghouse Solar Lights, a subsidiary of the International Development Company (headquartered in Abu Dhabi), sells a variety of solar-powered lighting products.[17]

It appears that most of these Westinghouse licensees are marketing operations with little or no captive manufacturing capability. It seems likely that George Westinghouse would not be pleased with the application of his name to such ephemeral ventures.

Companies Formed After the Death of George Westinghouse

The death of George Westinghouse in 1914 did not halt the creativity of the talented people who had worked for him. They continued to innovate, grow existing companies, and form new enterprises. We shall look briefly at two of the more notable additions to the Westinghouse family.

KDKA

In 1916, Frank Conrad, Westinghouse Electric & Manufacturing General Engineer, started experimenting with radio transmission. Conrad's pioneering work led Westinghouse Electric to establish a broadcasting service which sent election results to the local area on November 2, 1920. The call letters for that first commercial radio station, which still exists today, were KDKA.

In 1921, Westinghouse, GE, AT&T, and others founded the Radio Corporation of America (RCA) from the former American Marconi Company.[18] RCA formed the

National Broadcasting Company (NBC) in 1926, and all of the Westinghouse radio stations became affiliates of the NBC network.

Westinghouse continued to add radio stations across the country[a] and entered the TV market in 1948. The operation, officially known as Westinghouse Broadcasting and Cable, Inc., took the nickname "Group W" on May 20, 1963.[19]

By 1995, broadcasting became the tail that wagged the Westinghouse dog, and, as discussed previously, Westinghouse bought CBS and two years later became the CBS Corporation.[20]

Bettis Atomic Power Laboratory

Bettis Atomic Power Laboratory was founded in 1949 on a 207-acre site in West Mifflin, Pennsylvania, near Pittsburgh. Westinghouse Electric operated the Bettis Labs for the U.S. Government from its founding until 1998 to develop nuclear reactors for Navy ships and commercial power generation. Products included propulsion plants for:

- The USS *Nautilus*, the world's first operational nuclear-powered submarine;
- The USS *George Washington*, the U.S.'s first ballistic missile submarine;
- The USS *Long Beach*, a guided-missile cruiser and the world's first nuclear-powered surface combatant vessel; and
- The USS *Enterprise*, the world's first nuclear-powered aircraft carrier.[21]

Bettis also built the nuclear reactor that powered the world's first atomic electric power plant devoted to peacetime uses at Shippingport, Pennsylvania, about 25 miles west of Pittsburgh. The Shippingport Plant reached criticality on December 2, 1957, and remained in operation until October 1982.[22]

Ashes to Ashes, or Rising from the Ashes?

The industrial enterprises created by George Westinghouse

Historical marker for KDKA in downtown Pittsburgh (author's photograph).

(a) WJZ, originally licensed in Newark, NJ; WBZ, first located in Springfield, MA; KYW, initially in Chicago; WBZA in Boston; and WOWO in Fort Wayne, IN.

Chapter 28. The Next Century

USS *Enterprise*, in active service from 1961 to 2012 (Wikimedia Commons).

have adapted and morphed into companies with different names and different purposes. Some, such as the manufacturing plants in the Turtle Creek valley, have shrunk or disappeared. Others have transformed into service industries, such as broadcasting.

But focus on the creation part of the cycle.

Creativity is at the heart of the Westinghouse Arts Academy's objectives. How fitting that it is now located in the former Westinghouse Air Brake General Office Building, "The Castle," in Wilmerding.

David Bear and his Westinghouse Park 2nd Century Coalition (WP2CC) and Westinghouse 175 initiatives hope to restore Westinghouse Park, enhance the Westinghouse legacy, and inspire a new generation of Pittsburghers by the enormous contributions of George Westinghouse.

Westinghouse, by his effort and perseverance, created a livelihood for thousands of workers for over a century. But his most enduring contributions are not measured in industrial plants or even in millions of dollars paid in wages to his employees. Westinghouse advanced technology and literally made the world a better place to live.

Train transportation as we know it, both freight and passenger, could not exist without the Westinghouse air brake and other railroad-related inventions.

In the United States in 2011, the per capita consumption of electricity was 13 megawatt-hours.[23] Virtually all of that electricity was conveyed to us on high-voltage AC transmission lines. Westinghouse, through his foresight and against massive opposition, won the Battle of the Currents and made AC the standard throughout the world.

Westinghouse said, "If someday they say of me that in my work I have contributed something to the welfare and happiness of my fellow man, I shall be satisfied."

George Westinghouse has every right to be satisfied.

Appendix I—Westinghouse Family Genealogy

Describing family relationships in words can be confusing, especially when several people bear the same given name. So genealogists have developed several semi-pictorial formats to illustrate how people across many generations are related. The most familiar such format is the family tree, which starts with the earliest generation at the top and proceeds down the page, showing succeeding generations. Such an arrangement quickly expands beyond the limits of the printed page, but siblings of less interest are sometimes pruned to reduce space requirements.

A modified family tree, called an hourglass chart, includes ancestors of the person of interest and also his or her descendants. The hourglass chart, with selective branch pruning, suits the objective of capturing the family relationships of George Westinghouse, Jr., in a compact form. The next page contains an hourglass chart for the Westinghouse family.

This chart has been derived from multiple partial family trees in the files of the Westinghouse Collection at the Detre Library and Archives Division of the Senator John Heinz History Center in Pittsburgh, Pennsylvania. Genealogy research sites such as Wikitree.com and FindaGrave.com added more information.

The Hourglass Chart includes six of George Westinghouse, Jr.'s, great-grandparents; all four of his grandparents; his parents; all of his siblings; his wife, Marguerite; their son, George III and his wife, Evelyn Violet Brocklebank Westinghouse; and their six children.

Appendix I—Westinghouse Family Genealogy

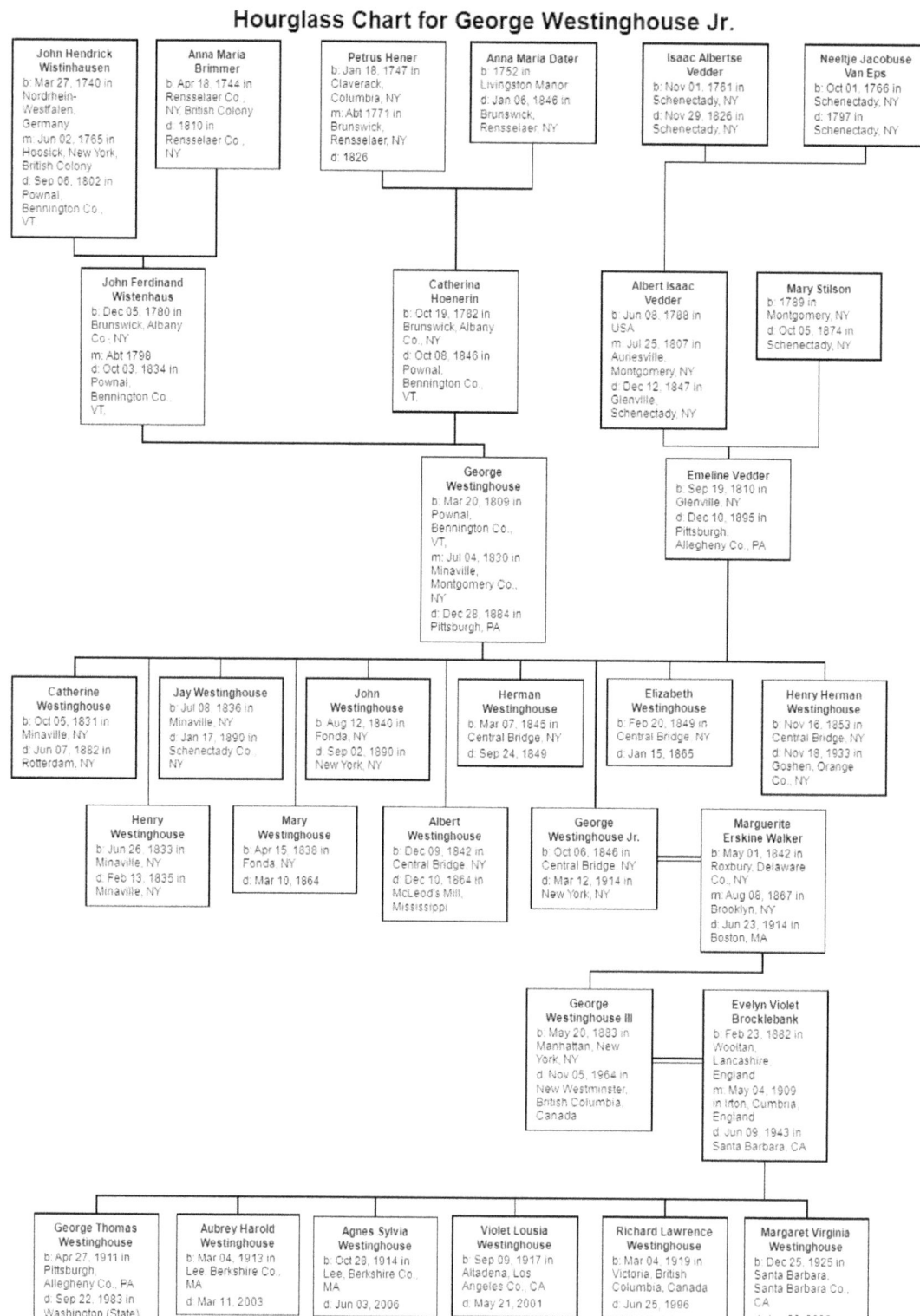

Appendix II—Automatic Air Brake Operation

How do pneumatic automatic air brakes for trains work? The following explanation and diagrams are adapted from a website titled "Brakes" by Dr. Piers Connor of PRC Rail Consulting Ltd., www.railway-technical.com.[1]

Consider first the "Brakes Released" condition shown in the diagram on the next page:

In the Brakes Released condition, the engineer sets his brake valve to send high-pressure air through the Brake Pipe to the TRIPLE VALVE, thus causing the Slide Valve to move all the way to the left. In this position, air from the Brake Pipe passes through the Feed Groove and recharges the air pressure in the Auxiliary Reservoir. At the same time, the Slide Valve opens the path for air to flow from the BRAKE CYLINDER to the atmosphere (Exhaust), thus reducing the pressure in the BRAKE CYLINDER and allowing the Spring to push back the Piston and release the Brake Block from the Wheel.

In the Brakes Applied condition, the train engineer sets his brake valve to release air from the Brake Pipe. The reduced pressure in the Brake Pipe allows the higher pressure in the Auxiliary Reservoir to push the Slide Valve in the TRIPLE VALVE all the way to the right, thus permitting high-pressure air from the Auxiliary Reservoir to flow to the BRAKE CYLINDER. The high pressure in the BRAKE CYLINDER overcomes the Spring and pushes the Piston to the left, thus forcing the Brake Block into the Wheel and stopping the train. Note that a leaking Brake Pipe or accidentally decoupled car would have the same effect—reducing the pressure in the Brake Pipe and applying the brakes. Thus, the system is fail-safe.

The TRIPLE VALVE was the heart of the Westinghouse Automatic Air Brake. The drawing shows the triple valve described in U.S. Patent 220,556 issued on October 14, 1879.

Starting in 1868 and through subsequent decades, Westinghouse developed and patented many improvements to his automatic air brake. Each new patent created an additional barrier to competition, allowing Westinghouse Air Brake to maintain its virtual monopoly on railroad brakes.

The actual Westinghouse air brake system is somewhat more complicated than the previous illustration indicates. A full pictorial diagram is shown on the next page.[2]

The same principles described above are used in train braking systems today, but of course, over 140 years of experience have led to numerous improvements.

Appendix II—Automatic Air Brake Operation

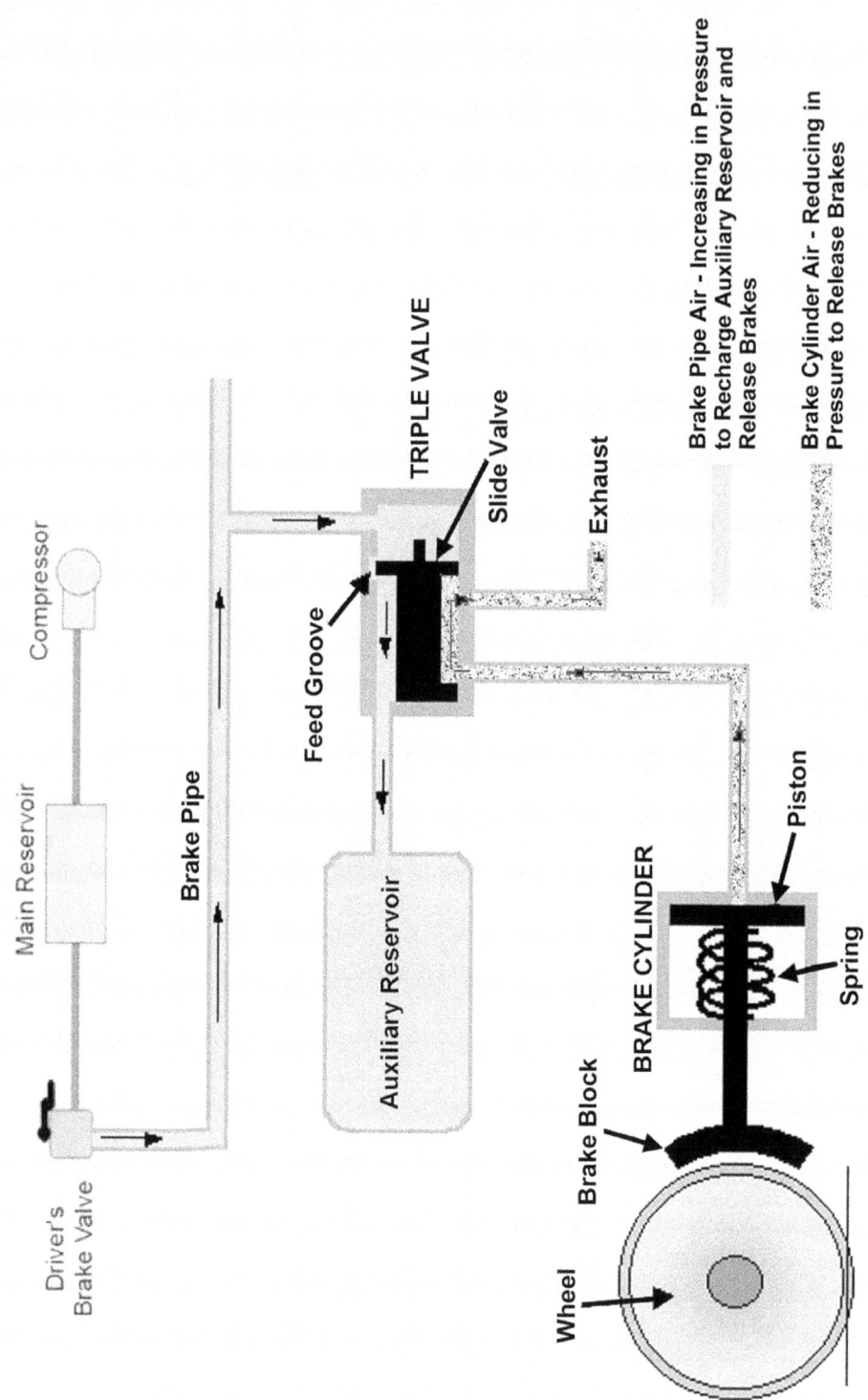

Westinghouse automatic air brake system with brakes released (drawing by P. Connor, www.railway-technical.com).

Appendix II—Automatic Air Brake Operation

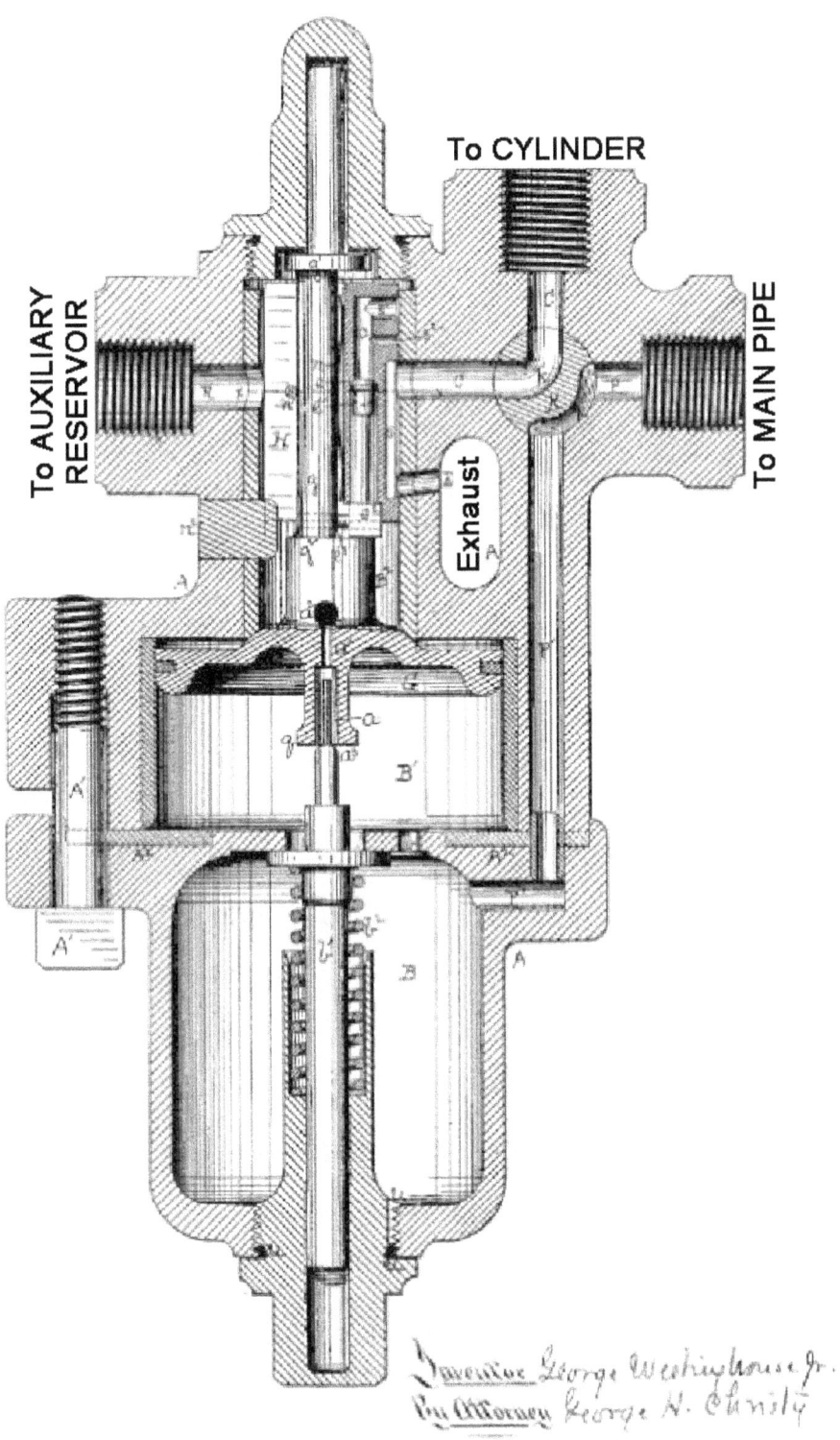

Westinghouse triple valve from October 14, 1879—drawing from U.S. Patent 220,556.

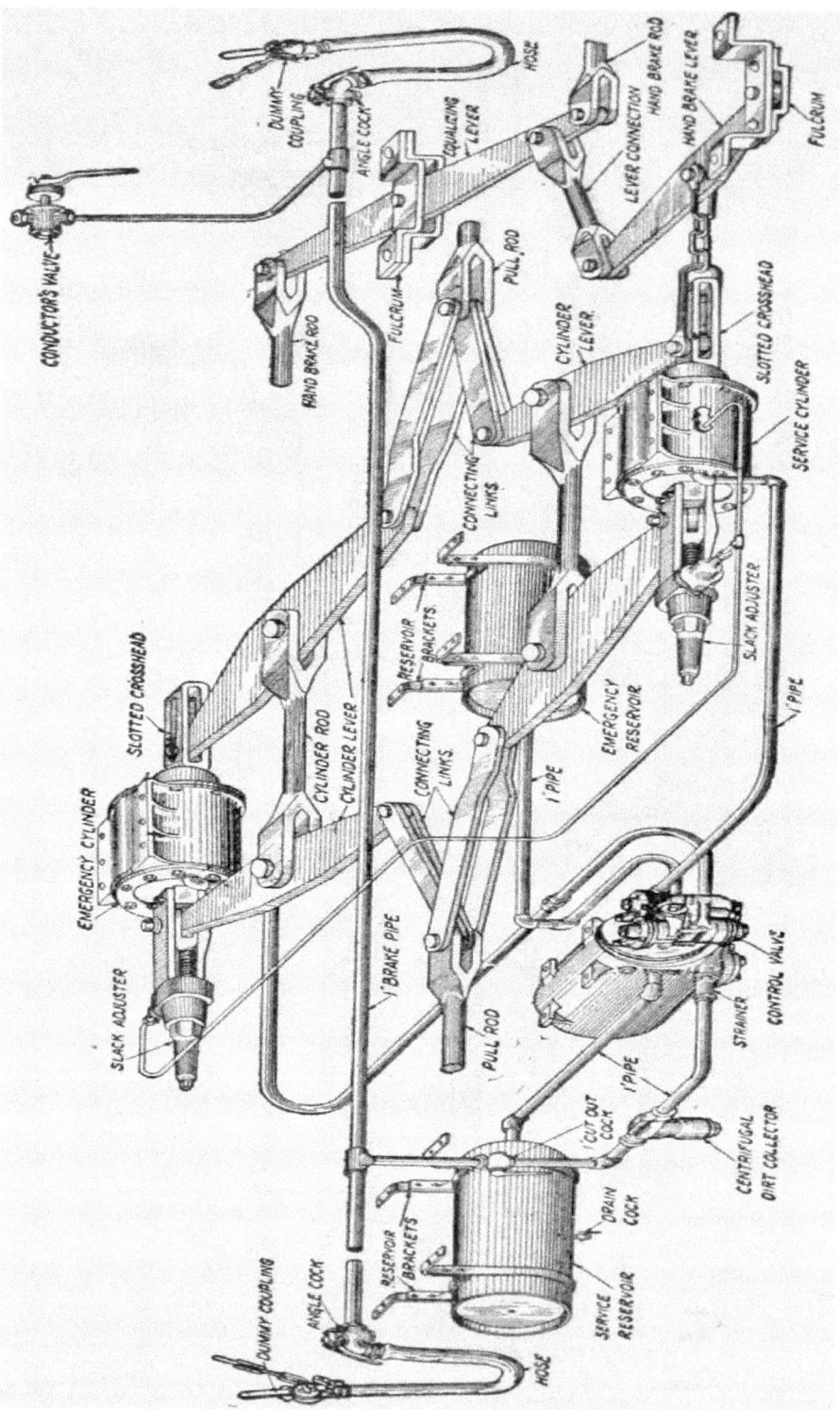

Westinghouse air brake as designed for passenger and freight service—1918 diagram (Wikimedia Commons).

Appendix III—Electrical Engineering 101

What is voltage in an electrical circuit? Electrons in an electrical circuit are invisible, so the analogy of a hydraulic (water) system, which is easier to visualize, is often used.

The pressure created by the pump in the hydraulic system corresponds to the electrical pressure, called voltage, provided by the battery or generator in the electrical system. The impediment to flow caused by the narrow pipe in the hydraulic system corresponds to the electrical resistance in the electrical system. And the rate of flow in the hydraulic system corresponds to the rate of electron flow, which is called current, in the electrical system. To summarize the corresponding elements of the hydraulic and electrical systems:

Hydraulic System	Electrical System	
	Name	*Symbol*
Pressure	Voltage	V
Impediment	Resistance	R
Flow	Current	I

In 1827, a German named Georg Simon Ohm published a paper describing the relationship among voltage, resistance, and current.[1]

Ohm's law is deceptively simple:

Voltage (Volts) = Current (Amperes) times Resistance (Ohms)

$V = I \times R$ (1)

So, the voltage across the electrical resistance equals the current through the resistance times its resistance.

The power dissipated in the resistance is given by voltage times current,

$P = V \times I$ (2)

Combining equations (1) and (2), we can easily show that the power dissipated in the resistance can also be written as

$P = I^2 \times R$ (3)

Now consider what happens if we have two separate resistors in the circuit, such as shown in the diagram.

It is evident that the same amount of current must flow through both resistors so that the power dissipated by the resistors is

230 Appendix III—Electrical Engineering 101

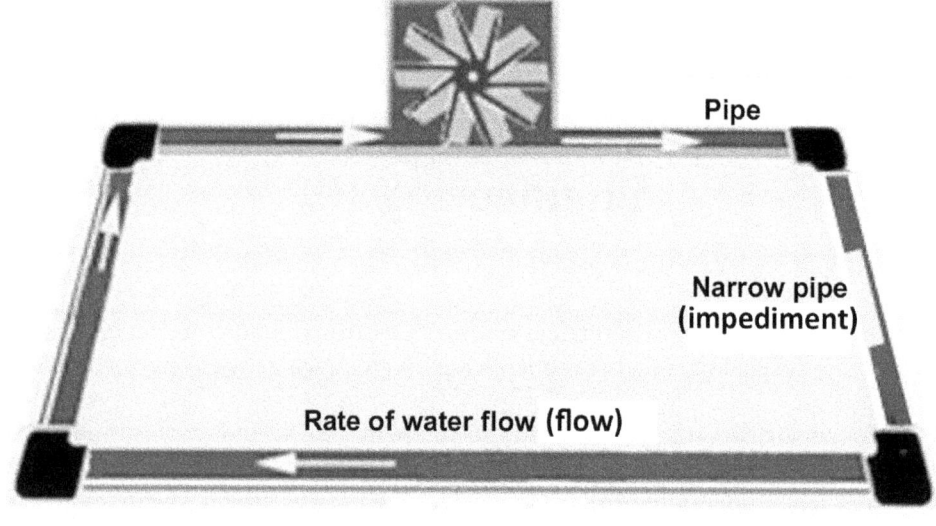

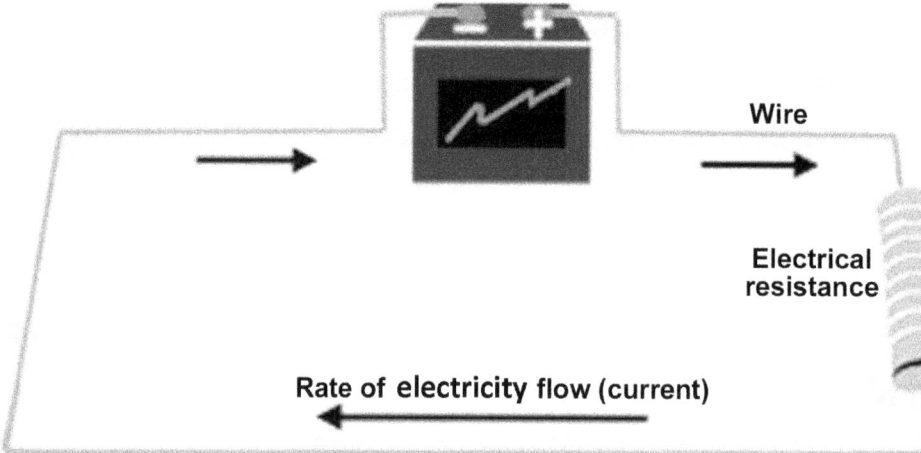

Analogous Systems—Hydraulic and Electrical.

$P_1 = I^2 \times R1$ (4), and
$P_2 = I^2 \times R2$ (5)

To apply this circuit to a real-life situation such as providing power to your home, think of R1 as the resistance of the wire from the generating station to your home, and R2 as the equivalent resistance of the items to be powered in your home (air conditioner, clothes dryer, etc.). So the power P_1 dissipated in the resistance R1 of the wire is wasted (called "loss"), and the power P_2 dissipated in the resistance R2 is productive. Applying Ohm's Law and a bit of algebra yields the following equation, which defines the percentage of power lost in terms of power delivered (P_2), wire resistance (R1), and voltage (V):

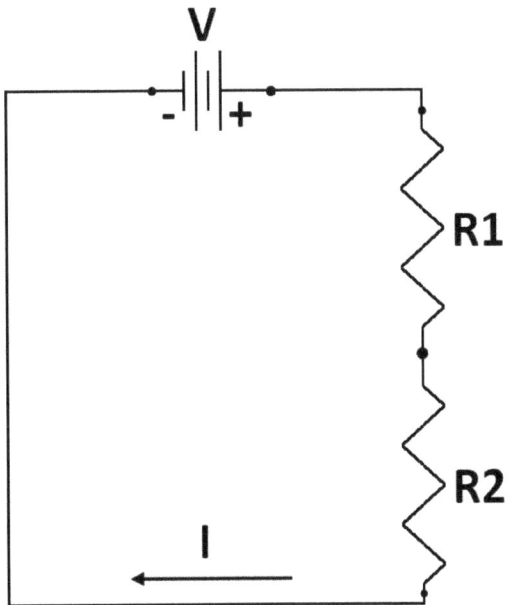

Circuit with two resistors in series.

P_{loss}/P_2 = percent Loss = $100 \times P_2 \times R1/V^2$ (6)

Although there was little initial response to Ohm's 1827 paper, by 1841, he received the Royal Society's highest award, the Copley Medal, for his pioneering work. The unit of resistance was named the "Ohm" in 1872.

DC Power Distribution

We will use Equation (6) to highlight the critical difference between Edison's DC power system and Westinghouse's AC system.

In the late 1800s, there was no practical way to change the voltage in a DC system, so DC power had to be distributed to the end-user at the same voltage at which it was generated.

For comparison purposes, we will assume the DC system has the following parameters:

- The generated DC voltage is 120 volts;
- The average household uses 1000 watts of power on a continuous basis (today's U.S. homes actually use about 1250 watts);
- The wire used to distribute Edison's DC power is very heavy, with a resistance of 0.025 Ohms per mile[a]; and
- All of Edison's customers were clustered at one mile from his generating plant.

(a) If implemented in copper, this low resistance would require ten strands of AWG 4/0 wire. Each strand would be almost ½ inch in diameter. Assuming such heavy wire for Edison's DC system allows a meaningful comparison to a modern AC distribution network, which uses similar wire for a 100-mile transmission line.

Inserting these numbers into Equation (6), and remember that two wires are required to carry the power to the customer:

Percent Loss$_{(DC)}$ = 0.347 × N (7)

where N = number of households served.

Plotting this equation:

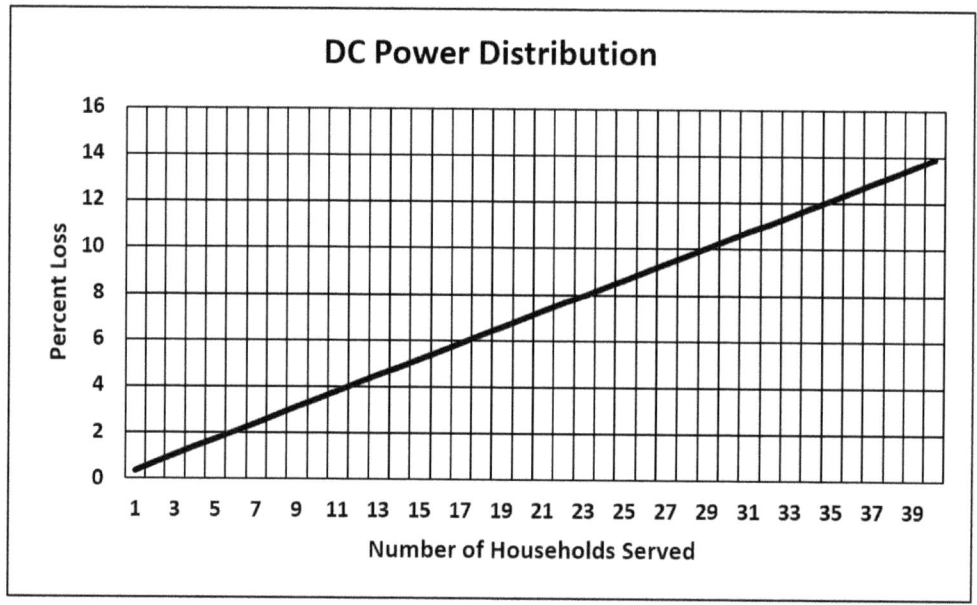

If we define 10 percent as an acceptable level of loss (100 watts would be wasted for every 1000 watts delivered to the users), then just 29 households could be served by Edison's DC system. All of those users would have to be within one mile of the generating plant.

AC Power Distribution

Now consider Westinghouse's AC distribution system. Remember, the overwhelming advantage of AC is that its voltage can easily be changed either up or down. Go back to Equation (6), which applies to both AC and DC systems:

P_{loss}/P_2 = percent Loss = 100 × P_2 × R1/V^2 (6)

For the AC distribution system, we will assume the following parameters:

- The generated AC voltage is 120 volts, but it can easily be increased to any reasonable level (probably 10,000 volts in the late 1800s; over 700,000 volts today);
- The average household uses 1000 watts of power on a continuous basis (again, today's U.S. homes actually use about 1250 watts);
- The wire used to distribute Westinghouse's AC power is the same as we assumed for Edison's DC power, with a resistance of 0.025 Ohms per mile; and
- The substation serving Westinghouse's customers is 100 miles from his generating plant.

This time we will use a fixed value of 10 percent for the loss percentage and calculate the voltage required as a function of the number of households served. Inserting these numbers into Equation (6) and solving for the transmission voltage,

Transmission Voltage $V = 224 \times N^{0.5}$ (8)

where N = number of households served.

Plotting this equation:

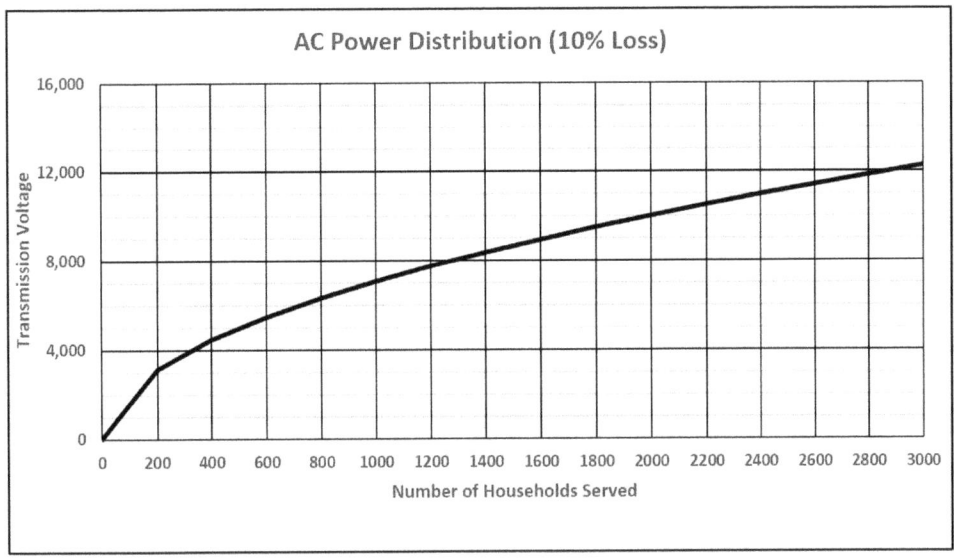

Several conclusions can be drawn from the AC Power Distribution graph:

- At an AC transmission voltage of 1,200 volts, the 29 households that had to be within one mile of Edison's DC generating station could now be up to 100 miles away from Westinghouse's AC generator.
- At an AC transmission voltage of 11,000 volts (the voltage used to transmit power from the Niagara Falls generating plant), over 2,400 households at a distance of 100 miles from Westinghouse's generator could be served.
- High-voltage AC transmission is clearly superior to low-voltage DC for long-distance transmission of electricity.

How Can AC Voltage be Increased or Decreased?

Lucien Gaulard and John Dixon Gibbs first exhibited a transformer-like device with an open iron core in London in 1882, then sold the patent for the idea to George Westinghouse. In 1885, three Hungarian engineers, Károly Zipernowsky, Ottó Bláthy, and Miksa Déri, developed an improved version and filed patents describing the ZBD transformer.

Edison held the U.S. rights to the ZBD transformer patents, so Westinghouse and his associates developed a different design that was easier to manufacture and did not infringe the ZBD patents. Westinghouse patented his new design in 1887.

So how does a transformer work? The diagram shows a simplified ideal transformer with primary and secondary windings around a shared iron core. A magnetic

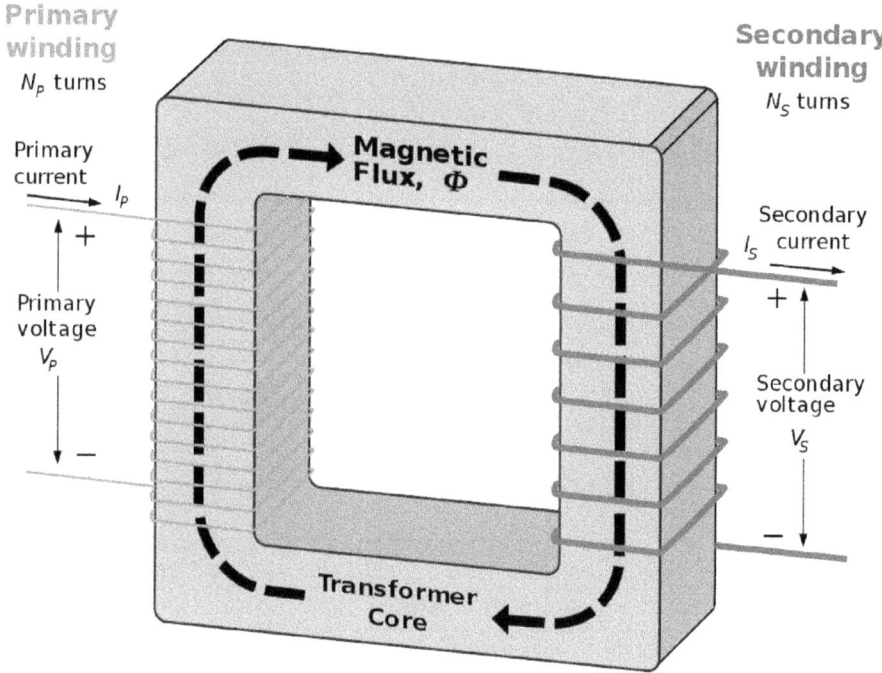

Ideal transformer (Wikimedia Commons).

flux is induced in the core by AC electricity flowing in the primary winding. The alternating flux induces a voltage in the secondary winding.

The magnitude of the secondary voltage is determined by the ratio of turns between the two windings. The equation governing the operation of this ideal transformer is:

Turns ratio = $a = N_p/N_s = V_p/V_s = I_s/I_p$

From this equation, we see that if a<1, the primary voltage V_p is less than the secondary voltage, V_s, and it is a step-up transformer. Conversely, if a>1, V_p is higher than V_s, and it becomes a step-down transformer.

In an AC power transmission system, the low voltage from a generator is applied to the primary of a step-up transformer, thus resulting in a much higher secondary voltage, which can be transmitted with low loss over hundreds of miles. At the far end of the transmission line, this high voltage is connected to the primary of a step-down transformer, resulting in a much lower voltage (typically 120 or 240 volts) for use in a house. Transformers used in AC transmission systems are a common sight at the top of poles along roadsides everywhere.

Appendix IV—How Does an Induction Motor Work?

The transformer from Appendix III provides a starting point for an explanation of how an induction motor operates. As shown previously, the diagram shows a simplified ideal transformer with primary and secondary windings around a shared iron core.

A magnetic flux is induced in the core by AC flowing in the primary winding. By Faraday's Law, the alternating flux induces a voltage in the secondary winding.[1]

Now let us change the transformer into a conceptual AC induction motor. First, cut apart the transformer to separate the right leg, as shown in the drawing. As long as the cuts are thin, magnetic flux will continue to flow all around the core and will still induce a voltage in the secondary winding.

Next, short out the secondary winding by connecting its two ends. The voltage

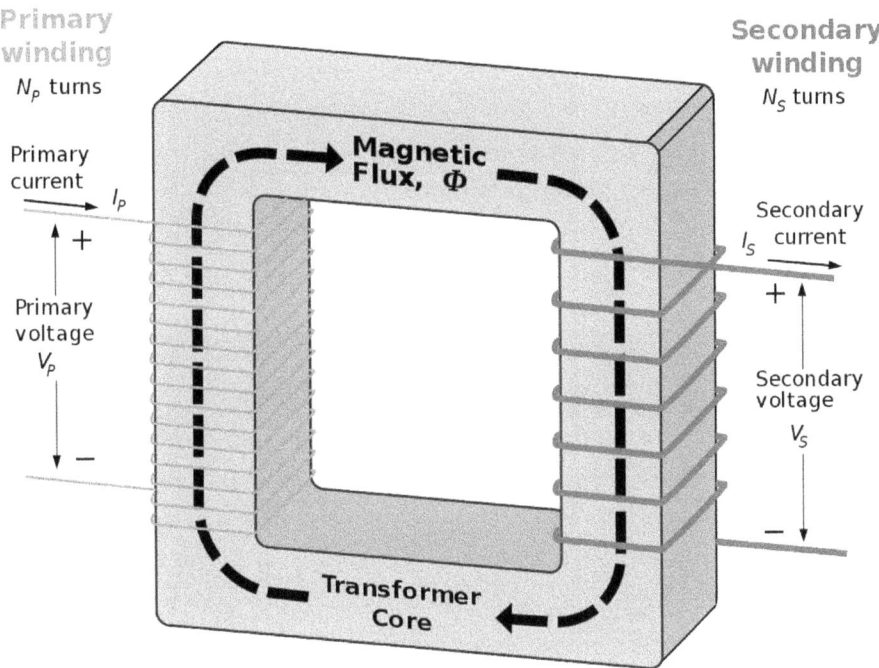

Ideal transformer (Wikimedia Commons).

Appendix IV—How Does an Induction Motor Work?

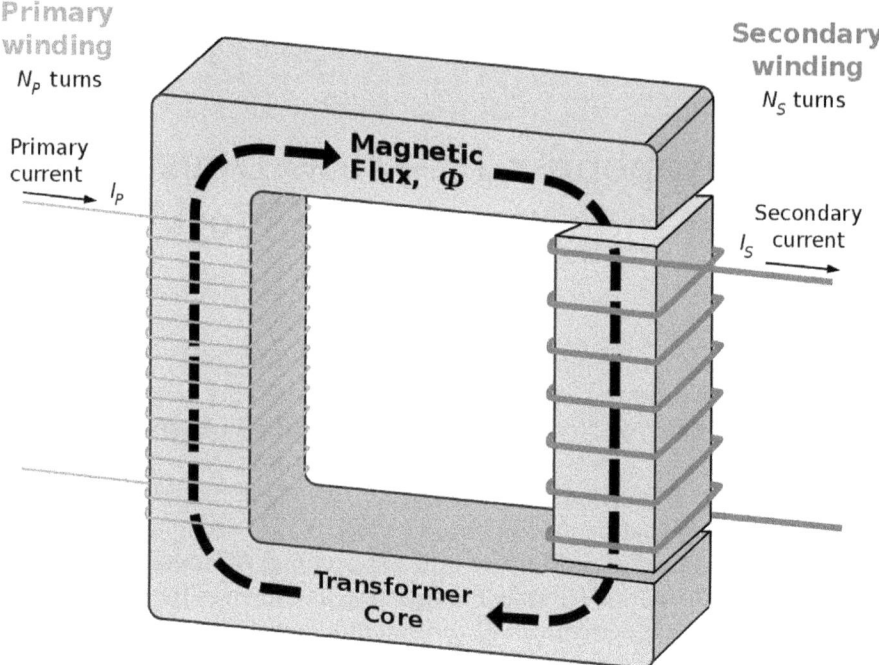

Transformer with right leg separated.

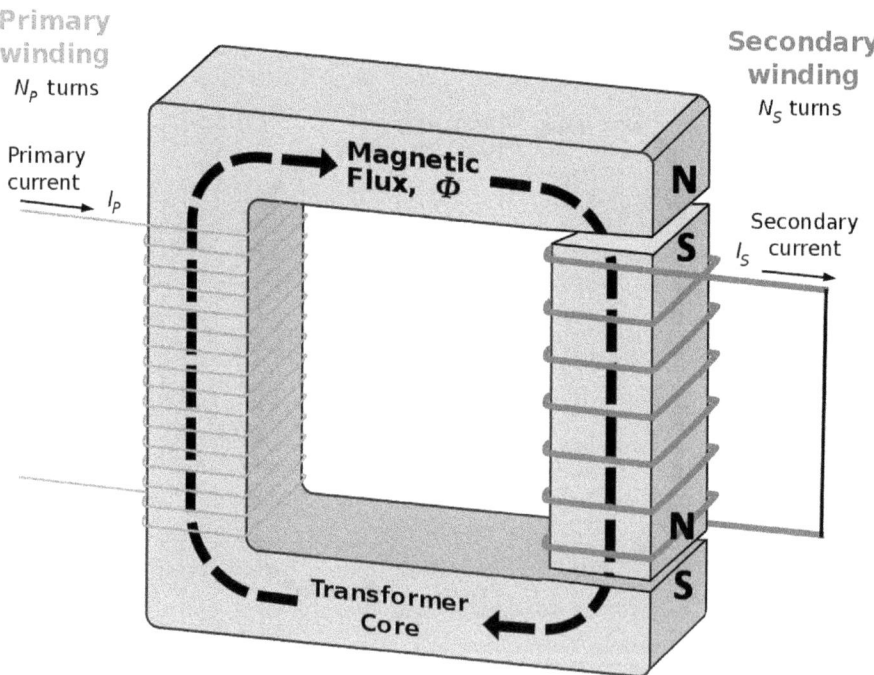

Transformer with right leg separated and secondary shorted.

Appendix IV—How Does an Induction Motor Work?

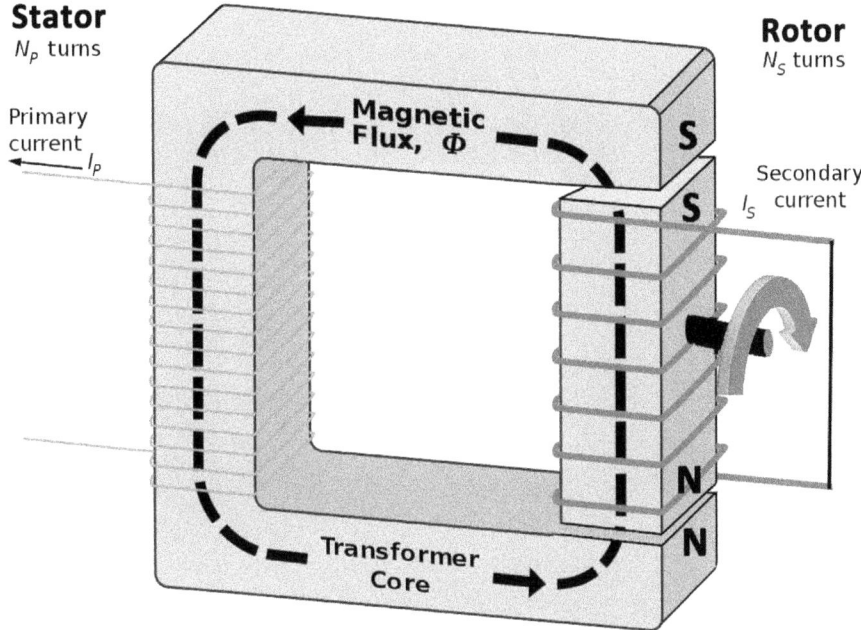

Conceptual induction motor with rotor ready to rotate.

induced in the secondary will cause a current to flow in the shorted winding. By Lenz's Law, that current will create a magnetic flux of its own in the separated right leg of the core.[2]

If nothing else changes in the system, then the total magnetic flux will not change. The North and South poles on the diagram reflect a stable system.

Now reverse the current flow in the primary, as would happen during the next half-cycle of an AC input. As a result, the magnetic flux in the C-shaped part of the core reverses, and the magnetic poles switch to the condition shown in the final drawing. But that condition is unstable as the two S poles repel and the two N poles repel. If the right leg is free to rotate around the black shaft, it will do so until the N at its bottom moves to the top and aligns with the S pole of the C-shaped portion. We would call the C-shaped portion the stator, and the right leg would be called the rotor.

With significant physical modifications, we would have an induction motor.

Tesla's original patent (U.S. Patent 381,968; issued May 1, 1888) provides a much more complete and accurate description of his induction motor.

Appendix V—How Does a Turbine Operate?

The operation of a steam turbine might be as unfamiliar to a non-technical audience today as it was a century ago. Around 1915, Herbert T. Herr, vice president of Westinghouse Electric & Manufacturing Company, wrote an explanation that remains helpful. The following discussion is adapted, with minimal changes, from his report.[1]

A turbine is essentially a machine for developing large amounts of power. It is different from the ordinary steam engine in that it converts the energy in steam into mechanical work by utilizing the velocity resulting from the steam expansion, either by action or reaction of a steam jet on the blades, as opposed to the conversion of steam into energy in reciprocating engines by direct pressure of the steam on a piston.

Multiple stages (of turbine blades) are necessary for the turbine to fractionally extract the energy of steam in its expansion from boiler pressure at the input to the condenser at the output. No known material can stand the stresses and speed necessary to extract in one stage the energy of a jet of steam expanding from 200 pounds per square inch (200 psi) pressure to atmospheric pressure (14.7 psi), as the steam speed under these conditions would be 4,300 feet per second.

A booklet written in 1910 by Westinghouse Machine Company engineers and titled, "The Westinghouse-Parsons Steam Turbine,"[2] provides further details of turbine operation. Once again, what follows is adapted from that document.

Any steam turbine depends for its operation upon the effect of steam expanding through suitably formed passages, thereby attaining a velocity. The steam then impinging upon suitable "buckets" gives up the energy due to velocity and thus gives motion to the rotating element of the turbine. In some cases, the steam expands through passages that are themselves capable of movement, in which case the effect of velocity is to give motion to the rotating element because of reaction. The Westinghouse-Parsons turbine uses both of the two effects. The following diagram illustrates the general structure of the turbine blades.

Both stationary and moving blades are shown in their respective location to one another.

The steam, in passing through Row 1, expands and falls from pressure P to pressure P1. By expanding, the steam does work upon itself and attains a velocity, thus providing energy to the moving blades in Row 2. Again, in the passage of steam through the blades of Row 2, the pressure falls from P1 to P2. This expansion again

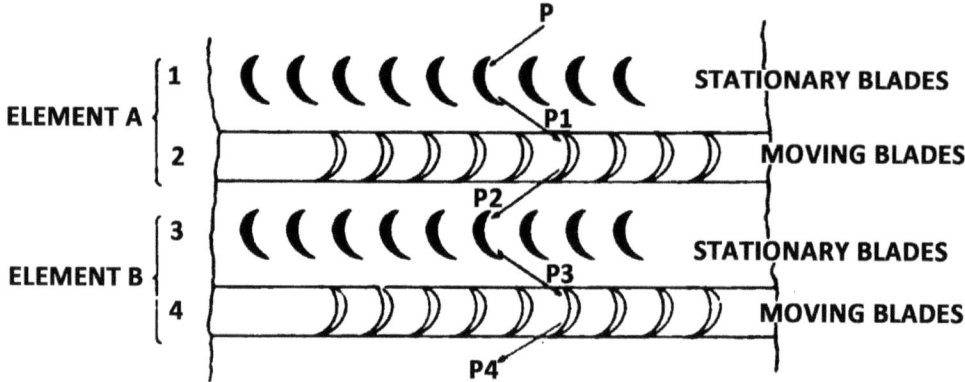

Conceptual model of a steam turbine.

produces a velocity, but this time its effect is to cause the shaft carrying the blades on Row 2 to rotate. This cycle is repeated several times until exhaust pressure P4 is attained.

The above descriptions make it clear that steam under high pressure is essential to the operation of a steam turbine. For efficient operation, that high-pressure steam must impose its force on the turbine blades and not leak through any spaces between the blades and the housing of the structure. In short, close spacing and tight tolerances are essential for efficient operation. When we consider the high temperatures and pressures involved, combined with high rotational speeds, material selection and turbine construction are complex.

A more detailed, but still understandable explanation of steam turbine structure and operation can be found in an article titled "Steam Turbine Electricity Generation Plants" at https://www.mpoweruk.com/steam_turbines.htm.

Appendix VI—Patent Law Primer

Brief History of the U.S. Patent System

In the U.S. colonies, the first patent was granted under Massachusetts law in 1641 to Samuel Winslow for his new method of making salt. Later, the U.S. Constitution formed the basis of federal patent law, which was codified in 1790. This initial patent law established a Patent Board which could grant to inventors the "sole and exclusive right and liberty of making, constructing, using and vending to others" their work.[1]

By 1836, 10,000 U.S. patents had been issued, and the combined volume of new applications and dissatisfaction of applicants who had been denied patents led to the formation of the U.S. Patent Office within the State Department.

In 1849, the Department of the Interior took over management of the Patent Office, and the definition of an invention was modified to "all new, useful, and non-obvious works."[2] The 1849 revision formed the basis for patent applications, rights, and litigation in the time of George Westinghouse.

Patent Law Glossary

Only those aspects of patent law helpful in understanding the content of this book are included here. Patent attorneys spend many years of specialized study to master the intricacies of law in general and patent law in particular, so this summary can but begin to cover the most relevant points.

Claim: A patent claim is the part of a patent or patent application that defines the scope of protection granted by the patent. The claims define, in technical terms, the extent of the protection conferred by a patent, or the protection sought in a patent application. The claims are of the utmost importance both during prosecution (the process of applying for and obtaining a patent) and during litigation.

A patent is a right to exclude others from making, using, selling, or offering for sale the subject matter defined by the claims. To exclude someone from using a patented invention in a court, the patent owner, or patentee, needs to demonstrate that what the other person is using falls within the scope of a claim of the patent.[3]

Divisional patent applications: When a patent application describes more than one invention, the applicant is required to split the parent application into one or more divisional applications, each claiming only a single invention. Even though a divisional application is filed later than the parent application, it will generally retain the same priority date. Thus, public disclosures after that earliest filing date cannot serve as prior art to threaten the validity of the claims. To be entitled to the benefit of the earliest filing date of the parent application, claims within the divisional application must be fully supported by the technical disclosure found within the parent application. Charles Van Depoele's 1887 patent application was divided into two applications, which later issued at patents in 1890 and 1893.

Double Patenting: Double patenting is the granting of two patents for a single invention, to the same proprietor and in the same country or countries. It is an accepted principle in most patent systems that two patents cannot be granted to the same applicant for one invention.[4] The doctrine of double patenting seeks to prevent the unjustified extension of patent exclusivity beyond the term of a patent.[5]

In the case of Thomson-Houston vs. Winchester Avenue Railroad Company and Westinghouse Electric & Manufacturing Company, the initial ruling by Judge Townsend was correct. He ruled that one of Van Depoele's divisional patents did not invalidate the other, a position consistent with current patent law. Judge Wallace improperly reversed that ruling.

Infringe: A product (or method, service, and so on) is said to infringe a patent if it falls within one or more of the claims of the patent. To determine infringement requires "reading" a claim onto the accused product. If all of the claim's elements are found in the product, the claim is said to "read on" the product, and the product infringes. If even a single element from the claim is missing from the product, the claim does not literally read on the product, and the product does not infringe the patent with respect to that claim.

Prior art: Prior art constitutes all information that has been made available to the public in any form before the priority date of a patent or patent application. If an invention has been described in the prior art, a patent on that invention is not valid. To anticipate a patent claim, the prior art is generally expected to provide a description sufficient to inform an average worker in the field (one said to be of "ordinary skill in the art") of some subject matter falling within the scope of the claim.[6]

Valid: Patent validity is a complex concept. In its simplest form, a patent claim is valid if there is no single piece of prior art that discloses every element of the claim. If even one element of the claim is undisclosed by the single piece of prior art, the claim is valid. However, if elements undisclosed by a single piece of prior art would have been obvious to one of ordinary skill in the art, or are inherent given the disclosed elements, then the patent claim is invalid. Furthermore, if two or more pieces of prior art together disclose every element of the patent claim, *and* if there is motivation to combine these references, then the patent claim can be found to be invalid.

Chapter Notes

Acknowledgments

1. Leupp, Francis E. *George Westinghouse: His Life and Achievements*. Boston: Little, Brown, and Company, 1919.
2. Prout, Henry G. *A Life of George Westinghouse*. New York: The American Society of Mechanical Engineers, 1921.

Preface

1. "America's 20 Most Beautiful Skylines, Ranked." 1 December 2014. *Thrillist*. Ed. Matt Meltzer. https://www.thrillist.com/travel/nation/best-skylines-in-america-seattle-chicago-and-las-vegas-top-our-list.
2. Collier, Sean, et al. "The 50 Greatest Pittsburghers of All Time." 19 December 2018. *Pittsburgh Magazine*. https://www.pittsburghmagazine.com/the-50-greatest-pittsburghers-of-all-time/.
3. "Rege Cordic." n.d. *Wikipedia*. https://en.wikipedia.org/wiki/Regis_Cordic.
4. Schoenfield, David. "The greatest game ever played." 13 October 2010. *ESPN*. https://www.espn.com/mlb/playoffs/2010/columns/story?id=5676003.
5. "The Pittsburgh 'H.'" n.d. *Visit Pittsburgh*. https://www.visitpittsburgh.com/things-to-do/arts-culture/history/the-pittsburgh-h/.

Chapter 1

1. *Emeline Vedder Westinghouse*. n.d. https://www.findagrave.com/memorial/164303284/emeline-westinghouse.
2. Prout, Henry G. *A Life of George Westinghouse*. New York: The American Society of Mechanical Engineers, 1921.
3. O'Callaghan, E.B. *The Documentary History of the State of New York*. Vol. III. Albany: Weed, Parsons & Co., 1850. https://play.google.com/books/reader?id=7QxfAAAAcAAJ&hl=en&pg=GBS.PP13.
4. *Early Settlement and Improvement of Columbia County, NY*. 1900. Hudson Gazette. http://history.rays-place.com/ny/colu/early.htm; *Robert Hunter (governor)*. n.d. https://en.wikipedia.org/wiki/Robert_Hunter_(governor).
5. Leupp, Francis E. *George Westinghouse: His Life and Achievements*. Boston: Little, Brown, and Company, 1919.
6. *Westinghouse Family Papers*. n.d. https://nyheritage.org/collections/westinghouse-family-papers; Prout. Op. cit.; Johnson, Carl. The first George Westinghouse of Schenectady. 14 January 2016. http://hoxsie.org/2016/01/04/the-first-george-westinghouse-of-schenectady/; *Catherine Westinghouse Moore*. n.d. https://www.findagrave.com/memorial/142743956/catherine-moore; Leupp. Op. cit.; *Jay Westinghouse*. n.d. https://www.findagrave.com/memorial/164138905/jay-westinghouse; *John Westinghouse*. n.d. https://www.findagrave.com/memorial/19076828/john-westinghouse; *George Westinghouse*. n.d. https://www.findagrave.com/memorial/1091/george-westinghouse; *Elizabeth Westinghouse*. n.d. https://www.geni.com/people/Elizabeth-Westinghouse/6000000079948864970; *Henry Herman Westinghouse*. n.d. https://www.findagrave.com/memorial/4737/henry-herman-westinghouse.
7. Leupp. Op. cit.
8. *Ibid*.
9. Strunk, Joseph. "Letter from Joseph Strunk to George Westinghouse, Sr." 17 January 1865. *New York Heritage Digital Collections*. https://cdm16694.contentdm.oclc.org/digital/collection/p16202coll7/id/9074/rec/29.
10. Leupp. Op. cit.; Jordan, John Woolf. *Genealogical and Personal History of Western Pennsylvania*. Vol. 1. New York: Lewis Historical Publishing Company, 1915.
11. Prout, op. cit.
12. Abrams, Michael. "George Westinghouse." 29 June 2012. *American Society of Mechanical Engineers*. https://www.asme.org/topics-resources/content/george-westinghouse.

Chapter 2

1. Leupp, Francis E. *George Westinghouse: His Life and Achievements*. Boston: Little, Brown, and Company, 1919.
2. *Ibid*.
3. Vogel, Charity. *The Angola Horror: The 1867 Train Wreck that Shocked the Nation and

Transformed American Railroads. Ithaca: Cornell University Press, 2013.
   4. Leupp, op. cit.

Chapter 3

   1. Barcousky, Len. "Eyewitness 1855: Rail and factory workers died on average at age 27." *Pittsburgh Post-Gazette* 1 March 2014. https://www.post-gazette.com/local/pittsburgh-history/2014/03/02/Eyewitness-1855-Rail-and-factory-workers-died-on-average-at-age-27/stories/201403020066.
   2. U.S. Senate Committee on Interstate Commerce. *Automatic Couplers and Power-Brakes*. Washington, D.C.: Government Printing Office, 1890. https://play.google.com/books/reader?id=cIuAAAAMAAJ&hl=en&pg=GBS.PA1.
   3. "George Westinghouse Historical Marker." n.d. *Pennsylvania Historical Markers*. https://explorepahistory.com/hmarker.php?markerId=1-A-1A9.
   4. Leupp, Francis E. *George Westinghouse: His Life and Achievements*. Boston: Little, Brown, and Company, 1919.
   5. *Ibid.*
   6. "Fréjus Rail Tunnel." n.d. *Wikipedia*. https://en.wikipedia.org/wiki/Fr%C3%A9jus_Rail_Tunnel.
   7. Leupp, op. cit.
   8. *Ibid.*

Chapter 4

   1. Leupp, Francis E. *George Westinghouse: His Life and Achievements*. Boston: Little, Brown, and Company, 1919.
   2. *Ibid.*
   3. *Ibid.*
   4. *Ibid.*
   5. *Ibid.*
   6. Westinghouse, George. "History of the Air Brake." *The Electric Journal* (1910): 227–236.
   7. Leupp, op. cit.

Chapter 5

   1. "A Glimpse Into the Future of Westinghouse." *Review of Reviews* (1931).
   2. Leupp, Francis E. *George Westinghouse: His Life and Achievements*. Boston: Little, Brown, and Company, 1919; Prout, Henry G. *A Life of George Westinghouse*. New York: The American Society of Mechanical Engineers, 1921.
   3. Westinghouse, George. "History of the Air Brake." *The Electric Journal* (1910): 227–236.
   4. Wohleber, Curt. "'St. George' Westinghouse." *American Heritage's Invention & Technology* Winter 1997. https://www.inventionandtech.com/content/%E2%80%9Cst-george%E2%80%9D-westinghouse-1?page=full.
   5. Leupp, op. cit.
   6. *Ibid.*
   7. *Ibid.*
   8. *Ibid.*
   9. Banning, Hubert A. and Henry Arden. "WESTINGHOUSE V. GARDNER, ETC., AIR-BRAKE CO." April 1875. *YesWeScan: The FEDERAL CASES*. https://law.resource.org/pub/us/case/reporter/F.Cas/0029.f.cas/0029.f.cas.0798.pdf.
   10. *Ibid.*
   11. *Ibid.*
   12. *Ibid.*
   13. *Ibid.*
   14. Prout, op. cit.
   15. *Ibid.*

Chapter 6

   1. Prout, Henry G. *A Life of George Westinghouse*. New York: The American Society of Mechanical Engineers, 1921.
   2. "Pacific Railroad Acts." n.d. *Wikipedia*. https://en.wikipedia.org/wiki/Pacific_Railroad_Acts.
   3. Staff, Linda Hall Library. "Evolution of Couplers and Brakes on 19th Century Railroads." n.d. *The Transcontinental Railroad*. https://railroad.lindahall.org/essays/couplers-brakes.html.
   4. Schmucker, Kristine. "Tools from Our Collection." 11 April 2013. *Voices of Harvey County*. http://harveycountyvoices.blogspot.com/2013/04/tools-from-our-collection-link-and-pin.html.
   5. Westinghouse, George. "History of the Air Brake." *The Electric Journal* (1910): 227–236.
   6. Prout, op. cit.
   7. Westinghouse, op. cit.; Prout, op. cit.
   8. Prout, op. cit.
   9. *Ibid.*
   10. "Geo. Westinghouse Dies in 68th Year." *New York Times* 13 March 1914. https://westinghousepark.wordpress.com/2019/04/23/new-york-times-march-13-1914/.

Chapter 7

   1. Skrabec, Jr., Quentin. *George Westinghouse: Gentle Genius*. New York: Algora Publishing, 2007.
   2. Prout, Henry G. *A Life of George Westinghouse*. New York: The American Society of Mechanical Engineers, 1921.
   3. *Ibid.*
   4. "History of Wilmerding." n.d. *WilmerdingRenewed.org (Archived copy)*. https://web.archive.org/web/20110217141641/http://www.wilmerdingrenewed.org/history1.html.
   5. "The Town of Pullman." n.d. *Historic*

Pullman Foundation. http://www.pullmanil.org/town.htm; "Pullman Company." n.d. *Wikipedia*. https://en.wikipedia.org/wiki/Pullman_Company#Company_town.
  6. "Working Conditions." n.d. *Library of Congress*. https://www.loc.gov/collections/films-of-westinghouse-works-1904/articles-and-essays/the-westinghouse-world/working-conditions/?c=200&fa=subject:industrialism.
  7. Skrabec, Jr., Quentin. *William McKinley, Apostle of Protectionism*. New York: Algora Publishing, 2008. https://books.google.com/books?id=NVcI2rm1O1EC&pg.
  8. McDonald, Charles W. "100 Years of Safer Railroads." August 1993. *Web Archive*. https://web.archive.org/web/20090731183723/http://www.fra.dot.gov/downloads/safety/rail_safety_program_booklet_v2.pdf.
  9. Skrabec, Jr., Quentin. *George Westinghouse: Gentle Genius*. New York: Algora Publishing, 2007.
  10. *Ibid*. "Life in Wilmerding—The Ideal Home Town." 2 September 1904. *Inside an American Factory: Films of the Westinghouse Works, 1904*. https://www.loc.gov/collections/films-of-westinghouse-works-1904/articles-and-essays/the-westinghouse-world/life-in-wilmerding/.
  11. "Working Conditions." Op. cit.
  12. "Life in Wilmerding—The Ideal Home Town." op. cit.
  13. "National Register of Historic Places listings in Allegheny County, Pennsylvania." n.d. *Wikipedia*. https://en.wikipedia.org/wiki/National_Register_of_Historic_Places_listings_in_Allegheny_County,_Pennsylvania.
  14. "Westinghouse Castle goes for $100,000 at sheriff's sale." 7 June 2016. *Pocono Record*. https://www.poconorecord.com/article/20160607/NEWS/160609689.

Chapter 8

  1. Cited in Farnsworth, M. M. "The Union Switch and Signal Company: A Review of Its Predecessors, Formation, Developments, Growth, Activities, Acquisitions and Affiliates." Internal to Company. Union Switch and Signal Company, 1948.
  2. Balliet, Herbert S., Keith E. Kellenberger, Henry M. Sperry. *The History of Dr. William Robinson's Invention of the Track Circuit*. New York: Signal Section, American Railway Association, 1922.
  3. "Interlocking." n.d. *Wikipedia*. https://en.wikipedia.org/wiki/Interlocking.
  4. "Landmarks in Signaling History." *Railway Age Gazette* 28 July 1916; Farnsworth, op. cit.
  5. Union Switch and Signal Company. *Electropneumatic interlocking*. Pittsburgh, 1914. https://play.google.com/books/reader?id=kZrVAAAAMAAJ&hl=en&pg=GBS.PA3.
  6. *Ibid*.
  7. "The Ousting of Westinghouse—Legal Effort to Be Made to Annul the Recent Election." *New York Times* 12 March 1891: 1.
  8. "Union Switch & Signal Co." *Railroad Gazette* 13 March 1891: 186. https://play.google.com/books/reader?id=dkRKzCk5x2gC&hl=en&pg=GBS.PA186.
  9. "Tables Quickly Turned—Mr. Westinghouse Again In Control of the Switch Company." *New York Times* 14 March 1891: 1.

Chapter 9

  1. "Thomas Hutchins to John McKee." *Deed Book 1, Allegheny County*. Pittsburgh, 11 September 1785. 152.
  2. Riggs, Walter L. "Early History of McKeesport: Part 2." 31 July 2012. *The Tube City Almanac*. http://www.tubecityonline.com/almanac/entry_1967.php.
  3. "John McKee to Pollard McCormick." *Deed Book R2, Allegheny County*. Pittsburgh, 13 October 1835. 71.
  4. Pollard McCormick to Peter Schoenberger." *Deed Book, Allegheny County*. Vol. 139. Pittsburgh, 1 January 1842. 196.
  5. "Dr. Peter Schoenberger to Nathaniel Holmes." *Deed Book, Allegheny County*. Vol. 141. Pittsburgh, 10 September 1859. 180.
  6. "Nathaniel Holmes to Thomas Miller." *Deed Book, Allegheny County*. Vol. 190. Pittsburgh, 26 September 1865. 507.
  7. "Thomas M. Miller to James H. Hopkins." *Deed Book, Allegheny County*. Vol. 246. Pittsburgh, 31 July 1869. 256.
  8. "James H. Hopkins to George Westinghouse." *Deed Book, Allegheny County*. Vol. 265. Pittsburgh, 1 January 1871. 515.
  9. Leupp, Francis E. *George Westinghouse: His Life and Achievements*. Boston: Little, Brown, and Company, 1919.
  10. Voelker, Elizabeth Anne. *Solitude:Majestic Showplace for a Millionaire's Genius*. College Thesis. Carnegie Institute of Technology. Pittsburgh, 1950.
  11. "Pittsburg Greatly Shocked." *New York Times* 26 June 1906. https://www.nytimes.com/1906/06/26/archives/pittsburg-greatly-shocked-benjamin-thaw-too-ill-to-be-told-of-his.html.
  12. Scully, Donald C. "Historical Society Notes and Documents: Homewood At the Turn of the Century." *Western Pennsylvania History: 1918–2018* 1 October 1976: 493–502. https://journals.psu.edu/wph/article/viewFile/3446/3277.
  13. "A Peer of the Realm." *Pittsburgh Press* 16 September 1897: 7.
  14. VanTrump, James D. ""Solitude" and the Nether Depths: The Pittsburgh Estate of George Westinghouse and Its Gas Well." *The Western Pennsylvania Historical Magazine* June 1959: 155–172. https://journals.psu.edu/wph/article/viewFile/2641/2474.

15. "King of Belgium Here as Prince Albert in '98." *Pittsburgh Post-Gazette* 1 September 1914: 9.
16. "George Westinghouse III." n.d. *FamilySearch*. https://www.familysearch.org/tree/person/details/LH7Z-8HM.

Chapter 10

1. "Natural Gas is King in Pittsburgh." n.d. *American Oil & Gas Historical Society*. https://aoghs.org/oil-almanac/pennsylvania-natural-gas/.
2. Morris, Sue. "Keepers of the Eternal Flame." 19 April 2013. *The Historical Dilettante*. http://historicaldilettante.blogspot.com/2013/04/keepers-of-eternal-flame.html.
3. Ibid.
4. Waples, David A. *The Natural Gas Industry in Appalachia*. Second Edition. Jefferson, NC: McFarland, 2012.
5. Morris, op. cit. "Natural Gas is King..." op. cit.; Tarr, Joel and Karen Clay. "Boom and Bust in Pittsburgh Natural Gas History: Development, Policy, and Environmental Effects, 1878–1920." *The Pennsylvania Magazine of History and Biography* 2015 October: 323–342.
6. Waples, op. cit.
7. Leupp, Francis E. *George Westinghouse: His Life and Achievements*. Boston: Little, Brown, and Company, 1919.
8. Ibid.
9. Ibid.
10. Tarr and Clay, op. cit.
11. VanTrump, James D. "'Solitude' and the Nether Depths: The Pittsburgh Estate of George Westinghouse and Its Gas Well." *The Western Pennsylvania Historical Magazine* June 1959: 155–172. https://journals.psu.edu/wph/article/viewFile/2641/2474.
12. Tarr and Clay, op. cit.
13. Waples, op. cit. P. 51; Skrabec, Jr., Quentin. *Henry Clay Frick: The Life of the Perfect Capitalist*. Jefferson: McFarland, 2010.
14. Private communication from Dawn Reid Brean, Associate Curator of Decorative Arts, The Frick Pittsburgh.
15. "H.J. Heinz and Clifford in Carriage." 1905. *Historic Pittsburgh*. https://historicpittsburgh.org/islandora/object/pitt:MSP57.B008.I06.
16. Skrabec, Jr., Quentin. *George Westinghouse: Gentle Genius*. New York: Algora Publishing, 2007.
17. Tarr and Clay, op. cit.
18. "Gas Pipes." *Pittsburgh Post Gazette* 8 August 1884
19. "City of Pittsburgh v. Equitable Gas Co." *Atlantic Reporter* 1917: 1049. https://books.google.com/books?id=CjU8AAAAIAAJ&pg=PA1049&lpg=PA1049&dq=%22Westinghouse+Ordinance%22+1884&source=bl&ots=lV3QPYiAd2&sig=ACfU3U09z0x2NqiVqI_ZDnRktlZtVJAkeQ&hl=en&sa=X&ved=2ahUKEwiCgZPEsM7kAhUnnq0KHcaICtEQ6AEwAHoECAYQAQ#v=onepage&q&f=false.
20. Cited in Tarr and Clay. P. 330.
21. Lief, Alfred. *Metering for America: 125 years of the gas industry and American Meter Company*. New York: Appleton-Century-Crofts, 1961. https://babel.hathitrust.org/cgi/pt?id=uc1.$b667030&view=1up&seq=52.
22. Tarr and Clay, op. cit.
23. "Early Years of The Philadelphia Company." Unattributed, undated report found at the Detre Archives of the Heinz History Center. n.d.
24. Ibid.
25. Crum, A. R. and A. S. Dungan, *Romance of American Petroleum and Gas*. Vol. 2. New York: Romance of American Petroleum and Gas Co., 1911. 2 vols.
26. "Pittsburgh, the 'Smoky City.'" 11 February 2015. *Popular Pittsburgh*. https://popularpittsburgh.com/darkhistory/.
27. "Impressions of Early Travelers (quotation from Thomas's Travels through the western country in 1816)." 1916. *Pittsburgh in 1816 (archived)*. Carnegie Library of Pittsburgh. https://web.archive.org/web/20081202145839/http://www.clpgh.org/research/Pittsburgh/history/pgh1816.html.
28. *Harper's Weekly*. 27 February 1892. *Hathi Trust*. https://babel.hathitrust.org/cgi/pt?id=mdp.39015014126026&view=1up&seq=161.
29. "Westinghouse Sues Carnegie: The Two Millionaires Have a Dispute About a Large Gas Bill." *New York Times* 30 September 1891: 1. https://timesmachine.nytimes.com/timesmachine/1891/09/30/103339251.html?pageNumber=1.
30. Tarr and Clay, op. cit. P. 332.
31. "Bipartisan Effort to Rebuild Pittsburgh." n.d. *Alex C. Walker Foundation*. http://walker-foundation.org/pittsburgh.

Chapter 11

1. Leupp, Francis E. *George Westinghouse: His Life and Achievements*. Boston: Little, Brown, and Company, 1919; Prout, Henry G. *A Life of George Westinghouse*. New York: The American Society of Mechanical Engineers, 1921.
2. Drew, Bernard A. and Gerald Chapman. "William Stanley Lighted a Town and Powered an Industry." *Berkshire History* Fall 1985.
3. Cited in Leupp, op. cit.
4. "Gaulard and Gibbs secondary generator." n.d. *Museo Galileo Virtual Museum*. https://catalogue.museogalileo.it/object/GaulardGibbsSecondaryGenerator.html.

Chapter 13

1. Leupp, Francis E. *George Westinghouse: His Life and Achievements*. Boston: Little, Brown, and Company, 1919.
2. Prout, Henry G. *A Life of George Westinghouse*. New York: The American Society of Mechanical Engineers, 1921.
3. "On the Human Side: William Stanley and

Electric Power Distribution." April 2011. *IEEE NY-Monitor 59*. http://sites.ieee.org/ny-monitor/files/2011/04/4_PDF_MEL4.pdf.
  4. Blalock, Thomas J. "In the Berkshires, Part 1: William Stanley Started Something." *IEEE Power & Energy* July/August 2012. https://magazine.ieee-pes.org/julyaugust-2012/history-3/.
  5. Drew, Bernard A. and Gerald Chapman. "William Stanley Lighted a Town and Powered an Industry." *Berkshire History* Fall 1985.
  6. "IEEE Edison Medal." n.d. *IEEE*. https://www.ieee.org/about/awards/medals/edison.html.
  7. Blalock, op. cit.
  8. Josephson, Matthew. *Edison: A Biography*. New York: McGraw Hill, 1959.
  9. Leupp, op. cit.
  10. Prout, op. cit.
  11. Leupp and Prout, op. cit.

Chapter 14

  1. Carnegie, Dale. "Little Known Facts About Well Known People: Nikola Tesla." *WOR Radio Broadcast*. New York, 11 January 1945. Radio broadcast.
  2. Cheney, Margaret. *Tesla: Man Out of Time*. New York: Dorset Press, 1981.
  3. Tesla, Nikola. "My Inventions." *Electrical Experimenter* February-June and October 1919.
  4. O'Neill, John J. *Prodigal Genius: The Life of Nikola Tesla*. New York: Ives Washburn, 1944.
  5. Tesla, op. cit.
  6. Seifer, Marc J. *Wizard: The Life and Times of Nikola Tesla : Biography of a Genius*. New York: Citadel Press; Kensington Publishing Corp., 1998.
  7. Tesla, op. cit.
  8. Seifer, op. cit.
  9. Cheney, op. cit.; Citation from Tesla, op. cit.
  10. Tesla, Nikola. *The Autobiography of Nikola Tesla*. n.d. https://www.mcnabb.com/music/tesla/bio.pdf.
  11. Cheney, op. cit.
  12. Biography.com Editors. "Nikola Tesla Biography." 4 September 2019. *Biography.com*. A&E Television Network. https://www.biography.com/inventor/nikola-tesla; Barksdale, Nate. "9 Things You May Not Know About Nikola Tesla." 22 August 2018. *History.com*. https://www.history.com/news/9-things-you-may-not-know-about-nikola-tesla; "Inventions & Experiments of Nikola Tesla." n.d. *jimdo.com*. https://teslaresearch.jimdo.com/edison-s-continental-1882–1885/; Carlson, W. Bernard. *Tesla: Inventor of the Electrical Age*. Princeton: Princeton University Press, 2013. https://books.google.com/books?id=5I5c9j8BEn4C; "Nikola Tesla." n.d. *Wikipedia*. https://en.wikipedia.org/wiki/Nikola_Tesla#CITEREFCarlson2013.
  13. O'Neill, John J. *Prodigal Genius: The Life of Nikola Tesla*. New York: Ives Washburn, 1944.
  14. Byllesby, H. H. "Letter to George Westinghouse." New York, 21 May 1888.
  15. Westinghouse, George. "Letter to T. B. Kerr." Lenox, MA, 5 July 1888.
  16. Byllesby, H. H. "Letter to George Westinghouse." New York, 6 July 1888.
  17. Kerr, T. B. "Letter to George Westinghouse." Pittsburgh, 9 July 1888.
  18. Westinghouse, George. "Letter to T. B. Kerr." Lenox, MA, 11 July 1888.
  19. Byllesby, H. H. "Letter to Nikola Tesla." Pittsburgh, 13 December 1888.
  20. "Legal Agreement between Nikola Tesla and Westinghouse Electric Company." Pittsburgh, 2 August 1889.
  21. O'Neill, op. cit.
  22. McDonald, Mildred E., Librarian. "Letter to Mr. J. T. Morris, New York International." Pittsburgh, 1 December 1952.
  23. Private communication with David Bear
  24. "Toshiba launches 600V sine-wave PWM Driver IC for three-phase brushless motors." 28 October 2019. *NewElectronics*. http://www.newelectronics.co.uk/electronics/toshiba-sine-wave-brushless-motor-industrial-home-appliances/220711/.
  25. "Advantages & Disadvantages Induction Motor." n.d. *Instrumentation Tools*. https://instrumentationtools.com/advantages-disadvantages-induction-motor/.

Chapter 15

  1. Cep, Casey. "The Real Nature of Thomas Edison's Genius." 21 October 2019. *The New Yorker*. https://www.newyorker.com/magazine/2019/10/28/the-real-nature-of-thomas-edisons-genius.
  2. "Thomas Alva Edison I (1847–1931)." n.d. *WikiTree*. https://www.wikitree.com/wiki/Edison-1#_note-1860us.
  3. Maranzani, Barbara. "Thomas Edison's Near Death Set Him on the Road to Fame." 27 June 2019. *Biography.com*. https://www.biography.com/news/thomas-edison-train-accident-young-boy-saved-telegraph.
  4. Essig, Mark. *Edison & the Electric Chair: A Story of Light and Death*. New York: Walker & Company, 2003.
  5. Cep, op. cit.
  6. "Edison Files: Edison's Two Families." n.d. *Edison Museum*. https://edisonmuseum.org/content8625.html?pageCatID=2&pageID=2.
  7. Essig, op. cit.
  8. "Thomas Edison and Menlo Park." n.d. The Thomas Edison Center at Menlo Park. http://www.menloparkmuseum.org/history/thomas-edison-and-menlo-park/.
  9. Wilson, Wendell E. "Thomas Alva Edison (1847–1931)." 2013. *The Mineralogical Record*. https://archive.ph/20130415080355/http://www.minrec.org/labels.asp?colid=737#selection-389.0-393.11.
  10. Shulman, Seth. *Owning the Future*. Boston:

Houghton Mifflin, 1999. https://archive.org/details/owningfuture00shul/page/158.

11. "Carbon Transmitter (Archived)." 4 August 2009. *IEEE Global History Network*. https://web.archive.org/web/20100318043500/http://www.ieeeghn.org/wiki/index.php/Carbon_Transmitter.

12. Edison, Thomas A. *The Papers of Thomas A. Edison: Menlo Park: The Early Years, April 1876-December 1877*. Ed. Paul B. Israel, Keith Nier, Robert A. Rosenberg. Vol. 3. Johns Hopkins University Press, 1995. 3 vols.

13. Essig, op. cit.

14. "History of the Cylinder Phonograph." n.d. Library of Congress. https://www.loc.gov/collections/edison-company-motion-pictures-and-sound-recordings/articles-and-essays/history-of-edison-sound-recordings/history-of-the-cylinder-phonograph/.

15. Friedel, Robert and Paul Israel. *Edison's Electric Light: The Art of Invention*. Baltimore: The Johns Hopkins University Press, 2010. https://books.google.com/books?id=8U-Naf4DuzMC&pg=PA26&source=gbs_selected_pages&cad=2#v=onepage&q&f=false.

16. "Moses G. Farmer." n.d. *Wikipedia*. https://en.wikipedia.org/wiki/Moses_G._Farmer.

17. Friedel, op. cit.

18. "Edison's Newest Marvel." *New York Sun* 16 September 1878. http://edison.rutgers.edu/yearofinno/EL/Doc1439_NYSun_9-16-78.pdf.

19. "Edison's Miracle of Light: Enhanced Transcript." n.d. *PBS: The American Experience*. http://www.shoppbs.pbs.org/wgbh/amex/edison/filmmore/transcript/index.html.

20. Deffree, Suzanne. "Edison Electric Light Co begins operation, October 15, 1878." 15 October 2019. *EDN Network*. https://www.edn.com/electronics-blogs/edn-moments/4398588/Edison-Electric-Light-Co-begins-operation:October-15--1878.

21. Essig, op. cit.

22. Edison, T. A. "Electric Lamp." 27 January 1880. *Patent Images*. https://patentimages.storage.googleapis.com/d4/9b/62/aac68f7e65536c/US223898.pdf.

23. Spehr, Paul. *The Man Who Made Movies: W.K.L. Dickson*. New Barnet: John Libbey Publishing Ltd., 2008. https://books.google.com/books?id=hcWCDgAAQBAJ&pg=PA38&lpg=PA38&dq=Major+Sherbourne+Eaton&source=bl&ots=7dLf7j84y4&sig=ACfU3U2PtUgbRDYIdyWDgRKGkbci5i67cw&hl=en&sa=X&ved=2ahUKEwjP85v1rMnlAhVIx1kKHVSmDnAQ6AEwCXoECAsQAQ#v=onepage&q=Major%20Sherbourne%20Eaton.

24. Morrow, L.W.W. "The Father of the Central-Station Industry." *Electrical World* 9 September 1922: 529–530.

25. Essig, op. cit.

26. Morrow, op. cit.

27. Ibid.

28. "Edison's Miracle of Light." op. cit.

29. Lieb, John W. "The Birth of an Industry." *Electrical World* 9 September 1922: 523–528.

30. "Miscellaneous City News: Edison's Electric Light: 'The Times' Building Illuminated By Electricity." *New York Times* 5 September 1882. https://ethw.org/w/images/a/ae/Edison_and_Pearl_Street%2C_Text%2C_031410.pdf.

31. Jonnes, Jill. *Empires of Light: Edison, Tesla, Westinghouse, and the Race to Electrify the World*. New York: Random House, 2003. https://www.amazon.com/Empires-Light-Edison-Westinghouse-Electrify-ebook/dp/B000FBJDA2/ref=sr_1_1?gclid=Cj0KCQjwr-_tBRCMARIsAN413WSxneqqGRSKNB0kewjvAYgi_1Vmw7ZYGNe-Ob-z842lBUTZBbgwtpoaAgsGEALw_wcB&hvadid=378051177081&hvdev=c&hvlocphy=9009644&hvnetw=g&hvpo.

32. "Mina Miller Edison: A Valuable Partner to Thomas Edison." n.d. *Thomas Edison Muckers: Your Blog for Everything Edison, Everyday*. http://www.edisonmuckers.org/mina-miller-edison/.

33. "Mina Miller Edison." 26 February 2015. *Thomas Edison National Historical Park, New Jersey*. https://www.nps.gov/edis/learn/historyculture/mina-miller-edison.htm.

34. *Thomas Edison National Historical Park: Research*. n.d. NPS. https://www.nps.gov/edis/learn/historyculture/research.htm.

35. Thomas Edison National Historical Park." n.d. *Wikipedia*. https://en.wikipedia.org/wiki/Thomas_Edison_National_Historical_Park.

36. Lieb, op. cit.

37. "Edwin J. Houston." n.d. *Wikipedia*. https://en.wikipedia.org/wiki/Edwin_J._Houston.

38. "Elihu Thomson." n.d. *Wikipedia*. https://en.wikipedia.org/wiki/Elihu_Thomson.

39. "Elihu Thomson." 14 February 2019. *Engineering and Technology History*. https://ethw.org/Elihu_Thomson; "Thomson-Houston Electric Company." n.d. *Wikipedia*. https://en.wikipedia.org/wiki/Thomson-Houston_Electric_Company.

40. Leupp, Francis E. *George Westinghouse: His Life and Achievements*. Boston: Little, Brown, and Company, 1919.

41. Ibid.

42. "Edison's Miracle of Light." op. cit.

Chapter 16

1. "John Pierpont Morgan and the American Corporation." n.d. *Annenberg Learner*. https://www.learner.org/series/a-biography-of-america/capital-and-labor/.

2. Flynn, John Thomas. *Men of Wealth: The Story of Twelve Significant Fortunes from the Renaissance to the Present Day*. New York: Simon and Schuster, 1941. https://books.google.com/books?id=6H1IS8CjYgMC&pg=PA528&lpg=PA528&dq=Amelia+Sturgis&source=bl&ots=Z_zSXW9Xh9&sig=ACfU3U1ehlmitPqnTnFP8QY08oAI4dadzw&hl=en&sa=X&ved=2ahUKEwiI7aPb_9XlAhUjUt8KHZ6HBUUQ6AEwD3oECAsQAQ#v=onepage&q=Amelia%20Sturgis&f=false.

3. "J.P. Morgan, The Robber Baron With the

Disturbing Facial Feature." n.d. *New England Historical Society.* https://www.newenglandhistoricalsociety.com/j-p-morgan-robber-baron-disturbing-facial-feature/.

4. Chernow, Ron. *The House of Morgan: An American Banking Dynasty and the Rise of Modern Finance.* New York: Grove Press, 1990.

5. "Amelia 'Memie' Morgan." 15 August 2018. *Geni.* https://www.geni.com/people/Amelia-Memie-Morgan/6000000003681536368.

6. "Marriage to Frances Louisa Tracy." n.d. *The Morgan Library & Museum.* https://www.themorgan.org/about/pierpont-morgan-banker/3.

7. Kenton, Will. "Morganization." 25 June 2019. *Investopedia.* https://www.investopedia.com/terms/m/morganization.asp; Beattie, Andrew. "The Kingpin of Wall Street: J.P. Morgan." 14 August 2019. *Investopedia.* https://www.investopedia.com/articles/economics/08/jp-morgan-kingpin-wall-street.asp.

8. The Editors of Encyclopaedia Britannica. "J.P. Morgan: American Financier." 5 September 2019. *Encyclopædia Britannica.* https://www.britannica.com/biography/J-P-Morgan; "J.P. Morgan Biography." 15 April 2019. *Biography.* https://www.biography.com/business-figure/jp-morgan.

9. Ford, Henry. *My Life and Work.* London: William Heinemann Ltd., 1923. https://www.gutenberg.org/cache/epub/7213/pg7213.html.

10. Newton, James D. *Uncommon Friends.* New York: Harcourt Brace Jovanovich, 1987.

Chapter 17

1. Edison, Thomas A. "The Dangers of Electric Lighting." *The North American Review* November 1889: 625–634. https://www.jstor.org/stable/25101896?seq=1#metadata_info_tab_contents.

2. Brandon, Craig. *The Electric Chair: An Unnatural American History.* Jefferson: McFarland, 1999; Essig, Mark. *Edison & the Electric Chair: A Story of Light and Death.* New York: Walker & Company, 2003.

3. Brandon, op. cit.

4. Beals, Gerry. "THOMAS EDISON 'QUOTES'." 1996. *Thomas Edison.com.* https://www.thomasedison.com/quotes.html.

5. Essig, op. cit.

6. Reynolds, Terry and Theodore Bernstein. "Edison and 'The Chair.'" *IEEE Technology and Society Magazine* March 1989: 19–28. <simson.net/ref/1989/Edison_and_The_Chair.pdf.

7. Ibid.

8. "Thomas Edison Biography." 26 August 2019. *Biography.com.* https://www.biography.com/inventor/thomas-edison.

9. Reynolds, op. cit.

10. Tighe, Janet A. "The New York Medico-Legal Society: Legitimating the Union of Law and Psychiatry (1867- 1918)." *International Journal of Law and Psychiatry* 9 (1986): 231–243. https://www.sciencedirect.com/sdfe/pdf/download/eid/1-s2.0-0160252786900488/first-page-pdf.

11. "More Information Touching Underground Matters and Alternating Currents Brought Out at Meeting of Board of Electrical Control." *Electrical Review: A Weekly Journal of Electric Light, Telephone Telegraph and Scientific Progress* 12.22 (1888). https://books.google.com/books?id=9bMuUkqmDMsC&pg=RA16-PA8&lpg=RA16-PA8&dq=June+5,+1888+letter+to+the+New++York+by+Harold+Brown&source=bl&ots=w41lpCseEi&sig=ACfU3U2MwTcvl-I1HveIO3Yc_hNFfV4Bt9A&hl=en&sa=X&ved=2ahUKEwjam_eLiOrlAhWMTt8KHVMDAJwQ6AEwDHoECAYQAQ#.

12. Reynolds, op. cit.

13. Green, Samantha. *Thomas Alva Edison and Nikola Tesla.* New York: Enslow Publishing, 2019. https://books.google.com/books?id=1Q6DDwAAQBAJ&pg=PA49&lpg=PA49&dq=June+5,+1888+letter+to+the+New++York+by+Harold+Brown&source=bl&ots=_3_1XlsaDu&sig=ACfU3U1PSUuco32z4IoGIgzb0we78dttUQ&hl=en&sa=X&ved=2ahUKEwiw-oaEh-rlAhUrZN8KHbLfBpUQ6AEwCnoECAkQAQ#v=onepag.

14. Ibid.

15. Ross, Dr. Irwin. "The Chair of Death: A Product of the A.C.-D.C. Controversy." *Consulting Engineer* April 1967: 102–104.

16. Sarat, Austin. *Gruesome Spectacles: Botched Executions and America's Death Penalty.* Stanford: Stanford Law Books, 2014. https://books.google.com/books?id=Glt_AwAAQBAJ&pg=PA70&lpg=PA70&dq=John+Debella+Kemmler&source=bl&ots=soATrSOCgZ&sig=ACfU3U2sNKGj9qOpPlfb9U7VJv-ZL8x-Dw&hl=en&sa=X&ved=2ahUKEwjLnMTd0-rlAhUMHqwKHYqHCpkQ6AEwAHoECAUQAQ#v=onepage&q=John%20Debella%20Kemmler&f=f; Gray, J. "People v. Kemmler." 21 March 1890. *Casetext.* https://casetext.com/case/people-v-kemmler.

17. Sarat, op. cit.

18. Reynolds, op. cit.

19. Bernard, Allen. "The First Electric Chair." *True: The Man's Magazine* June 1956: 71–72.

20. "William Kemmler." n.d. *Murderpedia (includes New York Herald article of August 7, 1890).* https://murderpedia.org/male.K/k/kemmler-william.htm.

Chapter 18

1. "The Club and the Dam." n.d. *Johnstown Area Heritage Association.* https://www.jaha.org/attractions/johnstown-flood-museum/flood-history/the-club-and-the-dam/.

2. McCullough, David. *The Johnstown Flood.* New York: Simon and Schuster, 1968.

3. "The Club and the Dam." op. cit.

4. Shappee, Nathan Daniel. 1940. "History of

Johnstown." Unpublished Doctoral Dissertation, University of Pittsburgh. https://www.jaha.org/edu/flood/why/dam_club_era.html#_ftnref11. p. 323.

5. "The Relief Effort." n.d. *Johnstown Area Heritage Association.* https://www.jaha.org/attractions/johnstown-flood-museum/flood-history/the-relief-effort/.

6. "Johnstown Flood." n.d. *Wikipedia.* https://en.wikipedia.org/wiki/Johnstown_Flood.

7. History.com Editors. "Homestead Strike." 7 June 2019. *History.com.* https://www.history.com/topics/industrial-revolution/homestead-strike.

8. Adamczyk, Joseph. "Homestead Strike." 25 September 2019. *ENCYCLOPÆDIA BRITANNICA.* https://www.britannica.com/event/Homestead-Strike.

9. *Ibid.*

10. "1892 Homestead Strike." n.d. *AFL-CIO: America's Unions.* https://aflcio.org/about/history/labor-history-events/1892-homestead-strike.

11. Standiford, Les. *Meet You in Hell: Andrew Carnegie, Henry Clay Frick, and the Bitter Partnership That Transformed America.* New York: Crown Publishers, 2005.

12. Fitch, John A. *The Steel Workers.* Ed. Paul Underwood Kellogg. Six vols. Philadelphia: William F. Fell Co., 1911. https://archive.org/stream/steelworkers00fitcrich#page/n9/mode/2up.

13. "Andrew Carnegie's steel workers, hours worked per week?" n.d. *History.* https://history.stackexchange.com/questions/42424/andrew-carnegies-steel-workers-hours-worked-per-week.

14. "How Did Andrew Carnegie Treat His Workers?" n.d. *Reference.com.* https://www.reference.com/business-finance/did-andrew-carnegie-treat-his-workers-de36d945a374a10f.

15. Skrabec, Jr., Quentin. *Benevolent Barons: American Worker-Centered Industrialists, 1850–1910.* Jefferson: McFarland, 2015. https://books.google.com/books?id=CLnwCQAAQBAJ&pg=PA159&lpg=PA159&dq=contrasting+carnegie+and+westinghouse&source=bl&ots=zTCdwvABH7&sig=ACfU3U2hlNQUkM2bb8SxxUo_NwKCGBzMcw&hl=en&sa=X&ved=2ahUKEwiN75Xknvzl AhXsmeAKHcvtA1gQ6AEwDnoECA4QAQ#v=onepage&q=contrasti.

16. Reis, Ed. "A Man for His People." *Mechanical Engineering Magazine* October 2008: 32–35. https://watermark.silverchair.com/me-2008-oct3.pdf?token=AQECAHi208BE49Ooan9kkhW_Ercy7Dm3ZL_9Cf3qfKAc485ysgAAA50wggOZBgkqhkiG9w0BBwagggOKMIIDhgIBADCCA38GCSqGSIb3DQEHATAeBglghkgBZQMEAS4wEQQMjRhQ4FGq—aaNLlDAgEQgIIDUC2xQLp_nZaSMsjwzQyMgcrlAid1yshBHVd_h5tCxp.

17. McCollester, Charles. "The Next Page: Dark Days in the Electric Valley." *Pittsburgh Post-Gazette* 7 June 2014. <https://www.post-gazette.com/opinion/Op-Ed/2014/06/08/The-Next-Page-Dark-days-in-the-Electric-Valley-Charles-McCollester/stories/201406080068.

18. *Westinghouse Memorial.* n.d. http://www.westinghousememorial.org/story.htm.

19. Heinrichs, E.H. "Anecdotes and Reminiscences of Westinghouse." Unpublished manuscript. October 1931.

20. "The Westinghouse Electric & Manufacturing Company." *The Westinghouse Company.* n.d. 165–202. From Westinghouse Collection, Detre Library and Archives Division, Senator John Heinz History Center, Pittsburgh, PA.

21. *Ibid.*

22. Skrabec, Jr., Quentin. *George Westinghouse: Gentle Genius.* New York: Algora Publishing, 2007; "Westinghouse On Top." *The Pittsburgh Press* 10 May 1891: 7.

23. Leupp, Francis E. *George Westinghouse: His Life and Achievements.* Boston: Little, Brown, and Company, 1919.

24. "Westinghouse On Top." op. cit.

25. "The Westinghouse Electric & Manufacturing Company." op. cit.

26. "Thomas Edison Feels That the Newly Formed General Electric, Under J.P. Morgan, Is Overcharging Him for His Lighting and Motor Supplies." n.d. *RAAB Collection.* https://www.raabcollection.com/science-autographs/edison-trust.

27. Tesla, Nikola. "My Inventions." *Electrical Experimenter* February-June and October 1919.

Chapter 19

1. "World's Columbian Exposition." n.d. *Encyclopedia of Chicago.* https://encyclopedia.chicagohistory.org/pages/1386.html.

2. "The House of Representatives' Selection of the Location for the 1893 World's Fair." n.d. *History, Art & Archives, U.S. House of Representatives.* https://history.house.gov/Historical-Highlights/1851–1900/The-House-of-Representatives--selection-of-the-location-for-the-1893-World-s-Fair/.

3. Leupp, op. cit.

4. "Daniel Burnham." n.d. *Wikipedia.* https://en.wikipedia.org/wiki/Daniel_Burnham.

5. "Ferris Wheel." n.d. *Wikipedia.* https://en.wikipedia.org/wiki/Ferris_wheel#The_original_Ferris_Wheel.

6. "Beijing Great Wheel." 29 May 2013. *Observation Wheel Directory.* http://www.observationwheeldirectory.com/wheels/beijing-great-wheel/; "Beijing to get Ferris wheel... finally." 16 March 2016. *China Daily News.* https://www.chinadaily.com.cn/china/2016-03/16/content_23901029.htm.

7. "World's Columbian Exposition." n.d. *Wikipedia.* https://en.wikipedia.org/wiki/World%27s_Columbian_Exposition#Great_Buildings; "World's Columbian Exposition." n.d. *Encyclopedia of Chicago.* https://encyclopedia.chicagohistory.org/pages/1386.html.

8. "List of world expositions." n.d. *Wikipedia.* https://en.wikipedia.org/wiki/List_of_world_expositions.

9. "Art Palace." n.d. *Chicagology*. https://chicagology.com/columbiaexpo/fair039/.

10. *EXPO: Magic of the White City*. Dir. Mark Bussler. Perf. Gene Wilder. Inecom Entertainment Company. 2005. https://www.amazon.com/EXPO-Inecom-Entertainment-Company-Bussler/dp/B01GWDHEXO.

11. Larson, Erik. *The Devil in the White City: Murder, Magic, and Madness at the Fair That Changed America*. New York: Crown Publishers (Random House), 2003. https://www.amazon.com/Devil-White-City-Madness-Changed/dp/0375725601.

12. Johnson, Rossiter, ed. *A History of the World's Columbian Exposition Held in Chicago in 1893*. Vol. 1. New York: D. Appleton and Company, 1897. 4 vols. http://livinghistoryofillinois.com/pdf_files/History%20of%20the%20Worlds%20Columbian%20Exposition%20Held%20in%20Chicago%20in%201893%20vol-1.pdf.

13. Skinner, Charles E. "Lighting the World's Columbian Exposition." Historical Society of Western Pennsylvania, 1934.

14. Johnson, op. cit.

15. "A Most Dangerous Trust: Some Allegations as to the General Electric Company." *New York Times* 19 November 1892: 5. https://timesmachine.nytimes.com/timesmachine/1892/11/19/106089238.html?pageNumber=5.

16. Leupp, op. cit.

17. Johnson, op. cit.

18. Sajna, Mike. "Westinghouse lights up night at Columbian Exposition." *FOCUS* 2 May 1993.

19. *Westinghouse Electric & Manufacturing Company Annual Report*. Detre Library and Archives, Senator John Heinz History Center. Pittsburgh, PA, 1894.

Chapter 20

1. "Niagara Falls: History of Power." n.d. Niagara Frontier. http://www.niagarafrontier.com/power.html#Sch.

2. *Jacob F. Schoellkopf*. n.d. https://en.wikipedia.org/wiki/Jacob_F._Schoellkopf; *Niagara Falls Hydraulic Power and Manufacturing Company*. n.d. https://en.wikipedia.org/wiki/Niagara_Falls_Hydraulic_Power_and_Manufacturing_Company.

3. Niagara Falls: History of Power, op. cit.

4. "Francis Lynde Stetson." n.d. *Wikipedia*. https://en.wikipedia.org/wiki/Francis_Lynde_Stetson.

5. Forbes, George. "The Electrical Transmission of Power from Niagara Falls: Part III." *The Electrical Engineer: A Weekly Review of Theoretical and Applied Electricity* 3 January 1894: 14–16.

6. Stetson, Francis Lynde. "The Use of the Niagara Water Power." *Cassier's Magazine* June 1895.

7. Prout, Henry G. *A Life of George Westinghouse*. New York: The American Society of Mechanical Engineers, 1921.

8. Leupp, Francis E. *George Westinghouse: His Life and Achievements*. Boston: Little, Brown, and Company, 1919.

9. 2015 Gold King Mine waste water spill." n.d. *Wikipedia*. https://en.wikipedia.org/wiki/2015_Gold_King_Mine_waste_water_spill.

10. Britton, Charles C. "An Early Electric Power Facility in Colorado." *The Colorado Magazine* 1972: 185–195. https://www.historycolorado.org/sites/default/files/media/document/2018/ColoradoMagazine_v49n3_Summer1972.pdf.

11. *Ibid*.

12. Nunn, Paul N. "We Did Not Know What Watts Were." *General Electric Review* September 1956: 43. Cited in Britton. *Ibid*.

13. *Colorado Experience: Hydro Power*. Rocky Mountain PBS. 2015. Video. https://www.youtube.com/watch?v=PXhP45lsVcI.

14. Britton, op. cit.

15. "Ames Hydroelectric Generating Plant." n.d. *Wikipedia*. https://en.wikipedia.org/wiki/Ames_Hydroelectric_Generating_Plant.

16. Stetson, op. cit.

17. Prout, op. cit.

18. Stillwell, Lewis Buckley. "Electric Power Generation at Niagara." *Cassier's Magazine—Engineering Illustrated* 1 June 1895: 253–304.

19. Prout, op. cit.

20. Adams, Edward Dean. *Niagara Power: History of The Niagara Falls Power Company 1886–1918: Evolution of its Central Power Station and Alternating Current System*. Vol. II. Niagara Falls: The Niagara Falls Power Company, 1927. II vols. https://archive.org/details/niagarapowerhist00adam_0.

21. Stillwell, op. cit.

22. Prout, op. cit.

23. "Momentous Decision Reached." *Buffalo Courier Express* October 1946.

24. Tesla, Nikola. "Nikola Tesla and George Westinghouse built the first hydro-electric power plant in 1895 in Niagara Falls and started the electrification of the world." 12 January 1897. *Tesla Memorial Society of New York*. https://www.teslasociety.com/exhibition.htm.

25. Froehlich, Fritz E., ed. *The Froehlich/Kent Encyclopedia of Telecommunications*. Vol. 17. New York: Marcel Dekker, Inc., 1999.

26. "Projects Worked On By the Westinghouse Electric & Manufacturing Company During This Period." n.d. *Library of Congress*. https://www.loc.gov/collections/films-of-westinghouse-works-1904/articles-and-essays/the-westinghouse-world/projects.

27. Seifer, Marc J. "Nikola Tesla: The Lost Wizard." Jan/Feb/Mar 2006. *ttmagazine*. http://teslatech.info/ttmagazine/v4n1/seifer.htm.

28. F.A. Merrick, Company President. Pittsburgh, 2 January 1934. Archives Publication VP 3477.

Chapter 21

1. *Along the Braddock Road.* 9 July 2020. Braddock Road Preservation Association, Braddock Battlefield Preservation Center, National Park Service, Braddock Battlefield History Center. https://www.facebook.com/fortnecessity/videos/278253620051712.
2. Preston, David L. "Ten questions about Braddock's Defeat." 9 July 2015. *history-braddocks-defeat.* https://blog.oup.com/2015/07/history-braddocks-defeat/.
3. *The Pittsburg Electrical Handbook.* The American Institute of Electrical Engineers, 1904. https://archive.org/details/pittsburgelectri00amer/page/45.
4. "Picture of Ben G. Lamme Given to Engineering Dept." *The Tiger (Clemson College)* 12 January 1927: 5.
5. *Ibid.*
6. Skrabec, Jr., Quentin. *George Westinghouse: Gentle Genius.* New York: Algora Publishing, 2007.
7. "Grand Army of the Republic." n.d. *Wikipedia.* https://en.wikipedia.org/wiki/Grand_Army_of_the_Republic.
8. Davidson, G. S. "Westinghouse and Pittsburgh; Grand Army Reception." *Anecdotes and Reminiscences of George Westinghouse 1846–1914 Contributed by His Former Associates.* Ed. Dr. Charles F. Scott. Pittsburgh: Westinghouse Air Brake Company, 1939.

Chapter 22

1. Prout, Henry G. *A Life of George Westinghouse.* New York: The American Society of Mechanical Engineers, 1921.
2. "Turbinia." n.d. *Wikipedia.* https://en.wikipedia.org/wiki/Turbinia.
3. Prout, op. cit.
4. *Ibid.*
5. *Ibid.*
6. National Electrical Manufacturers Association. *A chronological history of electrical development from 600 B.C.* New York: National Electrical Manufacturers Association, 1946. https://archive.org/details/chronologicalhis00natirich/page/96/mode/2up/search/turbine.
7. General Electric. "Steam Turbines." 2020. *GE Power.* https://www.ge.com/power/steam/steam-turbines.
8. "Electricity Explained: How Electricity is Generated." 5 November 2019. *U.S. Energy Information Administration.* https://www.eia.gov/energyexplained/electricity/how-electricity-is-generated.php.
9. Prout, op. cit.
10. "New Invention To Transform Shipbuilding." *New York Times* 3 October 1909: 1. https://www.nytimes.com/1909/10/03/archives/new-invention-to-transform-shipbuilding-device-of-melville.html.
11. Prout, op. cit.

Chapter 23

1. "Streetcars in North America." n.d. *Wikipedia.* https://en.wikipedia.org/wiki/Streetcars_in_North_America; "Horsecar." n.d. *Encyclopaedia Britannica.* https://www.britannica.com/technology/horsecar.
2. "Streetcars in North America." op. cit.
3. "Milestones: Richmond Union Passenger Railway, 1888." 10 August 2016. *Engineering and Technology History Wiki.* https://ethw.org/Milestones:Richmond_Union_Passenger_Railway,_1888; Sprague, John L. "Frank J. Sprague invents the constant-speed dc electric motor." *IEEE power & energy magazine* 1540–7977/16©2016IEEE (2016): 80–96.
4. Sprague, John L., op. cit.
5. Prentice, John. "Charles J. Van Depoele." n.d. *Tramway Information.* https://www.tramwayinfo.com/Tramframe.htm?https://www.tramwayinfo.com/tramways/Articles/Depoele.htm.
6. "The Early History of the DC Traction Motor." n.d. *The Railway Technical Website.* http://www.railway-technical.com/trains/rolling-stock-index-l/electric-locomotives/the-early-history-of-the-dc.html.
7. Prout, op. cit.
8. "Pittsburgh Trolley Pole Co." *Street Railway Journal* January 1896: 59. https://archive.org/stream/streetrailwayjo121896newy/streetrailwayjo121896newy#page/n938/mode/1up.
9. J. Aspinwall Hodge, Jr., and George L. Shearer. "Van Depoele Trolley Litigation." *Street Railway Journal* (1897): 482–483. https://books.google.com/books?id=IwJBAQAAIAAJ&pg=PA482&lpg=PA482&dq=U.S.+District+Court,+New+Haven,+U.S.+Patent+%22424695%22+5&source=bl&ots=9iR1VBGtzL&sig=ACfU3U1HYgUYQRrt7CAAlU1_95o9cNNtQw&hl=en&sa=X&ved=2ahUKEwjN-8rR6LrnAhUJyFkKHTeCByAQ6AEwCnoECAgQAQ#v=onepage&q=U.S.%20District%20Court%2C%20New%20Haven%2C%20US%20Patent%20%22424695%22%205&f=false.
10. "For the Present the Deal Is Off." *Pittsburgh Daily Post* 14 August 1895: 1.
11. Bain. "General Electric-Westinghouse Combination: A Mutual Working Agreement Made." *ELECTRICAL REVIEW-A Journal of Scientific and Electrical Progress* 14 March 1896: 141–143. https://play.google.com/books/reader?id=NuVwaqKNN9QC&hl=en&pg=GBS.PA141.
12. Prout, op. cit.
13. Hodge, op. cit.
14. "Pittsburgh Railways." n.d. *Wikipedia.* https://en.wikipedia.org/wiki/Pittsburgh_Railways.
15. Farley, Ren. *St. Clair River International Railway Tunnel.* February 2011. http://www.detroit1701.org/St.%20Clair%20River%20Railroad%20Tunnel.html; "Grand Trunk Western Railroad." n.d. *Wikipedia.* https://en.wikipedia.org/wiki/Grand_Trunk_Western_Railroad;

"St. Clair Tunnel." n.d. *Wikipedia.* https://en.wikipedia.org/wiki/St._Clair_Tunnel; "Tunnelling shield." n.d. *Wikipedia.* https://en.wikipedia.org/wiki/Tunnelling_shield.

16. *CN/GTW St. Clair Tunnels at Port Huron, MI.* 3 February 2019. https://industrialscenery.blogspot.com/2019/02/cngtw-st-clair-tunnels.html.

17. "Baldwin Locomotive Works." n.d. *Wikipedia.* https://en.wikipedia.org/wiki/Baldwin_Locomotive_Works#Electric_locomotives.

18. Wright, Roy V., ed. *Locomotive Cyclopedia of American Practice.* 6. New York: Simmons-Boardman Publishing Company, 1922. https://play.google.com/books/reader?id=oMY1AQAAMAAJ&hl=en&pg=GBS.PP5.

19. *CN/GTW St. Clair Tunnels at Port Huron, op.* cit.

20. "Railroad electrification in the United States." n.d. *Wikipedia.* https://en.wikipedia.org/wiki/Railroad_electrification_in_the_United_States.

21. Wadsworth, G. R. "Terminal Improvements of the New York Central & Hudson River in New York." *The Railroad Gazette* 20 October 1905: 366–369. https://play.google.com/books/reader?id=yJxMAAAAYAAJ&hl=en&pg=GBS.PA366.

22. "Railroad electrification in the United States." Op. cit.

23. "List of Westinghouse Locomotives." n.d. *Wikipedia.* https://en.wikipedia.org/wiki/List_of_Westinghouse_locomotives#Electric_locomotives.

Chapter 24

1. Tacitus. *The Annals.* Vol. VI. Rome, 109 AD. XVI vols. http://classics.mit.edu/Tacitus/annals.6.vi.html.

2. Wicker, Elmus. "Banking Panics in the U.S.: 1873–1933." n.d. *EH.net.* Economic History Association. https://eh.net/encyclopedia/banking-panics-in-the-us-1873-1933/; "List of banking crises." n.d. *Wikipedia.* https://en.wikipedia.org/wiki/List_of_banking_crises#19th_century.

3. Tallman, Ellis W. and Jon R. Moen. "Lessons from the Panic of 1907." *Economic Review* May/June 1990: 2–13.

4. Heinrichs, E.H., Personal press representative for George Westinghouse. "The 1907 Panic." n.d.

5. "Panic of 1907." n.d. *Wikipedia.* https://en.wikipedia.org/wiki/Panic_of_1907#cite_note-27.

6. Tallman, op. cit.

7. Dewing, Ph.D., Arthur S. *Corporate Promotions and Reorganizations.* Cambridge: Harvard University Press, 1914. https://play.google.com/books/reader?id=JU8tAAAAIAAJ&hl=en&pg=GBS.PP8.

8. *Ibid.*

9. Heinrichs, op. cit.

10. Skrabec, Jr., Quentin. *George Westinghouse: Gentle Genius.* New York: Algora Publishing, 2007.

11. "New President for Westinghouse Electric." *Electrical Review and Western Electrician* 6 August 1910: 256. https://books.google.com/books?id=pEY_AQAAMAAJ&pg=PA256&lpg=PA256&dq=Edwin+F.+Atkins+Westinghouse&source=bl&ots=PX622raad7&sig=ACfU3U0wFcVfa9HnjUkSpNaaOXlwWgvnUw&hl=en&sa=X&ved=2ahUKEwj93_Xd8NvnAhVILK0KHa1wDEgQ6AEwA3oECAYQAQ#v=onepage&q=Edwin%20F.%20Atkins Westinghouse&f=false.

12. "Westinghouse Electric: Mr. Westinghouse Appeals for Proxies." *Industrial World* 24 July 1911: 888–889. https://play.google.com/books/reader?id=aLBAAQAAMAAJ&hl=en&pg=GBS.RA3-PA888.

13. *Edwin Musser Herr.* n.d. https://www.geni.com/people/Edwin-Herr/6000000024008644052; "Results of Westinghouse Annual Meeting: E. M. Herr Chosen President of the Electric." *Industrial World* 7 August 1911: 944–945. https://play.google.com/books/reader?id=aLBAAQAAMAAJ&hl=en&pg=GBS.RA5-PA944.

14. "The Death Record: Robert Mather, Chairman of Westinghouse Electric." *Industrial World* 30 October 1911: 1321. https://play.google.com/books/reader?id=aLBAAQAAMAAJ&hl=en&pg=GBS.RA17-PA1321.

15. Jonnes, Jill. *Empires of Light: Edison, Tesla, Westinghouse, and the Race to Electrify the World.* New York: Random House, 2003. https://www.amazon.com/Empires-Light-Edison-Westinghouse-Electrify-ebook/dp/B000FBJDA2/ref=sr_1_1?gclid=Cj0KCQjwr-_tBRCMARIsAN413WSxneqqGRSKNB0kewjvAYgi_1Vmw7ZYGNe-Ob-z842lBUTZBbgwtpoaAgsGEALw_wcB&hvadid=378051177081&hvdev=c&hvlocphy=9009644&hvnetw=g&hvpo.

16. Dewing, op. cit. 185–186.

17. Carpenter, Jr., Niles. "The Westinghouse Electric & Manufacturing Company, the General Electric Company, and the Panic of 1907." *Journal of Political Economy* (1916): 230–253. https://www.jstor.org/stable/1819438.

Chapter 25

1. Nevin, Adelaide Mellier. *The Social Mirror: A Character Sketch of the Women of Pittsburg and Vicinity During the First Century of the Country's Existence.* Pittsburg: T. W. Nevin, 1888. https://archive.org/details/socialmirrorchar00nevi/page/93/mode/1up/search/Westinghouse.

2. Voelker, Elizabeth Anne. *Solitude: Majestic Showplace for a Millionaire's Genius.* College Thesis. Carnegie Institute of Technology. Pittsburgh, 1950.

3. "George Westinghouse, Sr. Pennsylvania, Pittsburgh City Deaths, 1870–1905." n.d.

Family Search. https://www.familysearch.org/ark:/61903/1:1:XZQB-XP8.

4. "Emeline (Vedder) Westinghouse (1810–1895)." n.d. *WikiTree.* https://www.wikitree.com/wiki/Vedder-329.

5. Leupp, Francis E. *George Westinghouse: His Life and Achievements.* Boston: Little, Brown, and Company, 1919.

6. Bear, David. "Finding Solitude in Westinghouse Park." *Pittsburgh Quarterly* 26 August 2019: 132–134.

7. Blalock, Thomas J. "In the Berkshires, Part 1: William Stanley Started Something." *IEEE Power & Energy* July/August 2012. https://magazine.ieeepes.org/julyaugust-2012/history-3/.

8. Leupp. op. cit.

9. Skrabec, Jr., Quentin. *George Westinghouse: Gentle Genius.* New York: Algora Publishing, 2007; "The Westinghouse Set: An American parcel-gilt silver and mixed metals 'Japanese style' seven-piece tea and coffee set with tray, Tiffany & Co., New York, circa 1885." 22 January 2006. *Sotheby's.* http://www.sothebys.com/en/auctions/ecatalogue/2006/americana-n08158/lot.135.html.

10. Blalock, op. cit.

11. Leupp. op. cit.

12. Drew, Bernard A. "Our Berkshires: 'George,' Mrs. Westinghouse said..." *Brattleboro Reformer* 5 May 2017. https://www.reformer.com/stories/george-mrs-westinghouse-said,506488?; Jennings, Jr., J.L. Sibley. *Massachusetts Avenue Architecture.* Ed. The Commission of Fine Arts. Vol. II. Washington: U.S. Government Printing Office, 1975. https://play.google.com/books/reader?id=bNRPAAAAMAAJ&hl=en&pg=GBS.PA124.

13. Bilis, Madeline. "On the Market: A Playhouse in the Berkshires." 3 April 2017. *Boston Magazine.* https://www.bostonmagazine.com/property/2017/04/03/westinghouse-playhouse-berkshires-otm/.

14. "Lenox as a Resort—Plunkett, Lee Road." 30 January 2016. *Lenox History.* https://lenoxhistory.org/tag/erskine-park/; Consolati, Florence. *See All the People.* Pittsfield: Berkshire Family History Association, 1978; Townes, John A. "A Second Spa for the Berkshires." *New York Times* 4 June 1989.

15. Jennings. op. cit.

16. "Biographies of the Secretaries of State: James Gillespie Blaine (1830–1893)." n.d. *U.S. Department of State, Office of the Historian.* https://history.state.gov/departmenthistory/people/blaine-james-gillespie.

17. Leupp. op. cit.

18. Slauson, Allen B., ed. *A History of the City of Washington: Its Men and Institutions.* Washington: The Washington Post Company, 1903.

19. Jennings. op. cit.

20. Leupp. op. cit.

21. Skrabec, Jr., Quentin. *The World's Richest Neighborhood: How Pittsburgh's East Enders Forged American Industry.* New York: Algora Publishing, 2010.

22. "Leon Czolgosz." n.d. *Wikipedia.* https://en.wikipedia.org/wiki/Leon_Czolgosz#Assassination_of_President_William_McKinley.

23. Skrabec, Jr., Quentin. *William McKinley, Apostle of Protectionism.* New York: Algora Publishing, 2008. https://books.google.com/books?id=NVcI2rm1O1EC&pg.

24. "Dauphin County Pennsylvania News, 1892 Harrisburg Train Wreck (compilation of newspaper reports)." 26 June 1892. *Genealogy Trails.* http://genealogytrails.com/penn/dauphin/news/1892wreck.html; Baer, Christopher T. *A General Chronology of the Pennsylvania Railroad Company.* 1892. http://www.prrths.com/newprr_files/Hagley/PRR_hagley_intro.htm.

25. Beck, Maggie. "Making tracks." *The Union Democrat* 20 February 2020: 1. https://www.pressreader.com/usa/the-union-democrat/20200220/page/1.

26. Voelker, op. cit.

27. "George Westinghouse III." n.d. *Family Search.* New York, New York City Births, 1846–1909. https://www.familysearch.org/ark:/61903/1:1:2WMM-JF4?from=lynx1UIV8&treeref=LH7Z-8HM.

28. "David McCullough '51 Archival Gallery." n.d. *Shady Side Academy.* https://www.shadysideacademy.org/about/history/david-mccullough-51-archival-gallery.

29. Yale University. *Directory of the Living Non-Graduates ff Yale University.* New Haven: Yale University, 1914. https://play.google.com/books/reader?id=WUobAAAAYAAJ&hl=en&pg=GBS.PA51.

30. Yale University. *University Catalogue, 1906.* New Haven: Yale University Publications, 1906.

31. "Society." *Pittsburgh Post-Gazette* 5 March 1909: 10.

32. "Mr. G. Westinghouse and Miss Violet Brocklebank." *Carlisle Express* 6 March 1909; "Society Weddings." *Pittsburgh Daily Post* 5 March 1909: 10.

33. Council of the City of Pittsburgh. *Municipal Record.* Vol. 47. Pittsburgh: Devine & Co., 1911. https://play.google.com/books/reader?id=D7pEAQAAMAAJ&hl=en&pg=GBS.PA104.

34. "United States Census." Vol. Roll 1304. Pittsburgh, Allegheny County, 15 April 1910.

35. *Pittsburgh and Allegheny Directory, 1907.* Pittsburgh: R.L. Polk & Co., 1907. https://historicpittsburgh.org/islandora/object/pitt%3A31735058193818#page/1/mode/2up/search/%22201\+N\+Murtland%22.

36. Consolati, op. cit.

37. "936 Mt. Newton Cross Road." n.d. *Andrew Mara.* Pemberton Homes. http://andrewmara.com/source/?225434.

38. "Westinghouse Family Moves to Arizona." *Pittsburgh Press* 10 March 1933.

39. Henry, Bonnie. "Westinghouse home has had a storied life." *Arizona Daily Star* 9 August 2009. https://tucson.com/lifestyles/bonnie-henry-westinghouse-home-has-had-a-storied-life/

article_d02877d5-d6e9-5065-b28a-6a7aca3781a7.html.
  40. "California Death Index, 1940–1997; Evelyn Violet Brocklebank." n.d. *FamilySearch*. https://www.familysearch.org/tree/person/sources/9VC4-QMC.
  41. "Hollywood Hospital (New Westminster, B.C.)." 8 April 2013. *Royal BC Museum*. https://search-bcarchives.royalbcmuseum.bc.ca/hollywood-hospital-new-westminster-b-c.
  42. "Return $200,000, Woman Told." *San Francisco Call* 16 July 1959: 1.

Chapter 26

  1. "A database of 50 years of FORTUNE's list of America's largest corporations ." n.d. *Fortune 500*. https://archive.fortune.com/magazines/fortune/fortune500_archive/letters/C.html.
  2. "Geo. Westinghouse Dies in 68th Year." *New York Times* 13 March 1914. https://westinghousepark.wordpress.com/2019/04/23/new-york-times-march-13–1914/.
  3. Dietrich, William. "George Westinghouse: The Mystery." Spring/Summer 2006. *Pittsburgh Quarterly*. Douglas Heuck.
  4. Prout, Henry G. *A Life of George Westinghouse*. New York: The American Society of Mechanical Engineers, 1921.
  5. Moore, Graham. *The Last Days of Night: A Novel*. New York: Random House, 2016.
  6. Cravath, Paul. "George Westinghouse the Man." *Mechanical Engineering* March 1937: 156–158.
  7. Quoted in Quentin Skrabec, Jr. *George Westinghouse: Gentle Genius*. New York: Algora Publishing, 2007.
  8. Reis, Ed. "George Westinghouse." *Mechanical Engineering* October 2008: 33–35.
  9. Garbedian, H. Gordon. *George Westinghouse: Fabulous Inventor*. New York: Dodd Mead & Company, 1943.
  10. "The New Westinghouse Air Spring." *New York Times* 11 February 1912; 46.
  11. Prout, op. cit.; "Westinghouse Air Spring." *The Power Wagon* 1 December 1912: 152–153. https://play.google.com/books/reader?id=I2HmAAAAMAAJ&hl=en&pg=GBS.PA153.
  12. Garbedian, op. cit.
  13. Reis, op. cit.
  14. "Death of Westinghouse." *Electrical World* 21 March 1914: 634–639. https://play.google.com/books/reader?id=C9tQAAAAYAAJ&hl=en&pg=GBS.PA639.
  15. "IEEE Edison Medal." n.d. *Wikipedia*. https://en.wikipedia.org/wiki/IEEE_Edison_Medal.
  16. "Honor for Westinghouse." *New York Times* 24 June 1913: 4. https://timesmachine.nytimes.com/timesmachine/1913/06/24/100630880.html?pageNumber=4.
  17. Leupp, Francis E. *George Westinghouse: His Life and Achievements*. Boston: Little, Brown, and Company, 1919.
  18. *Ibid*.
  19. "Westinghouse Will Filed." *New York Times* 19 March 1914.
  20. Leupp, op. cit.
  21. *Ibid*.
  22. "Mrs. Westinghouse's Gift." *New York Times* 31 December 1909: 1.
  23. Reis, op. cit.
  24. Leupp, op. cit.
  25. "Westinghouse's Estate." *New York Times* 22 November 2014; Westinghouse Will Filed, op. cit.
  26. "Westinghouse Widow Dies at Home in Lenox." *Pittsburgh Telegraph* 23 June 1914: 1, 8
  27. Leupp, op. cit.
  28. "Death of Westinghouse." *Electrical World* 21 March 1914: 634–639. https://play.google.com/books/reader?id=C9tQAAAAYAAJ&hl=en&pg=GBS.PA639.
  29. *Ibid*.

Chapter 27

  1. *Westinghouse Memorial*. n.d. http://www.westinghousememorial.org/story.htm.
  2. "Henry Hornbostel." n.d. *Wikipedia*. https://en.wikipedia.org/wiki/Henry_Hornbostel.
  3. "Eric Fisher Wood." n.d. *Wikipedia*. https://en.wikipedia.org/wiki/Eric_Fisher_Wood.
  4. "Daniel Chester French." n.d. *Wikipedia*. https://en.wikipedia.org/wiki/Daniel_Chester_French.
  5. "Paul Fjelde." n.d. *Wikipedia*. https://en.wikipedia.org/wiki/Paul_Fjelde.
  6. Detre Library and Archives Staff. "The Secret History of a Famous Westinghouse Photograph." 20 January 2016. Senator John Heinz History Center. https://www.heinzhistorycenter.org/blog/detre-library-archives/secret-history-famous-westinghouse-photograph.
  7. *Westinghouse Memorial*, op. cit.
  8. *Ibid*.
  9. *Ibid*.
  10. *Westinghouse Memorial*, op. cit.; Hornbostel, Henry. "The Westinghouse Memorial." from Westinghouse Collection, Detre Library and Archives Division, Senator John Heinz History Center, Pittsburgh, PA. 1930.
  11. Sawe, Benjamin Elisha. "Which U.S. City Is Known As "The City of Bridges"?" 22 January 2019. *World Atlas*. https://www.worldatlas.com/articles/which-us-city-is-known-as-the-city-of-bridges.html.
  12. Rosenblum, Charles. "A Monument Then and Now." *Pittsburgh Quarterly* 21 August 2016. https://pittsburghquarterly.com/articles/monument-then-and-now/.
  13. "George Westinghouse Bridge." n.d. *Bridge Browser*. https://historicbridges.org/bridges/browser/?bridgebrowser=pennsylvania/westinghouse/.
  14. Mellon, Steve. "The George Westinghouse

Bridge, Pittsburgh's engineering marvel." 10 October 2014. *The Digs.* https://newsinteractive.post-gazette.com/thedigs/2014/10/10/the-george-westinghouse-bridge-pittsburghs-engineering-marvel/.

Chapter 28

1. "H.H. Westinghouse Worth $2,100,145." *New York Times* 27 February 1935: 8. https://timesmachine.nytimes.com/timesmachine/1935/02/27/93456913.html?pageNumber=8.
2. Vondas, Jerry. "Man with a plan." 19 March 2006. *Tribune-Review.* https://phlf.org/2006/03/19/man-with-a-plan/.
3. Wabtec Corporation. n.d. https://en.wikipedia.org/wiki/Wabtec_Corporation.
4. "Westinghouse Air Brake Technologies." n.d. *MacroTrends.* https://www.macrotrends.net/stocks/charts/WAB/westinghouse-air-brake-technologies/revenue.
5. Friedmann, Nina. "WABCO to be Acquired by ZF Friedrichshafen for $136.50 per Share in Cash." 28 March 2019. *WABCO.* https://www.wabco-auto.com/en/media/media-center/press-releases/press-releases-single-view/news-article/wabco-to-be-acquired-by-zf-friedrichshafen-for-13650-per-share-in-cash/?cHash=3f700f12661faa4aa30c02f16fc7e530.
6. "Lloyd Groff Copeman." n.d. *Wikipedia.* https://en.wikipedia.org/wiki/Lloyd_Groff_Copeman.
7. "White Consolidated Industries Inc. History." 1996. *FundingUniverse.* http://www.fundinguniverse.com/company-histories/white-consolidated-industries-inc-history/; "Westinghouse Electric Corporation." n.d. *Wikipedia.* https://en.wikipedia.org/wiki/Westinghouse_Electric_Corporation.
8. Price, Gwilym A. "Westinghouse Expansion To Continue Into 1955." *Pittsburgh Post Gazette* 5 January 1954: 19.
9. "Westinghouse Plant Closing." *New York Times* 16 February 1985: 30. https://timesmachine.nytimes.com/timesmachine/1985/02/16/031887.html?pageNumber=30.
10. Gallagher, Jim. "WE closes East Pittsburgh plant." *Pittsburgh Post Gazette* 2 May 1987, Final ed.: 1.
11. Mochizuki, Takashi. "Toshiba Expects Write-Down of as Much as Several Billion Dollars." *The Wall Street Journal* 29 December 2016. https://www.wsj.com/articles/toshiba-expects-hefty-write-down-related-to-u-s-nuclear-unit-westinghouse-1482813231.
12. *Westinghouse Electric Company.* n.d. https://en.wikipedia.org/wiki/Westinghouse_Electric_Company.
13. *Westinghouse Licensing Program Signs Consumer Electronics Licensee.* 7 February 2017. http://lmca.net/news-and-insights/33rd-licensee-signed-westinghouse/.
14. *LMCA's Westinghouse Small Appliance Licensee Targets 5% Market Share in India.* 9 October 2017. https://www.licensing.org/press-release/lmcas-westinghouse-small-appliance-licensee-targets-5-market-share-in-india/.
15. *Westinghouse—Innovation You Can Be Sure Of.* n.d. http://www.westinghouselighting.com/.
16. Palenchar, Joseph. *TWICE: The Business of Consumer Electronics.* 7 November 2017. https://www.twice.com/product/ncc-plans-home-automation-hub-56132.
17. Outdoor Solar Lighting. n.d. http://westinghousesolarlights.com/.
18. "RCA." n.d. *Wikipedia.* https://en.wikipedia.org/wiki/RCA.
19. "Westinghouse Broadcasting." n.d. *Wikipedia.* https://en.wikipedia.org/wiki/Westinghouse_Broadcasting.
20. Ibid.
21. Bettis Atomic Power Laboratory. n.d. https://en.wikipedia.org/wiki/Bettis_Atomic_Power_Laboratory.
22. Shippingport Atomic Power Station. n.d. https://en.wikipedia.org/wiki/Shippingport_Atomic_Power_Station#cite_note-nrc-history-4.
23. "Electric energy consumption." n.d. *Wikipedia.* https://en.wikipedia.org/wiki/Electric_energy_consumption.

Appendix II

1. Connor, Dr. Piers. *Brakes.* n.d. http://www.railway-technical.com/trains/rolling-stock-index-l/train-equipment/brakes/.
2. Ludy, Llewellyn V. *AIR BRAKES: An Up-to-Date Treatise on the Westinghouse Air Brake.* Chicago: American Technical Society, 1918. https://archive.org/stream/airbrakesuptodat00ludy/airbrakesuptodat00ludy#page/135/mode/1up.

Appendix III

1. Ohm, Dr. Georg S. *Die galvanische Kette, mathematisch bearbeitet.* Berlin, 1827. https://web.archive.org/web/20090326094110/http://www.ohm-hochschule.de/bib/textarchiv/Ohm.Die_galvanische_Kette.pdf.

Appendix IV

1. "Electromagnetic Induction." n.d. *Electronics Tutorials.* https://www.electronics-tutorials.ws/electromagnetism/electromagnetic-induction.html.
2. Ibid.

Appendix V

1. Prout, Henry G. *A Life of George Westinghouse.* New York: The American Society of Mechanical Engineers, 1921.

2. Westinghouse Machine Company. *The Westinghouse-Parsons Steam Turbine*. East Pittsburgh, 1910. https://babel.hathitrust.org/cgi/pt?id=uc1.b4530455&view=1up&seq=4.

Appendix VI

1. "A Brief History of the U.S. Patent System." 19 April 2013. *Global Patent Solutions*. https://www.globalpatentsolutions.com/blog/brief-history-us-patent-system/.

2. "Patent History in the United States." 23 December 2019. *Laws*. https://patent.laws.com/patent-history/patent-history-in-the-united-states.

3. "Patent claim." n.d. *Wikipedia*. https://en.wikipedia.org/wiki/Patent_claim.

4. "Double patenting." n.d. *Wikipedia*. https://en.wikipedia.org/wiki/Double_patenting.

5. "804 Definition of Double Patenting [R-08.2017]." n.d. *uspto.gov*. https://www.uspto.gov/web/offices/pac/mpep/s804.html.

6. "Prior art." n.d. *Wikipedia*. https://en.wikipedia.org/wiki/Prior_art.

Bibliography

Abrams, Michael. "George Westinghouse." 29 June 2012. *American Society of Mechanical Engineers.* <https://www.asme.org/topics-resources/content/george-westinghouse>.

Adamczyk, Joseph. "Homestead Strike." 25 September 2019. *Encyclopædia Britannica.* <https://www.britannica.com/event/Homestead-Strike>.

Adams, Edward Dean. *Niagara Power: History of The Niagara Falls Power Company 1886–1918: Evolution of its Central Power Station and Alternating Current System.* Vol. II. Niagara Falls: The Niagara Falls Power Company, 1927. II vols. <https://archive.org/details/niagarapowerhist00adam_0>.

"Advantages & Disadvantages Induction Motor." n.d. *Instrumentation Tools.* <https://instrumentationtools.com/advantages-disadvantages-induction-motor/>.

"Along the Braddock Road." 9 July 2020. Braddock Road Preservation Association, Braddock Battlefield Preservation Center, National Park Service, Braddock Battlefield History Center. <https://www.facebook.com/fortnecessity/videos/278253620051712>.

"Amelia 'Memie' Morgan." 15 August 2018. *Geni.* <https://www.geni.com/people/Amelia-Memie-Morgan/6000000003681536368>.

"America's 20 Most Beautiful Skylines, Ranked." 1 December 2014. *Thrillist.* Ed. Matt Meltzer. <https://www.thrillist.com/travel/nation/best-skylines-in-america-seattle-chicago-and-las-vegas-top-our-list>.

"Ames Hydroelectric Generating Plant." n.d. *Wikipedia.* <https://en.wikipedia.org/wiki/Ames_Hydroelectric_Generating_Plant>.

"Andrew Carnegie's steel workers, hours worked per week?" n.d. *History.* <https://history.stackexchange.com/questions/42424/andrew-carnegies-steel-workers-hours-worked-per-week>.

"Art Palace." n.d. *Chicagology.* <https://chicagology.com/columbiaexpo/fair039/>.

Arthur S. Dewing. *Corporate Promotions and Reorganizations.* Cambridge: Harvard University Press, 1914. <https://play.google.com/books/reader?id=JU8tAAAAIAAJ&hl=en&pg=GBS.PP8>.

"Baldwin Locomotive Works." n.d. *Wikipedia.* <https://en.wikipedia.org/wiki/Baldwin_Locomotive_Works#Electric_locomotives>.

Banning, Hubert A., and Henry Arden. "Westinghouse v. Gardner, Etc., Air-Brake Co." April 1875. *YesWeScan: The Federal Cases.* <https://law.resource.org/pub/us/case/reporter/F.Cas/0029.f.cas/0029.f.cas.0798.pdf>.

Barcousky, Len. "Eyewitness 1855: Rail and factory workers died on average at age 27." *Pittsburgh Post-Gazette* 1 March 2014. <https://www.post-gazette.com/local/pittsburgh-history/2014/03/02/Eyewitness-1855-Rail-and-factory-workers-died-on-average-at-age-27/stories/201403020066>.

Barksdale, Nate. "9 Things You May Not Know About Nikola Tesla." 22 August 2018. *History.com.* <https://www.history.com/news/9-things-you-may-not-know-about-nikola-tesla>.

Battista, Judy. "Debate over Immaculate Reception rages on 47 years later." 20 December 2019. *NFL.com.* <http://www.nfl.com/news/story/0ap3000001089669/article/debate-over-immaculate-reception-rages-on-47-years-later>.

Beals, Gerry. "Thomas Edison 'Quotes.'" 1996. *Thomas Edison.com.* <https://www.thomasedison.com/quotes.html>.

Bear, David. "Finding Solitude in Westinghouse Park." *Pittsburgh Quarterly* 26 August 2019: 132–134.

Beattie, Andrew. "The Kingpin of Wall Street: J.P. Morgan." 14 August 2019. *Investopedia.* <https://www.investopedia.com/articles/economics/08/jp-morgan-kingpin-wall-street.asp>.

Beck, Maggie. "Making tracks." *The Union Democrat* 20 February 2020: 1. <https://www.pressreader.com/usa/the-union-democrat/20200220/page/1/>.

"Beijing Great Wheel." 29 May 2013. *Observation Wheel Directory.* <http://www.observationwheeldirectory.com/wheels/beijing-great-wheel/>.

"Beijing to get Ferris wheel … finally." 16 March 2016. *China Daily News.* <https://www.chinadaily.com.cn/china/2016-03/16/content_23901029.htm>.

Bernard, Allen. "The First Electric Chair." *True: The Man's Magazine* June 1956: 71–72.

Bernard A. Drew, Gerald Chapman. "William Stanley Lighted a Town and Powered an Industry." *Berkshire History* Fall 1985.

"Bettis Atomic Power Laboratory." n.d. <https://en.wikipedia.org/wiki/Bettis_Atomic_Power_Laboratory>.

Bilis, Madeline. "On the Market: A Playhouse in the Berkshires." 3 April 2017. *Boston Magazine*. <https://www.bostonmagazine.com/property/2017/04/03/westinghouse-playhouse-berkshires-otm/>.

"Biographies of the Secretaries of State: James Gillespie Blaine (1830–1893)." n.d. U.S. Department of State, Office of the Historian. <https://history.state.gov/departmenthistory/people/blaine-james-gillespie>.

Biography.com Editors. "Nikola Tesla Biography." 4 September 2019. *Biography.com*. A&E Television Network. <https://www.biography.com/inventor/nikola-tesla>.

"Bipartisan Effort to Rebuild Pittsburgh." n.d. Alex C. Walker Foundation. <http://walker-foundation.org/pittsburgh>.

Blalock, Thomas J. "In the Berkshires, Part 1: William Stanley Started Something." *IEEE Power & Energy* July/August 2012. <https://magazine.ieee-pes.org/julyaugust-2012/history-3/>.

"Brakemen's Brotherhood." n.d. *Wikipedia*. <https://en.wikipedia.org/wiki/Brakemen%27s_Brotherhood>.

Brandon, Craig. *The Electric Chair: An Unnatural American History*. Jefferson, NC: McFarland, 1999.

"A Brief History of the U.S. Patent System." 19 April 2013. Global Patent Solutions. <https://www.globalpatentsolutions.com/blog/brief-history-us-patent-system/>.

Britton, Charles C. "An Early Electric Power Facility in Colorado." *The Colorado Magazine* 1972: 185–195. <https://www.historycolorado.org/sites/default/files/media/document/2018/ColoradoMagazine_v49n3_Summer1972.pdf>.

"Brotherhood of Railroad Trainmen." n.d. *Wikipedia*. <https://en.wikipedia.org/wiki/Brotherhood_of_Railroad_Trainmen>.

Byllesby, H H. "Letter to George Westinghouse." New York, 21 May 1888.

_____. "Letter to George Westinghouse." New York, 6 July 1888.

_____. "Letter to Nikola Tesla." Pittsburgh, 13 December 1888.

"California Death Index, 1940–1997; Evelyn Violet Brocklebank." n.d. *FamilySearch*. <https://www.familysearch.org/tree/person/sources/9VC4-QMC>.

"Carbon Transmitter (Archived)." 4 August 2009. IEEE Global History Network. <https://web.archive.org/web/20100318043500/http://www.ieeeghn.org/wiki/index.php/Carbon_Transmitter>.

Carlson, W. Bernard. *Tesla: Inventor of the Electrical Age*. Princeton: Princeton University Press, 2013. <https://books.google.com/books?id=5I5c9j8BEn4C>.

Carnegie, Dale. "Little Known Facts About Well Known People—Nikola Tesla." *WOR Radio Broadcast*. New York, 11 January 1945. Radio broadcast.

Carpenter, Niles Jr. "Electric Companies and the Panic of 1907." *Journal of Political Economy* 24.3 (1916): 230–253.

_____. "The Westinghouse Electric and Manufacturing Company, the General Electric Company, and the Panic of 1907." *Journal of Political Economy* (1916): 230–253. <https://www.jstor.org/stable/1819438>.

"Catherine Westinghouse Moore." n.d. <https://www.findagrave.com/memorial/142743956/catherine-moore>.

Cep, Casey. "The Real Nature of Thomas Edison's Genius." 21 October 2019. *The New Yorker*. <https://www.newyorker.com/magazine/2019/10/28/the-real-nature-of-thomas-edisons-genius>.

Cheney, Margaret. *Tesla: Man Out of Time*. New York: Dorset Press, 1981.

Chernow, Ron. *The House of Morgan: An American Banking Dynasty and the Rise of Modern Finance*. New York: Grove Press, 1990.

"City of Pittsburgh v. Equitable Gas Co." *Atlantic Reporter* 1917: 1049. <https://books.google.com/books?id=CjU8AAAAIAAJ&pg=PA1049&lpg=PA1049&dq=%22Westinghouse+Ordinance%22+1884&source=bl&ots=lV3QPYiAd2&sig=ACfU3U09z0x2NqiVqI_ZDnRktlZtVJAkeQ&hl=en&sa=X&ved=2ahUKEwiCgZPEsM7kAhUnnq0KHcaICtEQ6AEwAHoECAYQAQ#v=onepage&q&f=false>.

"The Club and the Dam." n.d. Johnstown Area Heritage Association. <https://www.jaha.org/attractions/johnstown-flood-museum/flood-history/the-club-and-the-dam/>.

"CN/GTW St. Clair Tunnels at Port Huron, MI." 3 February 2019. <https://industrialscenery.blogspot.com/2019/02/cngtw-st-clair-tunnels.html>.

Collier, Sean, et al. "The 50 Greatest Pittsburghers of All Time." 19 December 2018. *Pittsburgh Magazine*. <https://www.pittsburghmagazine.com/the-50-greatest-pittsburghers-of-all-time/>.

Colorado Experience: Hydro Power. Rocky Mountain PBS. 2015. Video. <https://www.youtube.com/watch?v=PXhP45lsVcI>.

Connor, Dr. Piers. "Brakes." n.d. <http://www.railway-technical.com/trains/rolling-stock-index-l/train-equipment/brakes/>.

Conot, Matthew and Josephson, Robert E. "Thomas Edison, American Inventor." 14 October 2019. *Encyclopedia Britannica*. <https://www.britannica.com/biography/Thomas-Edison>.

Consolati, Florence. *See All the People*. Pittsfield, MA: Berkshire Family History Association, 1978.

Council of the City of Pittsburgh. *Municipal Record.* Vol. 47. Pittsburgh: Devine & Co., 1911. <https://play.google.com/books/reader?id=D7pEAQAAMAAJ&hl=en&pg=GBS.PA104>.

Cravath, Paul. "George Westinghouse the Man." *Mechanical Engineering* March 1937: 156–158.

Crum, A. R. and A. S. Dungan, *Romance of American Petroleum and Gas.* Vol. 2. New York: Romance of American Petroleum and Gas Co., 1911. 2 vols.

"Daniel Burnham." n.d. *Wikipedia.* <https://en.wikipedia.org/wiki/Daniel_Burnham>.

"Daniel Chester French." n.d. *Wikipedia.* <https://en.wikipedia.org/wiki/Daniel_Chester_French>.

"A database of 50 years of FORTUNE's list of America's largest corporations ." n.d. *Fortune 500.* <https://archive.fortune.com/magazines/fortune/fortune500_archive/letters/C.html>.

"Dauphin County Pennsylvania News, 1892 Harrisburg Train Wreck (compilation of newspaper reports)." 26 June 1892. *Genealogy Trails.* <http://genealogytrails.com/penn/dauphin/news/1892wreck.html>.

"David McCullough '51 Archival Gallery." n.d. *Shady Side Academy.* <https://www.shadysideacademy.org/about/history/david-mccullough-51-archival-gallery>.

Davidson, G. S. "Westinghouse and Pittsburgh; Grand Army Reception." *Anecdotes and Reminiscences of George Westinghouse 1846–1914 Contributed by His Former Associates.* Ed. Dr. Charles F. Scott. Pittsburgh: Westinghouse Air Brake Company, 1939.

"Death of Westinghouse." *Electrical World* 21 March 1914: 634–639. <https://play.google.com/books/reader?id=C9tQAAAAYAAJ&hl=en&pg=GBS.PA639>.

"The Death Record: Robert Mather, Chairman of Westinghouse Electric." *Industrial World* 30 October 1911: 1321. <https://play.google.com/books/reader?id=aLBAAQAAMAAJ&hl=en&pg=GBS.RA17-PA1321>.

Deffree, Suzanne. "Edison Electric Light Co begins operation, October 15, 1878." 15 October 2019. *EDN Network.* <https://www.edn.com/electronics-blogs/edn-moments/4398588/Edison-Electric-Light-Co-begins-operation—October-15--1878>.

Detre Library and Archives Staff. "The Secret History of a Famous Westinghouse Photograph." 20 January 2016. *Senator John Heinz History Center.* <https://www.heinzhistorycenter.org/blog/detre-library-archives/secret-history-famous-westinghouse-photograph>.

Dietrich, William. "George Westinghouse: The Mystery." Spring/Summer 2006. *Pittsburgh Quarterly.* Douglas Heuck.

"Dr. Peter Schoenberger to Nathaniel Holmes." *Deed Book, Allegheny County.* Vol. 141. Pittsburgh, 10 September 1859. 180.

"Double patenting." n.d. *Wikipedia.* <https://en.wikipedia.org/wiki/Double_patenting>.

Drew, Bernard A. "Our Berkshires: 'George,' Mrs. Westinghouse said..." *Brattleboro Reformer* 5 May 2017. <https://www.reformer.com/stories/george-mrs-westinghouse-said,506488?>.

"The Early History of the DC Traction Motor." n.d. *The Railway Technical Website.* <http://www.railway-technical.com/trains/rolling-stock-index-l/electric-locomotives/the-early-history-of-the-dc.html>.

"Early Years of The Philadelphia Company." Unattributed, undated report found at the Detre Archives of the Heinz History Center. n.d.

Edison, T. A. "Electric Lamp." 27 January 1880. *Patent Images.* <https://patentimages.storage.googleapis.com/d4/9b/62/aac68f7e65536c/US223898.pdf>.

Edison, Thomas A. "The Dangers of Electric Lighting." *The North American Review* November 1889: 625–634. <https://www.jstor.org/stable/25101896?seq=1#metadata_info_tab_contents>.

_____. *The Papers of Thomas A. Edison: Menlo Park: The Early Years, April 1876-December 1877.* Ed. Paul B. Israel, Keith Nier Robert A. Rosenberg. Vol. 3. Johns Hopkins University Press, 1995. 3 vols.

"Edison Files—Edison's Two Families." n.d. *Edison Museum.* <https://edisonmuseum.org/content8625.html?pageCatID=2&pageID=2>.

"Edison's Miracle of Light—Enhanced Transcript." n.d. *PBS: The American Experience.* <http://www.shoppbs.pbs.org/wgbh/amex/edison/filmmore/transcript/index.html>.

"Edison's Newest Marvel." *New York Sun* 16 September 1878. <http://edison.rutgers.edu/yearofinno/EL/Doc1439_NYSun_9-16-78.pdf>.

The Editors of Encyclopaedia Britannica. "J.P. Morgan: American Financier." 5 September 2019. *Encyclopædia Britannica.* <https://www.britannica.com/biography/J-P-Morgan>.

"Edwin J. Houston." n.d. *Wikipedia.* <https://en.wikipedia.org/wiki/Edwin_J._Houston>.

Edwin Musser Herr. n.d. <https://www.geni.com/people/Edwin-Herr/6000000024008644052>.

"804 Definition of Double Patenting [R-08.2017]." n.d. *uspto.gov.* <https://www.uspto.gov/web/offices/pac/mpep/s804.html>.

"Electric energy consumption." n.d. *Wikipedia.* <https://en.wikipedia.org/wiki/Electric_energy_consumption>.

"Electric power transmission." n.d. *Wikipedia.* <https://en.wikipedia.org/wiki/Electric_power_transmission#Advantage_of_high-voltage_power_transmission>.

"Electricity Explained: How Electricity is Generated." 5 November 2019. U.S. Energy Information Administration. <https://www.eia.gov/energyexplained/electricity/how-electricity-is-generated.php>.

"Electromagnetic Induction." n.d. *Electronics Tutorials.* <https://www.electronics-tutorials.ws/electromagnetism/electromagnetic-induction.html>.

"Elihu Thomson." 14 February 2019. *Engineering and Technology History*. <https://ethw.org/Elihu_Thomson>.

"Elihu Thomson." n.d. *Wikipedia*. <https://en.wikipedia.org/wiki/Elihu_Thomson>.

"Elizabeth Westinghouse." n.d. <https://www.geni.com/people/Elizabeth-Westinghouse/6000000079948864970>.

Ellis W. Tallman, Jon R. Moen. "Lessons from the Panic of 1907." *Economic Review* May/June 1990: 2–13.

"Emeline (Vedder) Westinghouse (1810–1895)." n.d. *WikiTree*. <https://www.wikitree.com/wiki/Vedder-329>.

"Emeline Vedder Westinghouse." n.d. <https://www.findagrave.com/memorial/164303284/emeline-westinghouse>.

E*quitable Resources, Inc. History.* n.d. <http://www.fundinguniverse.com/company-histories/equitable-resources-inc-history/>.

"Eric Fisher Wood." n.d. *Wikipedia*. <https://en.wikipedia.org/wiki/Eric_Fisher_Wood>.

Essig, Mark. *Edison & the Electric Chair: A Story of Light and Death.* New York: Walker & Company, 2003.

EXPO: Magic of the White City. Dir. Mark Bussler. Perf. Gene Wilder. Inecom Entertainment Company. 2005. <https://www.amazon.com/EXPO-Inecom-Entertainment-Company-Bussler/dp/B01GWDHEXO>.

"F.A. Merrick, Company President." Pittsburgh, 2 January 1934. Archives Publication VP 3477.

Farley, Ren. "St. Clair River International Railway Tunnel." February 2011. <http://www.detroit1701.org/St.%20Clair%20River%20Railroad%20Tunnel.html>.

Farnsworth, M. M. "The Union Switch and Signal Company: A Review of Its Predecessors, Formation, Developments, Growth, Activities, Acquisitions and Affiliates." Internal to Company. Union Switch and Signal Company, 1948.

"Ferris Wheel." n.d. *Wikipedia*. <https://en.wikipedia.org/wiki/Ferris_wheel#The_original_Ferris_Wheel>.

Fitch, John A. *The Steel Workers.* Ed. Paul Underwood Kellogg. Six vols. Philadelphia: William F. Fell Co., 1911. <https://archive.org/stream/steelworkers00fitcrich#page/n9/mode/2up>.

Flynn, John Thomas. *Men of Wealth: The Story of Twelve Significant Fortunes from the Renaissance to the Present Day.* New York: Simon & Schuster, 1941. <https://books.google.com/books?id=6H1IS8CjYgMC&pg=PA528&lpg=PA528&dq=Amelia+Sturgis&source=bl&ots=Z_zSXW9Xh9&sig=ACfU3U1ehlmitPqnTnFP8QY08oAI4dadzw&hl=en&sa=X&ved=2ahUKEwiI7aPb_9XlAhUjUt8KHZ6HBUUQ6AEwD3oECAsQAQ#v=onepage&q=Amelia%20Sturgis&f=false>.

"For The Present The Deal Is Off." *Pittsburgh Daily Post* 14 August 1895: 1.

Forbes, George. "The Electrical Transmission of Power from Niagara Falls—Part III." *The Electrical Engineer: A Weekly Review of Theoretical and Applied Electricity* 3 January 1894: 14–16.

Ford, Henry. *My Life and Work.* London: William Heinemann Ltd., 1923. <https://www.gutenberg.org/cache/epub/7213/pg7213.html>.

"Francis Lynde Stetson." n.d. *Wikipedia*. <https://en.wikipedia.org/wiki/Francis_Lynde_Stetson>.

"Fréjus Rail Tunnel." n.d. *Wikipedia*. <https://en.wikipedia.org/wiki/Fr%C3%A9jus_Rail_Tunnel>.

Friedmann, Nina. "WABCO to be Acquired by ZF Friedrichshafen for $136.50 per Share in Cash." 28 March 2019. WABCO. <https://www.wabco-auto.com/en/media/media-center/press-releases/press-releases-single-view/news-article/wabco-to-be-acquired-by-zf-friedrichshafen-for-13650-per-share-in-cash/?cHash=3f700f12661faa4aa30c02f16fc7e530>.

Froehlich, Fritz E., ed. *The Froehlich/Kent Encyclopedia of Telecommunications.* Vol. 17. New York: Marcel Dekker, Inc., 1999.

Gallagher, Jim. "WE closes East Pittsburgh plant." *Pittsburgh Post Gazette* 2 May 1987, Final ed.: 1.

Garbedian, H. Gordon. *George Westinghouse: Fabulous Inventor.* New York: Dodd Mead & Company, 1943.

"Gas Pipes." *Pittsburgh Post Gazette* 8 August 1884.

"Gaulard and Gibbs secondary generator." n.d. Museo Galileo Virtual Museum. <https://catalogue.museogalileo.it/object/GaulardGibbsSecondaryGenerator.html>.

General Electric. "Steam Turbines." 2020. *GE Power.* <https://www.ge.com/power/steam/steam-turbines>.

"Geo. Westinghouse Dies in 68th Year." *New York Times* 13 March 1914. <https://westinghousepark.wordpress.com/2019/04/23/new-york-times-march-13-1914/>.

"George Westinghouse." n.d. <https://www.findagrave.com/memorial/1091/george-westinghouse>.

"George Westinghouse Bridge." n.d. *Bridge Browser.* <https://historicbridges.org/bridges/browser/?bridgebrowser=pennsylvania/westinghouse/>.

"George Westinghouse Historical Marker." n.d. *Pennsylvania Historical Markers.* <https://explorepahistory.com/hmarker.php?markerId=1-A-1A9>.

"George Westinghouse III." n.d. *FamilySearch.* <https://www.familysearch.org/tree/person/details/LH7Z-8HM>.

"George Westinghouse III." n.d. *FamilySearch.* New York, New York City Births, 1846–1909. <https://www.familysearch.org/ark:/61903/1:1:2WMM-JF4?from=lynx1UIV8&treeref=LH7Z-8HM>.

"George Westinghouse, Sr. Pennsylvania, Pittsburgh City Deaths, 1870–1905." n.d. *Family Search.* <https://www.familysearch.org/ark:/61903/1:1:XZQB-XP8>.

"A Glimpse Into the Future of Westinghouse." *Review of Reviews* (1931).
"Grand Army of the Republic." n.d. *Wikipedia*. <https://en.wikipedia.org/wiki/Grand_Army_of_the_Republic>.
"Grand Trunk Western Railroad." n.d. *Wikipedia*. <https://en.wikipedia.org/wiki/Grand_Trunk_Western_Railroad>.
Gray, J. "People v. Kemmler." 21 March 1890. *Casetext*. <https://casetext.com/case/people-v-kemmler>.
Green, Samantha. *Thomas Alva Edison and Nikola Tesla*. New York: Enslow Publishing, 2019. <https://books.google.com/books?id=1Q6DDwAAQBAJ&pg=PA49&lpg=PA49&dq=June+5,+1888+letter+to+the+New++York+by+Harold+Brown&source=bl&ots=_3_1XlsaDu&sig=ACfU3U1PSUuco32z4IoGIgzb0we78dttUQ&hl=en&sa=X&ved=2ahUKEwiw-oaEh-rlAhUrZN8KHbLfBpUQ6AEwCnoECAkQAQ#v=onepag>.
Halstead, Murat. "A World's Fair—Electricity at the Fair." *The Cosmopolitan* September 1893: 577–583. <https://babel.hathitrust.org/cgi/pt?id=ucl.b000929978&view=1up&seq=529>.
Harper's Weekly. 27 February 1892. *Hathi Trust*. <https://babel.hathitrust.org/cgi/pt?id=mdp.39015014126026&view=1up&seq=161>.
Heinrichs, E. H., personal press representative for George Westinghouse. "The 1907 Panic." n.d.
Heinrichs, E.H. "Anecdotes and Reminiscences of Westinghouse." Unpublished manuscript. October 1931.
Henry, Bonnie. "Westinghouse home has had a storied life." *Arizona Daily Star* 9 August 2009. <https://tucson.com/lifestyles/bonnie-henry-westinghouse-home-has-had-a-storied-life/article_d02877d5-d6e9-5065-b28a-6a7aca3781a7.html>.
"Henry Herman Westinghouse." n.d. <https://www.findagrave.com/memorial/4737/henry-herman-westinghouse>.
"Henry Hornbostel." n.d. *Wikipedia*. <https://en.wikipedia.org/wiki/Henry_Hornbostel>.
Herbert S. Balliet, Keith E. Kellenberger, and Henry M. Sperry. *The History of Dr. William Robinson's Invention of the Track Circuit*. New York: Signal Section, American Railway Association, 1922.
"H.H. Westinghouse Worth $2,100,145." *New York Times* 27 February 1935: 8. <https://timesmachine.nytimes.com/timesmachine/1935/02/27/93456913.html?pageNumber=8>.
History.com Editors. "Homestead Strike." 7 June 2019. *History.com*. <https://www.history.com/topics/industrial-revolution/homestead-strike>.
"History of the Cylinder Phonograph." n.d. Library of Congress. <https://www.loc.gov/collections/edison-company-motion-pictures-and-sound-recordings/articles-and-essays/history-of-edison-sound-recordings/history-of-the-cylinder-phonograph/>.
"History of Wilmerding." n.d. *WilmerdingRenewed.org (Archived copy)*. <https://web.archive.org/web/20110217141641/http://www.wilmerdingrenewed.org/history1.html>.
"H.J. Heinz and Clifford in Carriage." 1905. *Historic Pittsburgh*. <https://historicpittsburgh.org/islandora/object/pitt:MSP57.B008.I06>.
"Hollywood Hospital (New Westminster, B.C.)." 8 April 2013. Royal BC Museum. <https://search-bcarchives.royalbcmuseum.bc.ca/hollywood-hospital-new-westminster-b-c>.
"Honor for Westinghouse." *New York Times* 24 June 1913: 4. <https://timesmachine.nytimes.com/timesmachine/1913/06/24/100630880.html?pageNumber=4>.
Hornbostel, Henry. "The Westinghouse Memorial." From Westinghouse Collection, Detre Library and Archives Division, Senator John Heinz History Center, Pittsburgh, PA. 1930.
"Horsecar." n.d. *Encyclopaedia Britannica*. <https://www.britannica.com/technology/horsecar>.
"The House of Representatives' Selection of the Location for the 1893 World's Fair." n.d. History, Art & Archives, U.S. House of Representatives. <https://history.house.gov/Historical-Highlights/1851-1900/The-House-of-Representatives—selection-of-the-location-for-the-1893-World-s-Fair/>.
"How Did Andrew Carnegie Treat His Workers?" n.d. *Reference.com*. <https://www.reference.com/business-finance/did-andrew-carnegie-treat-his-workers-de36d945a374a10f>.
"IEEE Edison Medal." n.d. IEEE. <https://www.ieee.org/about/awards/medals/edison.html>.
"IEEE Edison Medal." n.d. *Wikipedia*. <https://en.wikipedia.org/wiki/IEEE_Edison_Medal>.
"Impressions of Early Travelers (quotation from Thomas's Travels through the western country in 1816)." 1916. *Pittsburgh in 1816 (archived)*. Carnegie Library of Pittsburgh. <https://web.archive.org/web/20081202145839/http://www.clpgh.org/research/Pittsburgh/history/pgh1816.html>.
"Interlocking." n.d. *Wikipedia*. <https://en.wikipedia.org/wiki/Interlocking>.
"Inventions & Experiments of Nikola Tesla." n.d. *jimdo.com*. <https://teslaresearch.jimdo.com/edison-s-continental-1882–1885/>.
"J. F. Schoellkopf, Dye Pioneer, Dead." *The New York Times* 11 September 1942: 21. <https://www.nytimes.com/1942/09/11/archives/j-f-schoellkopf-dye-pioneer-dead-leader-in-making-of-anilines-i-84.html>.
"Jacob F. Schoellkopf." n.d. <https://en.wikipedia.org/wiki/Jacob_F._Schoellkopf>.
"James H. Hopkins to George Westinghouse." *Deed Book, Allegheny County*. Vol. 265. Pittsburgh, 1 January 1871. 515.
"Jay Westinghouse." n.d. <https://www.findagrave.com/memorial/164138905/jay-westinghouse>.
Jennings, J.L. Sibley, Jr. *Massachusetts Avenue Architecture*. Ed. The Commission of Fine Arts. Vol. II. Washington: U.S. Government Printing

Office, 1975. <https://play.google.com/books/reader?id=bNRPAAAAMAAJ&hl=en&pg=GBS.PA124>.

"John McKee to Pollard McCormick." *Deed Book R2, Allegheny County*. Pittsburgh, 13 October 1835. 71.

"John Pierpont Morgan and the American Corporation." n.d. *Annenberg Learner*. <https://www.learner.org/series/a-biography-of-america/capital-and-labor/>.

"John Westinghouse." n.d. <https://www.findagrave.com/memorial/19076828/john-westinghouse>.

Johnson, Carl. *The First George Westinghouse of Schenectady*. 14 January 2016. <http://hoxsie.org/2016/01/04/the-first-george-westinghouse-of-schenectady/>.

Johnson, Rossiter, ed. *A History of the World's Columbian Exposition Held in Chicago in 1893*. Vol. 1. New York: D. Appleton and Company, 1897. 4 vols. <http://livinghistoryofillinois.com/pdf_files/History%20of%20the%20Worlds%20Columbian%20Exposition%20Held%20in%20Chicago%20in%201893%20vol-1.pdf>.

"Johnstown Flood." n.d. *Wikipedia*. <https://en.wikipedia.org/wiki/Johnstown_Flood>.

Jonnes, Jill. *Empires of Light: Edison, Tesla, Westinghouse, and the Race to Electrify the World*. New York: Random House, 2003. <https://www.amazon.com/Empires-Light-Edison-Westinghouse-Electrify-ebook/dp/B000FBJDA2/ref=sr_1_1?gclid=Cj0KCQjwr-_tBRCMARIsAN413WSxneqqGRSKNB0kewjvAYgi_1Vmw7ZYGNe-Obz842lBUTZBbgwtpoaAgsGEALw_wcB&hvadid=378051177081&hvdev=c&hvlocphy=9009644&hvnetw=g&hvpo>.

Jordan, John Woolf. *Genealogical and Personal History of Western Pennsylvania*. Vol. 1. New York: Lewis Historical Publishing Company, 1915.

Josephson, Matthew. *Edison: A Biography*. New York: McGraw-Hill, 1959.

"J.P. Morgan Biography." 15 April 2019. *Biography*. <https://www.biography.com/business-figure/jp-morgan>.

"J.P. Morgan, The Robber Baron With the Disturbing Facial Feature." n.d. *New England Historical Society*. <https://www.newenglandhistoricalsociety.com/j-p-morgan-robber-baron-disturbing-facial-feature/>.

Kennedy, Amelia. "A Brief History of the Footnote—from the Middle Ages to Today." 1 September 2019. *quetext blog*. <https://www.quetext.com/blog/a-brief-history-of-the-footnote>.

Kenton, Will. "Morganization." 25 June 2019. *Investopedia*. <https://www.investopedia.com/terms/m/morganization.asp>.

Kerr, T. B. "Letter to George Westinghouse." Pittsburgh, 9 July 1888.

"King of Belgium Here as Prince Albert In '98." *Pittsburgh Post-Gazette* 1 September 1914: 9.

"Landmarks in Signaling History." *Railway Age Gazette* 28 July 1916.

Larson, Erik. *The Devil in the White City: Murder, Magic, and Madness at the Fair That Changed America*. New York: Crown Publishers (Random House), 2003. <https://www.amazon.com/Devil-White-City-Madness-Changed/dp/0375725601>.

"Legal Agreement between Nikola Tesla and Westinghouse Electric Company." Pittsburgh, 2 August 1889.

"Lenox as a Resort—Plunkett, Lee Road." 30 January 2016. *Lenox History*. <https://lenoxhistory.org/tag/erskine-park/>.

"Leon Czolgosz." n.d. *Wikipedia*. <https://en.wikipedia.org/wiki/Leon_Czolgosz#Assassination_of_President_William_McKinley>.

Leupp, Francis E. *George Westinghouse: His Life and Achievements*. Boston: Little, Brown, and Company, 1919.

Lieb, John W. "The Birth of an Industry." *Electrical World* 9 September 1922: 523–528.

Lief, Alfred. *Metering for America: 125 years of the Gas Industry and American Meter Company*. New York: Appleton-Century-Crofts, 1961. <https://babel.hathitrust.org/cgi/pt?id=uc1.$b667030&view=1up&seq=52>.

"Life in Wilmerding—The Ideal Home Town." 2 September 1904. *Inside an American Factory: Films of the Westinghouse Works, 1904*. <https://www.loc.gov/collections/films-of-westinghouse-works-1904/articles-and-essays/the-westinghouse-world/life-in-wilmerding/>.

"List of banking crises." n.d. *Wikipedia*. <https://en.wikipedia.org/wiki/List_of_banking_crises#19th_century>.

"List of prolific inventors." n.d. *Wikipedia*. <https://en.wikipedia.org/wiki/List_of_prolific_inventors>.

"List of Westinghouse Locomotives." n.d. *Wikipedia*. <https://en.wikipedia.org/wiki/List_of_Westinghouse_locomotives#Electric_locomotives>.

"List of world expositions." n.d. *Wikipedia*. <https://en.wikipedia.org/wiki/List_of_world_expositions>.

"Lloyd Groff Copeman." n.d. *Wikipedia*. <https://en.wikipedia.org/wiki/Lloyd_Groff_Copeman>.

"LMCA's Westinghouse Small Appliance Licensee Targets 5% Market Share in India." 9 October 2017. <https://www.licensing.org/press-release/lmcas-westinghouse-small-appliance-licensee-targets-5-market-share-in-india/>.

Ludy, Llewellyn V. *Air Brakes: An Up-to-Date Treatise on the Westinghouse Air Brake*. Chicago: American Technical Society, 1918. <https://archive.org/stream/airbrakesuptodat00ludy/airbrakesuptodat00ludy#page/135/mode/1up>.

Maranzani, Barbara. "Thomas Edison's Near Death Set Him on the Road to Fame." 27 June

2019. *Biography.com*. <https://www.biography.com/news/thomas-edison-train-accident-young-boy-saved-telegraph>.

"Marriage to Frances Louisa Tracy." n.d. The Morgan Library & Museum. <https://www.themorgan.org/about/pierpont-morgan-banker/3>.

McCollester, Charles. "The Next Page: Dark Days in the Electric Valley." *Pittsburgh Post-Gazette* 7 June 2014. <https://www.post-gazette.com/opinion/Op-Ed/2014/06/08/The-Next-Page-Dark-days-in-the-Electric-Valley-Charles-McCollester/stories/201406080068>.

McCullough, David. *The Johnstown Flood*. New York: Simon & Schuster, 1968.

McDonald, Charles W. "100 Years of Safer Railroads." August 1993. *Web Archive*. <https://web.archive.org/web/20090731183723/http://www.fra.dot.gov/downloads/safety/rail_safety_program_booklet_v2.pdf>.

McDonald, Mildred E., Librarian. "Letter to Mr. J. T. Morris, New York International." Pittsburgh, 1 December 1952.

Mellon, Steve. "The George Westinghouse Bridge, Pittsburgh's engineering marvel." 10 October 2014. *The Digs*. <https://newsinteractive.post-gazette.com/thedigs/2014/10/10/the-george-westinghouse-bridge-pittsburghs-engineering-marvel/>.

"Milestones: Richmond Union Passenger Railway, 1888." 10 August 2016. *Engineering and Technology History Wiki*. <https://ethw.org/Milestones:Richmond_Union_Passenger_Railway,_1888>.

"Mina Miller Edison." 26 February 2015. Thomas Edison National Historical Park, New Jersey. <https://www.nps.gov/edis/learn/historyculture/mina-miller-edison.htm>.

"Mina Miller Edison: A Valuable Partner to Thomas Edison." n.d. *Thomas Edison Muckers: Your Blog for Everything Edison, Everyday*. <http://www.edisonmuckers.org/mina-miller-edison/>.

"Miscellaneous City News: Edison's Electric Light: 'The Times' Building Illuminated by Electricity." *New York Times* 5 September 1882. <https://ethw.org/w/images/a/ae/Edison_and_Pearl_Street%2C_Text%2C_031410.pdf>.

Mochizuki, Takashi. "Toshiba Expects Write-Down of as Much as Several Billion Dollars." *The Wall Street Journal* 29 December 2016. <https://www.wsj.com/articles/toshiba-expects-hefty-write-down-related-to-u-s-nuclear-unit-westinghouse-1482813231>.

"Momentous Decision Reached." *Buffalo Courier Express* October 1946.

Moore, Graham. *The Last Days of Night: A Novel*. New York: Random House, 2016.

"More Information Touching Underground Matters and Alternating Currents Brought Out at Meeting of Board of Electrical Control." *Electrical Review: A Weekly Journal of Electric Light, Telephone Telegraph and Scientific Progress* 12.22 (1888). <https://books.google.com/books?id=9bMuUkqmDMsC&pg=RA16-PA8&lpg=RA16-PA8&dq=June+5,+1888+letter+to+the+New++York+by+Harold+Brown&source=bl&ots=w4llpCseEi&sig=ACfU3U2MwTcvlI1HveIO3Yc_hNFfV4Bt9A&hl=en&sa=X&ved=2ahUKEwjam_eLiOrlAhWMTt8KHVMDAJwQ6AEwDHoECAYQAQ#>.

Morris, Sue. "Keepers of the Eternal Flame." 19 April 2013. *The Historical Dilettante*. <http://historicaldilettante.blogspot.com/2013/04/keepers-of-eternal-flame.html>.

Morrow, L. W. W. "The Father of the Central-Station Industry." *Electrical World* 9 September 1922: 529–530.

"Moses G. Farmer." n.d. *Wikipedia*. <https://en.wikipedia.org/wiki/Moses_G._Farmer>.

"A Most Dangerous Trust: Some Allegations as to the General Electric Company." *New York Times* 19 November 1892: 5. <https://timesmachine.nytimes.com/timesmachine/1892/11/19/106089238.html?pageNumber=5>.

"Nathaniel Holmes to Thomas Miller." *Deed Book, Allegheny County*. Vol. 190. Pittsburgh, 26 September 1865. 507.

National Electrical Manufacturers Association. *A Chronological History of Electrical Development from 600 B.C.* New York: National Electrical Manufacturers Association, 1946. <https://archive.org/details/chronologicalhis00natirich/page/96/mode/2up/search/turbine>.

"National Register of Historic Places listings in Allegheny County, Pennsylvania." n.d. *Wikipedia*. <https://en.wikipedia.org/wiki/National_Register_of_Historic_Places_listings_in_Allegheny_County,_Pennsylvania>.

"Natural Gas is King in Pittsburgh." n.d. *American Oil & Gas Historical Society*. <https://aoghs.org/oil-almanac/pennsylvania-natural-gas/>.

Nevin, Adelaide Mellier. *The Social Mirror: A Character Sketch of the Women of Pittsburg and Vicinity During the First Century of the Country's Existence*. Pittsburg: T. W. Nevin, 1888. <https://archive.org/details/socialmirrorchar00nevi/page/93/mode/1up/search/Westinghouse>.

"New Invention To Transform Shipbuilding." *New York Times* 3 October 1909: 1. <https://www.nytimes.com/1909/10/03/archives/new-invention-to-transform-shipbuilding-device-of-melville.html>.

"New President for Westinghouse Electric." *Electrical Review and Western Electrician* 6 August 1910: 256. <https://books.google.com/books?id=pEY_AQAAMAAJ&pg=PA256&lpg=PA256&dq=Edwin+F.+Atkins+Westinghouse&source=bl&ots=PX622raad7&sig=ACfU3U0wFcVfa9HnjUkSpNaaOXlwWgvnUw&hl=en&sa=X&ved=2ahUKEwj93_Xd8NvnAhVILK0KHalwDEgQ6AEwA3oECAYQAQ#v=onepage&q=Edwin%20F.%20Atkin>.

"The New Westinghouse Air Spring." *New York Times* 11 February 1912: 46.

Newton, James D. *Uncommon Friends*. New York: Harcourt Brace Jovanovich, 1987.

"Niagara Falls." n.d. *Wikipedia*. <https://en.wikipedia.org/wiki/Niagara_Falls#Hydroelectric_power>.

"Niagara Falls: History of Power." n.d. *Niagara Frontier*. <http://www.niagarafrontier.com/power.html#Sch>.

"Niagara Falls Hydraulic Power and Manufacturing Company." n.d. <https://en.wikipedia.org/wiki/Niagara_Falls_Hydraulic_Power_and_Manufacturing_Company>.

"Nikola Tesla." n.d. *Wikipedia*. <https://en.wikipedia.org/wiki/Nikola_Tesla#CITEREFCarlson2013>.

"936 Mt. Newton Cross Road." n.d. *AndrewMara*. Pemberton Homes. <http://andrewmara.com/source/?225434>.

Nunn, Paul N. "We Did Not Know What Watts Were." *General Electric Review* September 1956: 43.

O'Callaghan, E.B. *The Documentary History of the State of New York*. Vol. III. Albany: Weed, Parsons & Co., 1850. <https://play.google.com/books/reader?id=7QxfAAAAcAAJ&hl=en&pg=GBS.PP13>.

Ohm, Dr. Georg S. *Die galvanische Kette, mathematisch bearbeitet*. Berlin, 1827. <https://web.archive.org/web/20090326094110/http://www.ohm-hochschule.de/bib/textarchiv/Ohm.Die_galvanische_Kette.pdf>.

"On the Human Side—William Stanley and Electric Power Distribution." April 2011. *IEEE NY-Monitor 59*. <http://sites.ieee.org/ny-monitor/files/2011/04/4_PDF_MEL4.pdf>.

O'Neill, John J. *Prodigal Genius: The Life of Nikola Tesla*. Ives Washburn, 1944.

"The Ousting of Westinghouse: Legal Effort to Be Made to Annul the Recent Election." *New York Times* 12 March 1891: 1.

"Outdoor Solar Lighting." n.d. <http://westinghousesolarlights.com/>.

"Pacific DC Intertie." n.d. *Wikipedia*. <https://en.wikipedia.org/wiki/Pacific_DC_Intertie>.

"Pacific Railroad Acts." n.d. *Wikipedia*. <https://en.wikipedia.org/wiki/Pacific_Railroad_Acts>.

Palenchar, Joseph. *TWICE—The Business of Consumer Electronics*. 7 November 2017. <https://www.twice.com/product/ncc-plans-home-automation-hub-56132>.

"Panic of 1907." n.d. *Wikipedia*. <https://en.wikipedia.org/wiki/Panic_of_1907#cite_note-27>.

"Patent claim." n.d. *Wikipedia*. <https://en.wikipedia.org/wiki/Patent_claim>.

"Patent History in the United States." 23 December 2019. *Laws*. <https://patent.laws.com/patent-history/patent-history-in-the-united-states>.

"Paul Fjelde." n.d. *Wikipedia*. <https://en.wikipedia.org/wiki/Paul_Fjelde>.

"A Peer of the Realm." *Pittsburgh Press* 16 September 1897: 7.

"Picture of Ben G. Lamme Given to Engineering Dept." *The Tiger (Clemson College)* 12 January 1927: 5.

T*he Pittsburg Electrical Handbook*. The American Institute of Electrical Engineers, 1904. <https://archive.org/details/pittsburgelectri00amer/page/45>.

"Pittsburg Greatly Shocked." *New York Times* 26 June 1906. <https://www.nytimes.com/1906/06/26/archives/pittsburg-greatly-shocked-benjamin-thaw-too-ill-to-be-told-of-his.html>.

Pittsburgh and Allegheny Directory, 1907. Pittsburgh: R.L. Polk & Co., 1907. <https://historicpittsburgh.org/islandora/object/pitt%3A31735058193818#page/1/mode/2up/search/%22201\+N\+Murtland%22>.

"The Pittsburgh 'H.'" n.d. *Visit Pittsburgh*. <https://www.visitpittsburgh.com/things-to-do/arts-culture/history/the-pittsburgh-h/>.

"Pittsburgh Railways." n.d. *Wikipedia*. <https://en.wikipedia.org/wiki/Pittsburgh_Railways>.

"Pittsburgh, the "Smoky City."" 11 February 2015. *Popular Pittsburgh*. <https://popularpittsburgh.com/darkhistory/>.

"Pittsburgh Trolley Pole Co." *Street Railway Journal* January 1896: 59. <https://archive.org/stream/streetrailwayjo121896newy/streetrailwayjo121896newy#page/n938/mode/1up>.

"Pollard McCormick to Peter Schoenberger." *Deed Book, Allegheny County*. Vol. 139. Pittsburgh, 1 January 1842. 196.

Prentice, John. "Charles J. Van Depoele." n.d. *Tramway Information*. <https://www.tramwayinfo.com/Tramframe.htm?https://www.tramwayinfo.com/tramways/Articles/Depoele.htm>.

Preston, David L. "Ten questions about Braddock's Defeat." 9 July 2015. *history-braddocks-defeat*. <https://blog.oup.com/2015/07/history-braddocks-defeat/>.

Price, Gwilym A. "Westinghouse Expansion To Continue Into 1955." *Pittsburgh Post Gazette* 5 January 1954: 19.

"Prior art." n.d. *Wikipedia*. <https://en.wikipedia.org/wiki/Prior_art>.

"Projects Worked On By the Westinghouse Electric and Manufacturing Company During This Period." n.d. *Library of Congress*. <https://www.loc.gov/collections/films-of-westinghouse-works-1904/articles-and-essays/the-westinghouse-world/projects>.

Prout, Henry G. *A Life of George Westinghouse*. New York: The American Society of Mechanical Engineers, 1921.

"Pullman Company." n.d. *Wikipedia*. <https://en.wikipedia.org/wiki/Pullman_Company#Company_town>.

"Railroad electrification in the United States." n.d. *Wikipedia*. <https://en.wikipedia.org/wiki/Railroad_electrification_in_the_United_States>.

Ralph I. Hauser, Director of Engineering. "Letter

to J. M. Wallace, Vice President & General Manager." Pittsburgh, 26 August 1974.

"RCA." n.d. *Wikipedia.* <https://en.wikipedia.org/wiki/RCA>.

"Rege Cordic." n.d. *Wikipedia.* <https://en.wikipedia.org/wiki/Regis_Cordic>.

Reis, Ed. "George Westinghouse." *Mechanical Engineering* October 2008: 33–35.

____. "A Man for His People." *Mechanical Engineering Magazine* October 2008: 32–35. <https://watermark.silverchair.com/me-2008-oct3.pdf?token=AQECAHi208B E49Ooan9kkhW_Ercy7Dm3ZL_9Cf3qfKA c485ysgAAA50wggOZBgkqhkiG9w0BBw agggOKMIIDhgIBADCCA38GCSqGSIb3 DQEHATAeBglghkgBZQMEAS4wEQQM jRhQ4FGq—aaNl1DAgEQgIIDUC2xQLp_ nZaSMsjwzQyMgcrlAid1yshBHVd_h5tCxp>.

"The Relief Effort." n.d. *Johnstown Area Heritage Association.* <https://www.jaha.org/attractions/johnstown-flood-museum/flood-history/the-relief-effort/>.

"Results of Westinghouse Annual Meeting: E. M. Herr Chosen President of the Electric." *Industrial World* 7 August 1911: 944–945. <https://play.google.com/books/reader?id=aLBAAQAA MAAJ&hl=en&pg=GBS.RA5-PA944>.

"Return $200,000, Woman Told." *San Francisco Call* 16 July 1959: 1.

Reynolds, Terry S., and Theodore Bernstein. "Edison and 'The Chair.'" *IEEE Technology and Society Magazine* (1989): 19–28. <https://simson.net/ref/1989/Edison_and_The_Chair.pdf>.

Riggs, Walter L. "Early History of McKeesport: Part 2." 31 July 2012. *The Tube City Almanac.* <http://www.tubecityonline.com/almanac/entry_1967.php>.

Robert Friedel, Paul Israel. *Edison's Electric Light: The Art of Invention.* Baltimore: The Johns Hopkins University Press, 2010. <https://books.google.com/books?id=8U-Naf4DuzMC&pg=PA26&source=gbs_selected_pages&cad=2#v=onepage&q&f=false>.

Roscoe, William E. *History of Schoharie County, New York.* Syracuse: D. Mason & Co., 1882.

Rosenblum, Charles. "A Monument Then and Now." *Pittsburgh Quarterly* 21 August 2016. <https://pittsburghquarterly.com/articles/monument-then-and-now/>.

Ross, Dr. Irwin. "The Chair of Death—A Product of the A.C.-D.C. Controversy." *Consulting Engineer* April 1967: 102–104.

"St. Clair Tunnel." n.d. *Wikipedia.* <https://en.wikipedia.org/wiki/St._Clair_Tunnel>.

Sajna, Mike. "Westinghouse lights up night at Columbian Exposition." *FOCUS* 2 May 1993.

Sarat, Austin. *Gruesome Spectacles: Botched Executions and America's Death Penalty.* Stanford: Stanford Law Books, 2014. <https://books.google.com/books?id=Glt_AwAAQBAJ&pg=PA70&lpg=PA70&dq=John+Debella+Kemmler&source=bl&ots=soATrSOCgZ&sig=ACfU3U2 sNKGj9qOpPlfb9U7VJv-ZL8x-Dw&hl=en&sa=X&ved=2ahUKEwjLnMTd0-rlAhUMHqwKH YqHCpkQ6AEwAHoECAUQAQ#v=onepage&q=John%20Debella%20Kemmler&f=f>.

Sawe, Benjamin Elisha. "Which U.S. City Is Known As 'The City of Bridges'?" 22 January 2019. *World Atlas.* <https://www.worldatlas.com/articles/which-us-city-is-known-as-the-city-of-bridges.html>.

Schmucker, Kristine. "Tools from Our Collection." 11 April 2013. *Voices of Harvey County.* <http://harveycountyvoices.blogspot.com/2013/04/tools-from-our-collection-link-and-pin.html>.

Schoenfield, David. "The Greatest Game Ever Played." 13 October 2010. *ESPN.* <https://www.espn.com/mlb/playoffs/2010/columns/story?id=5676003>.

"Schoharie County, New York." n.d. <https://en.wikipedia.org/wiki/Schoharie_County,_New_York>.

Scully, Donald C. "Historical Society Notes and Documents: Homewood at the Turn of the Century." *Western Pennsylvania History: 1918–2018* 1 October 1976: 493–502. <https://journals.psu.edu/wph/article/viewFile/3446/3277>.

Sebak, Rick. "One Morning, the North Side Exploded (Archived)." November 2006. *WQED Pittsburgh.* <https://web.archive.org/web/20080110055014/http://www.wqed.org/mag/columns/sebak/2006/1106_explosion.shtml>.

Seifer, Marc J. "Nikola Tesla: The Lost Wizard." Jan/Feb/Mar 2006. *ttmagazine.* <http://teslatech.info/ttmagazine/v4n1/seifer.htm>.

____. *Wizard: The Life and Times of Nikola Tesla: Biography of a Genius.* New York: Citadel Press; Kensington Publishing Corp., 1998.

"Shippingport Atomic Power Station." n.d. <https://en.wikipedia.org/wiki/Shippingport_Atomic_Power_Station#cite_note-nrc-history-4>.

Shulman, Seth. *Owning the Future.* Boston: Houghton Mifflin, 1999. <https://archive.org/details/owningfuture00shul/page/158>.

Skinner, Charles E. "Lighting the World's Columbian Exposition." Historical Society of Western Pennsylvania, 1934.

Skrabec, Quentin, Jr. *Benevolent Barons—American Worker-Centered Industrialists, 1850–1910.* Jefferson, NC: McFarland, 2015. <https://books.google.com/books?id=CLnwCQAAQBAJ&pg=PA159&lpg=PA159&dq=contrasting+carnegie+and+westinghouse&source=bl&ots=zTCdw vABH7&sig=ACfU3U2hlNQUkM2bb8SxxUo_NwKCGBzMcw&hl=en&sa=X&ved=2ahUKEw iN75XknvzlAhXsmeAKHcvtA1gQ6AEwDnoE CA4QAQ#v=onepage&q=contrasti>.

____. *George Westinghouse: Gentle Genius.* New York: Algora Publishing, 2007.

____. *Henry Clay Frick: The Life of the Perfect Capitalist.* Jefferson: McFarland, 2010.<https://www.reddit.com/r/trains/comments/60p37p/

how_much_will_a_train_stretch_once_all_the_slack/>.

———. *William McKinley, Apostle of Protectionism*. New York: Algora Publishing, 2008. <https://books.google.com/books?id=NVcI2rm1O1EC&pg=>.

———. *The World's Richest Neighborhood: How Pittsburgh's East Enders Forged American Industry*. New York: Algora Publishing, 2010.

Slauson, Allen B., ed. *A History of the City of Washington—Its Men and Institutions*. Washington: The Washington Post Company, 1903.

"Society." *Pittsburgh Post-Gazette* 5 March 1909: 10.

"Society Weddings." *Pittsburgh Daily Post* 5 March 1909: 10.

Spehr, Paul. *The Man Who Made Movies: W.K.L. Dickson*. New Barnet, England: John Libbey Publishing Ltd., 2008. <https://books.google.com/books?id=hcWCDgAAQBAJ&pg=PA38&lpg=PA38&dq=Major+Sherbourne+Eaton&source=bl&ots=7dLf7j84y4&sig=ACfU3U2PtUgbRDYIdyWDgRKGkbci5i67cw&hl=en&sa=X&ved=2ahUKEwjP85v1rMnlAhVIx1kKHVSmDnAQ6AEwCXoECAsQAQ#v=onepage&q=Major%20Sherbourne%20Eaton>.

Sprague, John L. "Frank J. Sprague invents the constant-speed dc electric motor." *IEEE Power & Energy Magazine* 1540-7977/16©2016IEEE (2016): 80–96.

Standiford, Les. *Meet You in Hell: Andrew Carnegie, Henry Clay Frick, and the Bitter Partnership That Transformed America*. New York: Crown Publishers, 2005.

"Steam turbine." n.d. *Wikipedia*. <https://en.wikipedia.org/wiki/Steam_turbine#Reaction_turbines>.

Stetson, Francis Lynde. "The Use of the Niagara Water Power." *Cassier's Magazine* July 1895.

Stillwell, Lewis Buckley. "Electric Power Generation at Niagara." *Cassiers Magazine—Engineering Illustrated* 1 July 1895: 253–304.

"Streetcars in North America." n.d. *Wikipedia*. <https://en.wikipedia.org/wiki/Streetcars_in_North_America/>.

Strunk, Joseph. "Letter from Joseph Strunk to George Westinghouse, Sr." 17 January 1865. *New York Heritage Digital Collections*. <https://cdm16694.contentdm.oclc.org/digital/collection/p16202coll7/id/9074/rec/29>.

"Tables Quickly Turned: Mr. Westinghouse Again In Control of the Switch Company." *New York Times* 14 March 1891: 1.

Tacitus. *The Annals*. Vol. VI. Rome, 109 AD. XVI vols. <http://classics.mit.edu/Tacitus/annals.6.vi.html>.

Tarr, Joel, and Karen Clay. "Boom and Bust in Pittsburgh Natural Gas History: Development, Policy, and Environmental Effects, 1878–1920." *The Pennsylvania Magazine of History and Biography* 2015 October: 323–342.

Tesla, Nikola. *The Autobiography of Nikola Tesla*. n.d. <https://www.mcnabb.com/music/tesla/bio.pdf>.

———. "My Inventions." *Electrical Experimenter* February-June and October 1919.

———. "Nikola Tesla and George Westinghouse built the first hydro-electric power plant in 1895 in Niagara Falls and started the electrification of the world." 12 January 1897. *Tesla Memorial Society of New York*. <https://www.teslasociety.com/exhibition.htm>.

"Thomas Alva Edison I (1847–1931)." n.d. *Wiki Tree*. <https://www.wikitree.com/wiki/Edison-1#_note-1860us>.

"Thomas Edison and Menlo Park." n.d. The Thomas Edison Center at Menlo Park. <http://www.menloparkmuseum.org/history/thomas-edison-and-menlo-park/>.

"Thomas Edison Biography." 26 August 2019. *Biography.com*. <https://www.biography.com/inventor/thomas-edison>.

"Thomas Edison Feels That the Newly Formed General Electric, Under J.P. Morgan, Is Overcharging Him for His Lighting and Motor Supplies." n.d. *RAAB Collection*. <https://www.raabcollection.com/science-autographs/edison-trust>.

"Thomas Edison National Historical Park." n.d. *Wikipedia*. <https://en.wikipedia.org/wiki/Thomas_Edison_National_Historical_Park>.

"Thomas Edison National Historical Park—Research." n.d. NPS. <https://www.nps.gov/edis/learn/historyculture/research.htm>.

"Thomas Hutchins to John McKee." *Deed Book 1, Allegheny County*. Pittsburgh, 11 September 1785. 152.

"Thomas M. Miller to James H. Hopkins." *Deed Book, Allegheny County*. Vol. 246. Pittsburgh, 31 July 1869. 256.

"Thomson-Houston Electric Company." n.d. *Wikipedia*. <https://en.wikipedia.org/wiki/Thomson-Houston_Electric_Company>.

Tighe, Janet A. "The New York Medico-Legal Society: Legitimating the Union of Law and Psychiatry (1867- 1918)." *International Journal of Law and Psychiatry* 9 (1986): 231–243. <https://www.sciencedirect.com/sdfe/pdf/download/eid/1-s2.0-0160252786900488/first-page-pdf>.

"Toshiba launches 600V sine-wave PWM Driver IC for three-phase brushless motors." 28 October 2019. *NewElectronics*. <http://www.newelectronics.co.uk/electronics/toshiba-sine-wave-brushless-motor-industrial-home-appliances/220711/>.

"The Town of Pullman." n.d. Historic Pullman Foundation. <http://www.pullmanil.org/town.htm>.

Townes, John A. "A Second Spa for the Berkshires." *New York Times* 4 June 1989.

"Tunnelling shield." n.d. *Wikipedia*. <https://en.wikipedia.org/wiki/Tunnelling_shield>.

"Turbinia." n.d. *Wikipedia*. <https://en.wikipedia.org/wiki/Turbinia>.

"2015 Gold King Mine waste water spill." n.d.

Wikipedia. <https://en.wikipedia.org/wiki/2015_Gold_King_Mine_waste_water_spill>.

U*ncas, 1590–1683.* 2019. The Yale Indian Papers Project. <https://yipp.yale.edu/bio/bibliography/uncas-1590-1683>.

Union Switch and Signal Company. "Electro-pneumatic interlocking." Pittsburgh, 1914. <https://play.google.com/books/reader?id=kZrVAAAAMAAJ&hl=en&pg=GBS.PA3>.

"Union Switch & Signal Co." *Railroad Gazette* 13 March 1891: 186. <https://play.google.com/books/reader?id=dkRKzCk5x2gC&hl=en&pg=GBS.PA186>.

"United States Census." Vol. Roll 1304. Pittsburgh, Allegheny County, 15 April 1910.

U.S. Senate Committee on Interstate Commerce. *Automatic Couplers and Power-Brakes.* Washington, D.C.: Government Printing Office, 1890. <https://play.google.com/books/reader?id=-cIuAAAAMAAJ&hl=en&pg=GBS.PA1>.

VanTrump, James D. "'Solitude' and the Nether Depths: The Pittsburgh Estate of George Westinghouse and Its Gas Well." *The Western Pennsylvania Historical Magazine* June 1959: 155–172. <https://journals.psu.edu/wph/article/viewFile/2641/2474>.

Voelker, Elizabeth Anne. *Solitude—Majestic Showplace for a Millionaire's Genius.* College Thesis. Carnegie Institute of Technology. Pittsburgh, 1950.

Vogel, Charity. *The Angola Horror: The 1867 Train Wreck that Shocked the Nation and Transformed American Railroads.* Ithaca: Cornell University Press, 2013.

Vondas, Jerry. "Man with a plan." 19 March 2006. *Tribune-Review.* <https://phlf.org/2006/03/19/man-with-a-plan/>.

"Wabtec Corporation." n.d. <http://www.companyhistories.com/Wabtec-Corporation-Company-History.html>.

"Wabtec Corporation." n.d. <https://en.wikipedia.org/wiki/Wabtec_Corporation>.

Wadsworth, G. R. "Terminal Improvements of the New York Central & Hudson River in New York." *The Railroad Gazette* 20 October 1905: 366–369. <https://play.google.com/books/reader?id=yJxMAAAAYAAJ&hl=en&pg=GBS.PA366>.

Waples, David A. *The Natural Gas Industry in Appalachia.* Second Edition. Jefferson, NC: McFarland, 2012.

"Westinghouse Air Brake Company." n.d. *Wikipedia.* <https://en.wikipedia.org/wiki/Westinghouse_Air_Brake_Company>.

"Westinghouse Air Brake Technologies." n.d. *MacroTrends.* <https://www.macrotrends.net/stocks/charts/WAB/westinghouse-air-brake-technologies/revenue>.

"Westinghouse Air Spring." *The Power Wagon* 1 December 1912: 152–153. <https://play.google.com/books/reader?id=I2HmAAAAMAAJ&hl=en&pg=GBS.PA153>.

"Westinghouse Broadcasting." n.d. *Wikipedia.* <https://en.wikipedia.org/wiki/Westinghouse_Broadcasting>.

"Westinghouse Castle goes for $100,000 at sheriff's sale." 7 June 2016. *Pocono Record.* <https://www.poconorecord.com/article/20160607/NEWS/160609689>.

"The Westinghouse Electric and Manufacturing Company." n.d. *Library of Congress.* <https://www.loc.gov/collections/films-of-westinghouse-works-1904/articles-and-essays/the-westinghouse-world/the-westinghouse-electric-and-manufacturing-company/>.

____. *The Westinghouse Company.* n.d. 165–202.

W*estinghouse Electric and Manufacturing Company Annual Report.* Detre Library and Archives, Senator John Heinz History Center. Pittsburgh, PA, 1894.

W*estinghouse Electric Company.* n.d. <https://en.wikipedia.org/wiki/Westinghouse_Electric_Company>.

Westinghouse Electric Corporation. *George Westinghouse: 1846–1914.* Pittsburgh: Westinghouse Electric Corporation, 1946. <https://babel.hathitrust.org/cgi/pt?id=wu.89068792373&view=1up&seq=1>.

"Westinghouse Electric Corporation." n.d. *Wikipedia.* <https://en.wikipedia.org/wiki/Westinghouse_Electric_Corporation>.

"Westinghouse Electric: Mr. Westinghouse Appeals for Proxies." *Industrial World* 24 July 1911: 888–889. <https://play.google.com/books/reader?id=aLBAAQAAMAAJ&hl=en&pg=GBS.RA3-PA888>.

"Westinghouse Family Moves to Arizona." *Pittsburgh Press* 10 March 1933.

W*estinghouse Family Papers.* n.d. <https://nyheritage.org/collections/westinghouse-family-papers>.

"Westinghouse Family Papers." n.d. *New York Heritage Digital Collections.* <https://nyheritage.org/collections/westinghouse-family-papers>.

Westinghouse, George. "History of the Air Brake." *The Electric Journal* (1910): 227–236.

____. "Letter to T. B. Kerr." Lenox, MA, 11 July 1888.

____. "Letter to T. B. Kerr." Lenox, MA, 5 July 1888.

"Westinghouse Licensing Program Signs Consumer Electronics Licensee." 7 February 2017. <http://lmca.net/news-and-insights/33rd-licensee-signed-westinghouse/>.

Westinghouse Machine Company. *The Westinghouse-Parsons Steam Turbine.* East Pittsburgh, 1910. <https://babel.hathitrust.org/cgi/pt?id=uc1.b4530455&view=1up&seq=4>.

Westinghouse Memorial. n.d. <http://www.westinghousememorial.org/history.htm>.

"Westinghouse On Top." *The Pittsburgh Press* 10 May 1891: 7.

"Westinghouse Plant Closing." *New York Times* 16 February 1985: 30. <https://timesmachine.nytimes.com/timesmachine/1985/02/16/031887.html?pageNumber=30>.

"Westinghouse Romance." *Dundee Evening Telegraph* 20 February 1907: 6. <https://www.britishnewspaperarchive.co.uk/viewer/bl/0000563/19070220/102/0006>.

"The Westinghouse Set: An American parcel-gilt silver and mixed metals 'Japanese style' seven-piece tea and coffee set with tray, Tiffany & Co., New York, circa 1885." 22 January 2006. Sotheby's. <http://www.sothebys.com/en/auctions/ecatalogue/2006/americana-n08158/lot.135.html>.

"Westinghouse Sues Carnegie—The Two Millionaires Have a Dispute About a Large Gas Bill." *New York Times* 30 September 1891: 1. <https://timesmachine.nytimes.com/timesmachine/1891/09/30/103339251.html?pageNumber=1>.

"Westinghouse Widow Dies at Home in Lenox." *Pittsburgh Telegraph* 23 June 1914: 1, 8.

"Westinghouse Will Filed." *New York Times* 19 March 1914.

Westinghouse—Innovation You Can Be Sure Of. n.d. <http://www.westinghouselighting.com/>.

"Westinghouse's Estate." *New York Times* 22 November 2014.

"What is the purpose of the Federal Reserve System?" 3 November 2016. Board of Governors of the Federal Reserve System. <https://www.federalreserve.gov/faqs/about_12594.htm>.

"White Consolidated Industries Inc. History." 1996. *FundingUniverse*. <http://www.fundinguniverse.com/company-histories/white-consolidated-industries-inc-history/>.

Wicker, Elmus. "Banking Panics in the U.S.: 1873–1933." n.d. *EH.net*. Economic History Association. <https://eh.net/encyclopedia/banking-panics-in-the-us-1873-1933/>.

"William Kemmler." n.d. *Murderpedia* (includes *New York Herald* article of August 7, 1890). <https://murderpedia.org/male.K/k/kemmler-william.htm>.

Wilson, Wendell E. "Thomas Alva Edison (1847–1931)." 2013. *The Mineralogical Record.* <https://archive.ph/20130415080355/http://www.minrec.org/labels.asp?colid=737#selection-389.0–393.11>.

Wohleber, Curt. "'St. George' Westinghouse." *American Heritage's Invention & Technology* Winter 1997. <https://www.inventionandtech.com/content/%E2%80%9Cst-george%E2%80%9D-westinghouse-1?page=full>.

"Working Conditions." n.d. *Library of Congress.* <https://www.loc.gov/collections/films-of-westinghouse-works-1904/articles-and-essays/the-westinghouse-world/working-conditions/?c=200&fa=subject:industrialism>.

"World's Columbian Exposition." n.d. *Encyclopedia of Chicago.* <https://encyclopedia.chicagohistory.org/pages/1386.html>.

_____. n.d. *Wikipedia.* <https://en.wikipedia.org/wiki/World%27s_Columbian_Exposition#Great_Buildings>.

Wright, Roy V., ed. *Locomotive Cyclopedia of American Practice.* 6. New York: Simmona-Boardman Publishing Company, 1922. <https://play.google.com/books/reader?id=oMY1AQAAMAAJ&hl=en&pg=GBS.PP5>.

Yale University. *Directory Of The Living Non-Graduates Of Yale University.* New Haven: Yale University, 1914. <https://play.google.com/books/reader?id=WUobAAAAYAAJ&hl=en&pg=GBS.PA51>.

_____. *University Catalogue, 1906.* New Haven: Yale University Publications, 1906.

Index

AC 1, 78, 80, 81–82, 83–89, 93, 96, 98, 110–111, 118–125, 134, 139–143, 147–151, 157–158, 160, 175, 179, 190, 221, 231–234, 235–238; *see also* alternating current
Adams, Edward Dean 145, 151, 155
ADHD 92, 99
Aeolipile 166
air brake: automatic 37–43, 225–228; straight 31–36, 37–38, 42–43
air spring 204, 208
Akron (Ohio) 109
Albany (New York) 12, 19, 116, 120
Albany and Susquehanna Railroad 116
Albert, Prince (Belgium) 64–65
Albert, Prince (England) 135
ALCOA 154; *see also* Pittsburgh Reduction Company
Alexandria (Virginia) 40
Allegheny City (Pennsylvania) 44, 60, 140, 161
Allegheny County (Pennsylvania) 47, 51, 60, 126, 177, 214
Allegheny County Airport (Pennsylvania) 3
Allegheny Mountains 32
Allegheny River 4, 44, 60, 158
Altadena (California) 200
alternating current 81–82, 86, 87, 93–94, 118, 121–122, 124, 148, 150, 154, 156, 158, 192, 202, 214, 229–234; *see also* AC
alternator 147
Altoona (Pennsylvania) 28, 32, 40
Amalgamated Association of Iron and Steel Workers (AAISW) 128–129
Ambler, Augustine 24–25
Amerex 219
American Association of Engineering Societies 205
American Federation of Labor (AFL) 130
American Institute of Electrical Engineers (AIEE) 92, 93, 96, 205
American Legion 211

American Marconi Company 219
American Mutoscope Biograph Company 163
American Production and Inventory Control Society 52
American Society for the Prevention of Cruelty to Animals (ASPCA) 122
American Society of Mechanical Engineers (ASME) 193–194, 205
American Standard 59, 217
Ames Power Plant 146–148
ampere (electrical) 229
Anderson, Cook 27
Angelo Brothers Lighting 219
Angola (New York) 20,
Angola Horror 20
Animas River 146
Annals 181
Anne (Queen of England) 12, 192
Annual Report 142
Appleton (Wisconsin) 173
arc lamp 77, 88, 92, 103
Arlington National Cemetery 208
Armstrong, Thomas 63
Armstrong Cork Company 63
Arnold, William 93
Association of Master Mechanics 31
Astor, John Jacob 145
AT&T 219
Atkins, Edwin F. 185
Atlas Shrugged 90
Auburn State Prison (New York) 123, 195
Aurora Accomodation (train) 24
Austrian Polytechnic Institute 91
auxiliary reservoir 37, 225
Aviation Gas Turbine Division 218
Azores 113

Baggaley, Ralph 27–29, 31
Bahamas 200
Bainbridge Island (Washington) 200
Baldwin Locomotive Works 179
Baltimore & Ohio Railroad 116
Barings Bank (London) 132
Barney, Charles Tracy 182
Barton, Clara 127

Batchelor, Charles 92, 100–101
battery 55, 110, 229
Battle of the Currents 221
Battle of the Monongahela 159, 160
Bear, David 62, 190, 221
Beechwood Boulevard (Pittsburgh) 63
Beechwood Hall (Frew estate) 63
Behrend, Bernard 93
Beijing Great Wheel 137
Belgium 34, 64, 173, 205
Bell, Alexander Graham 101–102
Bell, Chichester 102
Bell, James 206
Bell Telephone Company 101
Bellerive (Switzerland) 113
Belmont, August 133
Belmont, August, Jr. 133–134, 161, 186
Bennett, Julia 200
Bennington County 11
Berkman, Alexander 129
Berkshire Courier 86–87
Berkshire Mountains 132, 190
Bern (Switzerland) 217
Bessemer Steel Works 20
Bettis Atomic Power Laboratory 220
Bidwell Plow Works 56
Big Sister Creek 20
Blaine, James Gillespie 193
Blaine House (Washington, D.C.) 193–194, 207
Blathy, Otto 233
Block Signal and Train Control Board 55
block system 53–54
Board of Patent Control 176
Bodwell, H.N. 87
Boston 58, 100, 113, 133, 178, 220
Boston & Providence Railroad 32
Brace Street (Buffalo) 154
Braddock (Pennsylvania) 160
Braddock, General Edward 158–160, 214
Braddock Avenue (Pittsburgh) 63,
Braddock Road 160
Bradford (Pennsylvania) 68
Blairsville (Pennsylvania) 218
brakeman (or brakemen) 22–23, 36, 39–40

272 Index

Brakes Applied 225
Brakes Released 225
Brewer, H.A. 87
Bridge, James Howard 130
Brimmer, Anna Maria 11
British Board of Trade 34
British Columbia 200
British Nuclear Fuels Limited (BNFL) 218–219
British Railways Accident Commission 34
British Westinghouse Electric and Manufacturing Company 145, 203
Brocklebank, Charles H. 198
Brocklebank, Evelyn Violet 198, 223; see also Westinghouse, Evelyn
Brocklebank, Lady 198
Brocklebank, Sylvia 198, 199
Brooklyn (New York) 21
Brown, Alfred 92–93
Brown, Harold P. 121–124
Brown, Mrs. James 108
Brown-Boveri 168
Brush Electric Light Company 118
brushes (motor) 96–97, 171
Brust, Anna 11
Buchanan, William 55–56
Budapest (Hungary) 92, 96
Budapest Telephone Exchange 92
Buffalo (New York) 118, 119, 123, 145, 148, 150, 154, 156, 195
Buffalo and Erie Railroad 20
Buffalo Society for the Prevention of Cruelty to Animals (BSPCA) 119
Burlington (Iowa) 41
Burnham, Daniel Hudson 135–136
Burnham and Root 136
Butte (Montana) 182
Butterworth, Mr. 93
Byllesby, H.H. 86, 93–94, 106

Caldwell, John, Jr. 127
Caledonian Railway 34
California 109, 145, 184, 196, 200
California State Railroad Museum (Sacramento) 196
Cambria County (Pennsylvania) 126
Cambridge University 148
Canada 99, 156, 178–179, 183, 200
Canadian Pacific Railroad 39
Canton (Ohio) 195
Captain of Industry 188
car replacer 19–20, 22, 26, 27
carbon brush 96
carbon electrode 77
carbon filament 103–105, 109, 139
carbon microphone 101
Card, W.W. 29, 31
Carman, William 100
Carnegie, Andrew 3, 64, 74, 126–131, 204, 206, 214–215

Carnegie Institute of Technology 211
Carnegie Library 63
Carnegie-Mellon University 211
Carnegie Steel Company 63, 128–129
Carnegie Tech 211
Carpenter, Niles 186
Carson, Rachel 3
Casino Building (Wilmerding) 163
Cassatt, Alexander Johnston 28, 31
Cassier's (magazine) 151, 174
cast iron 19–20
cast steel 20, 26
"Castle" 51, 221; see also General Office Building
Cataract Construction Company 145, 148, 150–151
CBS 218, 220
Central Bridge (New York) 14–15
Central High School (Philadelphia) 110
Central Pacific Railroad 86
Central Park (New York City) 207
Central Telegraph Office (Budapest) 92
Chafee, G.W. 119
chain brake 22, 24, 34
Chang, Li Hung 151
Charles II (King of Spain) 12
Chautauqua (New York) 109
Cheney, Margaret 90
Chernow, Ron 114
Chesapeake & Ohio Railroad 116
Chicago 24, 32, 46–47, 134–136, 138–139, 151, 158, 168, 195, 220
Chicago and North Western Railroad 32
Chicago, Burlington and Quincy Railroad 24, 41
Chicago, Detroit & Canada Grand Trunk Junction Rail Road Company 178
Chicago, Rock Island and Pacific Railway 185
Chicago World's Fair 1; see also Columbian Exposition and World's Columbian Exposition
Childs, Judge Henry 123
cholera
Christy, George 26
Churchill, Winston 124
Churchill, Pennsylvania 3
Circle Tree Ranch (Tucson) 200
"Circle-W" 3, 9
City Council (Pittsburgh) 73, 211
"City of Bridges" 214
Civil War (US) 17–18, 40, 115, 163, 182
claim (patent) 35, 104, 109, 241–242
Clayton (Frick estate) 63, 70, 96
Cleveland, Grover 68, 136, 193
Cleveland, Ohio 11
Cockran, W. Bourke 124
Coffin, Charles 110, 134, 176

Coffin, Lorenzo Stephen 22–23, 41, 49
Colorado 146, 148, 201
Columbia (steamship) 104
Columbia School of Mines 122
Columbian Exposition (1893) 136–138, 151, 192, 213; see also Chicago World's Fair and World's Columbian Exposition
Columbus, Christopher 135
Committee on Automatic Freight Car Brakes 41–42, 49
Committee on Interstate Commerce 22
Committee on Science and the Arts 38
commutator 92, 96–98
company town 5, 44–45, 50
Connor, Piers 38, 225–226
Conrad, Frank 219
Consolidated Electric Light 78
Continental Edison Company 92
Coolidge, Calvin 214
Cooper, Peter 209
Cooper Hewitt Electric Company 209
Cooper Union 209
Copeman Electric Stove Company 218
Cordic, Rege 3
Cordic & Company 3
core (transformer) 80, 83–84, 233–238
Cornell University 93
Corsair (ship) 116
Cortelyou, George 182
couplers (train car) 39–41
Court of Appeals (Second Circuit) 85, 176
Court of Honor 136–137, 192
Covell, Vernon R. 214
Cravath, Paul 203
Cravath, Swaine, and Moore 203
Croatia 90–91
Cromwell, Oliver 37, 198
Croushore Drug Store 47
Crystal Palace 135
current (electrical) 142, 146, 148, 150, 154–158, 192, 202, 214, 229–238
Current War 1, 80, 81, 98, 143
Cuyahoga River 11
Czolgosz, Leon 195

The Dalles (Oregon) 157
Davis-Monthan airfield (Tucson) 200
Dayton (Ohio) 205
DC 1, 77, 80, 81–82, 85, 87–88, 92, 106, 108–110, 118, 121–122, 125, 134, 139, 143, 144, 147–148, 157, 160, 173, 231–233; see also direct current
DeBella, John 123
de Moleyns, Frederick 103
Department of Justice (DoJ) 176
Deri, Miksa 233
Detre Library 93, 223
Detroit 46, 99–100, 178–179

Detroit Young Men's Society 100
"Devil in the White City" 138
"Diamond W" (Tucson) 200
direct current 81, 86, 103, 121–122, 146, 202; *see also* DC
District Court (District of Connecticut) 175
divisional (patent) 242–242
double patenting 175, 177, 242
Drake, Edwin 67
Drake stop (Pittsburgh) 177
drawing board 41, 62, 212, 213
Dredge, James 33–34, 36, 37, 43
Drigg (England) 198–199
Duncan, Sherman & Company 113–114
Dunfermline (Scotland) 129
Dupont Circle 193
Durston, Charles 124
Duryea, Charles 204
Duryea, Frank 204

East Conemaugh (Pennsylvania) 127
East End (Pittsburgh) 127
East End Gas Company 70
East Liberty (Pittsburgh) 68
East Liberty Valley (Pennsylvania) 61
East Pittsburgh 111, 160–163, 166, 188, 199, 218
East Pittsburgh Improvement Company 45, 47
East St. Louis (Illinois) 56
East Tanque Verde Road 200
Eaton, Sherbourne 105, 108
Edgar Thomson steelworks 74, 215
Edgewood Towne Center 58–59
Edinburgh (Scotland) 148
Edison, Marion Estelle ("Dot") 100, 110
Edison, Samuel Ogden 99
Edison, Thomas Alva 1, 5, 78, 81–82, 86–87, 92–93, 98, 99–112, 116–117, 118–124, 131, 134, 138–139, 143, 156, 171, 179, 202, 205, 231–233
Edison, Thomas, Jr. ("Dash") 100
Edison, William Leslie 104
Edison (Electric) Illuminating Company 105
Edison Electric Light Company 92
Edison General Electric Company 117, 134, 173
Edison Machine Works 92
Edison Manufacturing Company 122
Edison Medal 87, 205
Edison Phonograph Company 102
Ediswan 109
Eiffel, Gustave 136
Electric Club 162
Electric Club Journal 163
electric meter 88
"Electric Power Generation at Niagara" 151

Electrical World (magazine) 93, 209
Electricity Hall 142
Electro-Dynamic Light Company 139
electro-pneumatic 56–57
Elliott, Nancy 99
Elm Court 192
Emerson, Edward 68, 71
Emerson, Margaret 193
Emerson, Ralph Waldo 68
Engineering: An Illustrated Weekly Journal 33
Engineering and Science Hall of Fame 205
Engineers' Society of Western Pennsylvania 211
England 11–12, 23, 33–34, 37, 42–43, 44, 80, 84, 103, 145, 198, 200
English High School (Boston) 113
Enterprise (aircraft carrier) 220–221
Environmental Protection Agency 146
Equitable Gas Company 71, 73
Erie Canal 144
Erie Railroad 53
Erskine Park (Massachusetts) 132, 189–193, 198–200, 204, 206–208
Evershed, Thomas 144–145, 148
"EXPO: Magic of the White City" 138

fail-safe 37, 225
family tree 11, 223
Faraday, Michael 93
Faraday's Law 235
farm equipment 14, 17
Farmer, Moses 103
"Faust" 92
Federal Reserve System 182–183
Fell, George E. 119
Ferraris, Galileo 80, 96
Ferris, George Washington Gale, Jr. 136
Ferris Wheel 136–137
Field, Marshall 135
Fifth Avenue (Pittsburgh) 63
Fifth Avenue Presbyterian Church (New York City) 207
FindaGrave 223
Firestone, Harvey 117
First Methodist Church (Wilmerding) 47
Fjelde, Paul 211
flagman 196
"Flying W" Ranch (Tucson) 200
Forbes, George 148, 150–151
Forbes, John 8
Forbes Field 5
Ford, Henry 116–117
Fort Cumberland (Maryland) 158
Fort Duquesne (Pennsylvania) 158–159
Fort Meyers (Florida) 117
Fort Niagara 158
Fort Wayne 220

Fort Worth 201
Fortune 500 202, 218
Foxhollow School for Girls 193
France 25, 115, 145, 158
Franklin Institute 38
freight trains 32, 39, 42
French, Daniel Chester 211
French and Indian War 158, 214
French Legion of Honor 205
French Westinghouse Air Brake Company 151, 204
frequency 3, 43, 81, 97–98, 150
Frew, William 63
Frick, Henry Clay 63–64, 70, 72, 74, 96, 126–131, 197, 206
frog 19–20, 26–27

Galt, John 90
Galveston (Texas) 127
Gardner and Ranson Air Brake Company 35
Garfield, James 193
Garrison Alley (Pittsburgh) 56–57, 85–87, 158, 160
gas meter 73, 88
gas well 67–76, 133
gauge (track) 39, 160
Gaulard, Lucien 78, 80, 233
Gaulard and Gibbs transformer 79–80, 83–84, 233
GE Transportation 217
General Electric Company (GE) 14, 112, 117, 134, 139, 148, 149, 150–151, 168, 173, 175–177, 183, 186, 217, 219
General of Domestic Life 188
General Office Building (Wilmerding) 47, 51–52; *see also* "Castle"
General Robinson Street (Allegheny City) 44
generator 78, 86–87, 92, 108, 110, 119, 140, 142, 147–151, 166, 168, 171, 229, 233–234
George Washington (submarine) 220
Germany 12, 113, 205
Gerry, Elbridge 120, 122
Gibbs, John Dixon 78–80, 83–84, 233
Gilded Age 113
Gillespie Tool Company 69, 133
Gilliland, Ezra 109
Gimbels (department store) 177
"Girl in the Red Velvet Swing" 63
Glen Eyre 132–133, 157, 190, 194–196, 198
Glen Hotel 47, 49
Glenmont (Edison estate) 109–110
Goethe, Johann Wolfgang 92
Gold King Mine 146–148
Gompers, Samuel 130–131
Goodman, E.H. 59
Gottwals School (Wilmerding) 47
Gould, Jay 100, 116
graded-diameter pipe 72
Graf, John 52, 221

Index

Grand Army of the Republic (GAR) 163, 195
Grand Army Week 163–164
Grand Plaza 136
Grand Trunk Herald (newspaper) 100
Grand Trunk Railroad 99, 178–179
Grant Hill 30
Grashof Medal 205
Graz University of Technology 91
Great Barrington (Massachusetts) 190
Great Britain 109, 122, 178
Great Depression of 1929 217
Greenlawn (Heinz estate) 63, 70
Greensburg (Pennsylvania) 88, 214
"Group W" 220

Hale, Matthew 120
half-day work 44
Harper's Weekly (magazine) 71, 74, 106, 129
Harrisburg (Pennsylvania) 56–57, 195–196
Harrison, Benjamin 193
Hatch, Charles 124
Hayes, Rutherford B. 102
Haymaker, Michael 67–68
Haymaker, Obediah 67–68
Haymaker well 67–68
Heinrichs, Ernst H. 131, 184, 186
Heinz, H.J. 3, 44, 52, 63–64, 70
Heinze, F. Augustus 182
Henry, Joseph 102
Hero 165–166
Herr, Edwin Musser 185
Herr, Herbert Thacker 168, 185, 213, 239
Herr, Theodore Witmer 185
Hewitt, Abram 209
Hewitt, Peter Cooper 209
Higinbotham, Harlow Niles 142
Hill, David B. 120–121
Hodge, Nehemiah 43
Hoenerin, Catherina 11
Hollywood Sanitarium (British Columbia) 200
Holmes, Nathaniel 60
Holmwood 193
Homestead (Pennsylvania) 128–131
Homestead Strike 126, 128–131
Homewood (Pittsburgh) 62, 70
Homewood (Wilkins estate) 62
Hoosac Railway Company 176
Hopkins, James 60, 62
Hopkins, Mark 86
Hopkins, Mary Frances 86–87
Hopkinson (inventor) 80
Hornbostel, Henry 211
Horne, Joseph 63
Horrocks, Christopher 207
Horrocks School 47
Hort, John 123
Hotel Langham 207
Hotel New Yorker 157
Hourglass Chart 11, 223

House of Representatives (US) 39, 135, 195
Houston, Edwin 110
Hudson River 116
Hudson River Railroad 20, 55
Hunter, Elizabeth 11
Hunter, Robert 12
Hutchins, Thomas 60
HVDC 156–157
hysteresis 84

IEEE Milestone 87
incandescent lamp 77–78, 86–88, 92, 102–109, 121, 136, 138–140, 150, 154, 161
Indianapolis 32
induction coil 84
induction motor 92–93, 96–98, 142–143, 151, 156, 175, 235–238
infringe 25, 233, 242
injunction 176
Institute of Electrical and Electronics Engineers (IEEE) 87, 205
intellectual property 14, 78, 111, 175
Interlocking Switch and Signal Company 56
interlocking system 55–57, 77
Interstate Commerce 22, 55
Italy 25, 77, 78, 205
interference (patent) 78
International Development Company 219
International Niagara Commission 145–146, 148
Invention Factory 101–102
Irton Hall (England) 198

J. Pierpont Morgan 103–106, 108, 112, 113–117, 133–134, 135, 145, 157, 173, 182, 186
Jackson, C.H. 56
Jackson Manufacturing Company 56
Janney, Eli 40–41
Jarvie, James N. 184
Jersey City (New Jersey) 57
Jewett, Edgar B. 154
Jewett, Thomas 29
Jobard, Marcellin 103
John Fritz Medal 205
Johns Hopkins University 148
Johnstown (Pennsylvania) 126–128
Johnstown Flood 126–128, 131
Johnstown Flood Museum 127
Johnstown Library 127
Joncaire, Daniel 144
J.P. Morgan Library and Art Museum (New York City) 113

Karlovac (Croatia) 90
KDKA 3, 5, 219–220
Keller, Emil E. 199
Kemmler, William 123–125, 195
Kennelly, Arthur 122
Kerr, Thomas B. 94, 195

Keystone Gas Company 68
Keystone pickle factory 44
Khilkov, Prince Mikhail 63
Kilchenstein, Ed 213
King Albert I (Belgium) 64
King Charles II (Spain) 12
King Henry VI (England) 198
King Leopold I (Belgium) 205
King Leopold II (Belgium) 205
King Umberto I (Italy) 205
Kingston (New York) 21, 204
Knickerbocker Trust Company (New York City) 182
Knox, Philander Chase 63, 127, 128
Konigliche Technische Hochschule (Berlin) 205
Kruesi, John 100–101

Lake Conemaugh (Pennsylvania) 126–127
Lake Erie 178
Lake Geneva (Switzerland) 113
Lake Huron 178
Lake Michigan 137
Lamme, Benjamin 63, 149, 151, 156, 162
Lane, Robert 92
Lang Avenue (Pittsburgh) 62
Lange, Philip 88
Larson, Erik 138
"Last Days of Night" 203
Lathrop, Austin 122
La Tour-de-Peilz (Switzerland) 113
Laurel Lake 192
Lawrence, David L. 76
Lawrence, William Leonard 90
Lawrenceville (Pittsburgh) 68, 87
lawsuit 35, 74
Lee (Massachusetts) 190, 199
Leipzig (Germany) 205
Lenox (Massachusetts) 168, 190, 192
Lenz's Law 238
Lester (Pennsylvania) 218
Leupp, Francis 14, 61, 189, 206
Liberty Avenue (Pittsburgh) 31–32, 44
Liberty Street (New York City) 93
Lillie, C.H. 86–87
Lincoln, Abraham 115, 211
Lincoln Highway 3, 214
Lincoln Memorial 211
link-and-pin 23, 39–40
Lionel 39
Liverpool (England) 198
Living Age (magazine) 25
Lockport (New York) 156
Lockstadt, Charles F. 139
Logan, Meda 75–76
London 78, 113, 115, 132, 135, 145, 148, 189, 203, 233
London and North-Western (railroad) 34
London Chronicle (newspaper) 125
Long Beach (guided missile cruiser) 220

Long Island Railroad 53
Lorant, Stefan 76
Lord Kelvin 64, 82, 143, 145, 202–203; *see also* Thomson, William
Los Angeles 157
Lusitania (ship) 168, 193
Lyndhurst (Thaw estate) 63
Lynn (Massachusetts) 110, 173

Macalpine, John H. 169, 194
MacDonald, Carlos 123–124
Machinery Hall 140–142
Mackenzie, James 100
Mackenzie insurgents 99
Madison Avenue (New York City) 108, 113
Madison Square Garden 63, 157
magnetic flux 235, 238
Main Reservoir 37
Maitland Avenue (Pittsburgh) 199
Man (patent) 78, 104, 139
Man, Albon 78,
Mandic, Djuka 90
Manhattan Elevated Railroad Company 56, 171
Marguerite Avenue (Wilmerding) 49
Maribor (Slovenia) 91
Martin, Thomas Commerfield 93
Marx 39
Massachusetts 43, 86, 103, 110, 132, 171, 173, 189–191, 199, 241
Massachusetts Avenue (Washington, D.C.) 193
Master Car Builders Association 40–42
Mather, Robert 1895–186
Mauretania (ship) 168–169
Mazeroski, Bill 5
McCormick, Cyrus 135
McCormick, Pollard 60
McKee, John 60
McKeesport (Pennsylvania) 60, 159
McKinley, William 65, 195
McLeod's Mill (Mississippi) 17
Medico-Legal Society (New York) 121–122
Mellon, Andrew W. 63, 127, 214
Mellon, Richard Beatty 63
Mellon, Richard King 76
Mellon, Thomas A. 63
Melville, George Wallace 169, 194
Menlo Park (New Jersey) 100–105, 109–111, 171
Mercantile National Bank (New York City) 182
"Merchandise Creditors Modified Plan" 184
Metcalf, William 74, 76
Michigan Central Railroad 32
Midland Railway (England) 43
Milan (Ohio) 99
Miller, Lewis 109
Miller, Mina 109
Miller, Thomas 60

Mills, Darius Ogden ("D.O.") 109, 145
Milwaukee 24
Minaville (New York) 11, 14
Mineral Point (Pennsylvania) 127
Mohaques (Indian tribe) 12
Mohaques River 12
Mohawk River 12
Monongahela River 4, 30, 74, 128, 158–160
Mont Cenis Tunnel 25
Montgomery County (New York) 11
Montreal (Canada) 178
Montreal Board of Trade 178
Moore, Alexander Pollock 63
Moore, Graham 203
Morgan, Anne Tracy 115, 157
Morgan, J. Pierpont 103–106, 108, 112, 113–117, 133–134, 135, 145, 157, 173, 182, 186
Morgan, J. Pierpont, Jr. 115
Morgan, Juliet Pierpont 115
Morgan, Junius Spencer 113, 115
Morgan, Louisa Pierpont 115
"Morganization" 116, 186
Morrow, L.W.W. 105
Morse code 100, 109
MotivePower 217
motor 82, 88–89, 90, 92–98, 108, 142–143, 147–148, 150–151, 154, 156, 171, 173–175, 177, 235–238
Mt. Clemens (Michigan) 100
Murrysville (Pennsylvania) 67–69, 70, 71, 73
"Murrysville Freak" 67
Muscoota (ship) 18
Museum of Science and Industry 137
Musk, Elon 1, 89
Musser, Annie 185
Musser, Emma 185

Napoleon 25, 205
National Academy of Sciences 102
National Bank of Commerce (New York City) 182, 184
National Broadcasting Company (NBC) 220
National Historic Landmark 179
National Historical Park 110
National Inventors Hall of Fame 205
National Monument 46
National Park Service 110, 159
National Record Mart 177
National Registry of Historic Places 15, 52, 211
Natural Gas Commission (Pittsburgh) 73
Nautilus (submarine) 220
Navajo Nation 145
Naval Academy 171
Naval Consulting Board 121
Navy (US) 18, 170, 171, 220
"The NCC" 219
Negley, Joanna 45

Negley, William 45
Neptune (ship) 169–170
Nesbit, Evelyn 63
New England 113, 130
New Haven (Connecticut) 175, 197
New Jersey 92, 100, 109, 122
New Mexico 146
New Westminster (British Columbia) 200
New York (City) 63, 66, 69, 92, 100, 104, 106, 108, 113, 114–115, 121, 129, 130, 133, 135, 145, 157, 168, 171, 179, 182, 188, 195, 196–197, 207, 208, 210, 219
New York (State) 87, 109, 115, 118, 120–124, 145, 154, 176, 190, 195, 204
New York Cavalry 17
New York Central and Hudson River Railroad 55
New York Central Railroad 109, 116
New York Elevated Railroad Company 56
New York Evening Post (newspaper) 121
New York Express (train) 19
New York Herald (newspaper) 102, 125
New York National Bank of Commerce 184
New York National Guard 17
New York, New Haven, and Hartford Railroad 66, 168, 179
New York Stock Exchange 182
New York Sun (newspaper) 103
New York Times (newspaper) 90, 104, 106, 125, 169, 213, 218
New York, West Shore & Buffalo Railroad 116
New York Yankees 5
Newark (New Jersey) 100, 220
Newark Trials 43
News Reporting Telegraph Company 100
Niagara Falls 1, 65, 91, 125, 134, 143, 144–146, 148–157, 158, 160, 162, 213, 233
Niagara Falls Hydraulic Power & Manufacturing Company 144
Niagara Falls Power Company 145
Niagara River 144
Nice (France) 115
Nordrhein-Westfalen 11, 13
North Adams (Massachusetts) 43, 171
North America 144, 158
North American Review (magazine) 118
North Braddock (Pennsylvania) 159–160
North Murtland Street (Pittsburgh) 199
North Side (Pittsburgh) 44, 71
Northern District Court (New York) 176
Northern District Court (Ohio) 35

Index

Nottingham and Newark Division 43
Nunn, Lucien 146–147
Nunn, Paul 147–148

Ohio River 4, 158
Ohio State University 151
ohm (electrical) 229, 231–232
Ohm, Georg Simon 229, 231
Ohm's Law 229, 231
Oil City (Pennsylvania) 67
Old Colony Railroad 32
Olde Frothingslosh 3
Oliver, George T. 207
Olmstead, Frederick Law 136
O'Neill, John J. 93, 95
Order of Leopold 205
Order of the Crown of Italy 205
Oregon Railroad and Navigation Company 104–105
OSHA 49
Oslo (Norway) 211

Pacific DC Intertie 157
Pacific Express (train) 32–33
Pacific Railroad Act 39
Palace of Fine Arts 137
Palatines 12–13
Panhandle 29–31. 33
Panic of 1893 161, 182
Panic of 1907 182–187
Pantaleoni, Diomede 77
Pantaleoni, Guido 77–78, 84, 86, 96
Paris 92, 101, 115, 188
Paris Exposition (1889) 136–137
Parsons, Charles Algernon 165–166, 168
Parsons Marine Steam Turbine Company 165
Pattison, Robert E. 129
Peabody, Morgan & Company 113, 115
Pearl Street Station (New York City) 105–108
Peary, Robert E. 207
Peck, Charles F. 92–94
Pelton, Lester Allan 147
Pelton water wheel 147
Penn Avenue (Pittsburgh) 63
Penn Fuel Company 68, 71, 73
Penn-Lincoln Parkway (Pittsburgh) 3, 59
Pennsy 28–29; *see also* Pennsylvania Railroad; PRR
Pennsylvania (Commonwealth) 29, 31, 40, 44, 47, 56, 67–68, 70, 73, 85, 88, 126, 127, 128, 173, 177, 190, 195, 211, 220
Pennsylvania Humane Society 206
Pennsylvania Main Line Canal 126
Pennsylvania Railroad 28, 31, 33, 36, 38, 44, 51, 53, 57, 60–61, 63, 71, 74, 116, 127, 163, 188, 195; *see also* Pennsy; PRR
Penrose (Armstrong estate) 63
Peoples Natural Gas Company 68
Peterson, Frederick 122

Pew, Joseph 68, 71
Philadelphia (Pennsylvania) 110, 205, 218
Philadelphia & Reading Railroad 116
Philadelphia Company 71–74, 127, 151
phonograph 101–102
Piccirilli, Masaniello 213
Pierpont, John 113
Pierpont, Juliet 113
Pinkerton Detective Agency 129
Pitcairn, Robert 28, 31, 44, 63, 64, 85, 127, 196, 197
Pitt, William 8
Pittsborough 8
Pittsburg 8, 51, 74, 76, 188
Pittsburgh 3–5, 8, 26, 27–28, 30, 31–32, 33, 3843, 44, 52, 56, 58, 60–66, 67–69, 71–76, 77–78, 80, 85–86, 88, 93–96, 126–127, 128, 130–133, 136, 147–148, 158, 160, 161, 163–164, 168, 177, 186, 188–190, 192, 196, 197, 199, 206–207, 211, 213, 214, 218, 220, 223
Pittsburgh Brewing Company 3
Pittsburgh City Council 73, 211
Pittsburgh Magazine 3, 65
Pittsburgh Opera 32
Pittsburgh Parks Conservancy 214
Pittsburgh Pirates 5, 44
Pittsburgh Post-Gazette (newspaper) 8, 73, 190, 214, 218
Pittsburgh Press (newspaper) 132, 163
Pittsburgh Railway Company 177
Pittsburgh Reduction Company 148; *see also* ALCOA
Pittsburgh Stock Exchange 8, 182
"Pittsburgh: The Story of an American City" 76
"Plan for the Adjustment of Debt" 184
Playhouse (at Erskine Park) 193
PNC Park 44
pneumatic 18, 25, 56–57, 161, 225
Pocono Sandstone 67
Point Breeze (Pennsylvania) 33, 62–65, 70, 197
pollution 74, 179
polyphase 95, 142–143, 151
Polytechnic Institute (Troy) 14
Pompton (New Jersey) 20
Ponds at Foxhollow 193
Pope, Franklin 80, 84, 85
Port Huron (Michigan) 99–100, 178–179
Portland (Maine) 178
Post-Gazette (newspaper) 8, 73, 190. 214, 218
Poughkeepsie (New York) 32
power house 154–156, 179
Pownal (New Hampshire) 11, 158
Prague International Hygienic and Pharmaceutical Exposition 46

Presbyterian 207
Price, Gwilym A. 218
prior art (patent) 35, 242
priority (patent) 35, 96, 242
Priory Hospitality Group 52, 221
"*Prodigal Genius*" 95
prosecution (patent) 176, 241
Prout, Henry Goslee 17, 174, 202
PRR 28, 31–32, 45, 60, 62, 196; *see also* Pennsy; Pennsylvania Railroad
Prussia 11
Public Service Company 147
Pullman (passenger cars) 36, 45, 196
Pullman, George 36, 45–46
Pullman, Illinois 46
Pullman Company 36, 195

quadruplex telegraph 100
Queen Anne 12, 192
Queen Victoria 64, 135, 156

Radio Corporation of America (RCA) 219
"*Ragtime*" 63
Rahway (New Jersey) 92
Railroad Safety Appliance Act (1893) 40, 49, 195
Railway Companies' Association 43
Rankine, William B. 145
Readjustment Committee 184
Rebellion of 1837 99
Red Cross 127
Redfern Hotel 47
Reed, James Hay 128
Relief Department 49–50
remote control 157
Republican 76, 193, 195
resistance (electrical) 84, 104, 109, 229, 231–232
Rheinland-Palatinate 11, 13
rheumatic fever 113
rhinophyma 114
Rhodes, Godfrey 41
Richmond (Virginia) 171, 173
robber baron 9, 116, 131
Robinson, William 54–56
Rockefeller, John D. 113, 116, 182
Rockwell, A.D. 123–124
Rockwell, Colonel 119
Rodman, Hugh 111
Rodman Chemical Company 111
Rogers, Fred 3
Rome (Italy) 78, 181
Roosevelt, Theodore 182,
Roosevelt, Theodore, Jr. 211
Root, John Wellborn 136
rosacea 114
Roselle (New Jersey) 109
rotary engine 165
rotating field 151, 162
rotor (electrical) 96–98, 150, 237–238
rotor (turbine) 167
Rowland, A.T. 58–59
Rowland, Henry 148
Roxbury (New York) 21

Royal Aquarium (London) 78
Royal Commission on Railway Accidents 43
Rubber Factory 86
Ruff, Benjamin 126
RULES FOR STATION LIGHTING 121
Russell, Lillian 63

St. Clair River 178–179
St. Clair Tunnel 179
St. Clair Tunnel Company 179
St. George's Church (New York City) 114
St. Lawrence River 178
St. Paul's Church (Irton, England) 198
St. Louis 32, 56, 129, 135
Salk, Jonas 3
San Juan Mountains 147
San Miguel River 146–147
Santa Barbara (California) 200
Sarnia (Ontario) 178–179
Sawyer, William 78, 109, 139
Sawyer-Man lamp 78, 104, 139
scarlet fever 99
Schenectady (New York) 11, 14, 18–19, 21, 22, 25–26, 60
Schenley Park (Pittsburgh) 5, 211–212, 214, 216
Schmid, Albert 80, 83, 148, 151, 156
Schneck, Henry de Bois 190, 192
Schoellkopf, Jacob 144–145
Schoenberger, Peter 60
Schoharie (County) 12, 14
Schoharie (creek) 12
Schoharie (town) 190
Schoharie Junction 116
Scotland 34, 44, 129, 148
Scott, Charles 149
Scott, Thomas 71
Scott Legacy Medal 38, 205
Scranton (Pennsylvania) 173
Scully, Donald 63
Sellers, Coleman 148
semaphore 54–55
Senator John Heinz History Center 52, 223
Serbian Orthodox Church 90, 156
Serrell, Lemuel 92
Shady Side Academy 197
Shallenberger, Oliver 63, 80, 83, 88–89, 94, 96, 98, 148
Sheffield Scientific School ("Sheff") 197
Sherman Antitrust Act 176
Shippingport (Pennsylvania) 220
shock absorbers 9
Shulman, Seth 101
Siemens 86, 218
Siemens-Halske 139
Simmons, Doc 32–33
Skrabec, Quentin 46
slip rings 96–98
Sloane, William D. 192
Smiljan (Croatia) 90–91
Smith (property) 190, 192

Smith, George Lemuel 118–119
Smith, Grace Margaret 200
Smith, John Y. 43
Smithsonian Institution 102
"Smoky City" 3, 27, 60, 74,
Socialists 46
Solitude (Westinghouse estate) 5, 33, 60–66, 67–72, 96, 133, 157, 188–190, 193, 195–199, 206, 211
South Bend (Indiana) 173
South-Eastern Railway 34
South Fork (Pennsylvania) 126–128
South Fork Dam 126
South Fork Fishing and Hunting Club 126–128
Southern Telecom 219
Southside Metal and Machine Works 139
Southwick, Alfred Porter 119–120
Spanish flu 190, 211
"Spirit of American Youth" 211–212
Sprague, Robert C. 171
Sprague Electric Company 171
Sprague Electric Railway & Motor Company (SERM) 171
Sprague, Frank Julian 171, 173, 178
Springfield (Illinois) 163
Squirrel Hill Tunnels (Pittsburgh) 3
stamp mill 146–147
Stanley, William 63, 78, 83–84, 86–87, 190
Stanley Theater (Pittsburgh) 177
Stanwood, Harriet 193
Stars and Stripes (ship) 18
State Savings Bank (Butte, MT) 182
stator 96, 98, 150, 238
steam turbine 165–167, 213, 218, 239–240; *see also* turbine
Stetson, Francis Lynde 145
Steubenville (Ohio) 29–30
Stevens Institute of Technology 148
Stewart, Dr. William A. 206–207
Stillwell, Lewis B. 63, 147, 148, 151, 156
Stillwell, Mary Jane 100
Stilson, Mary (grandmother) 11
Stockbridge (Massachusetts) 190
Stockbridge Road (Massachusetts) 199
stopper lamp 78, 104, 139–140, 192
streetcar 171–174, 177, 180; *see also* trolley
Sturges, Amelia "Memie" 114–115
Sturges, Jonathan 114–115
Sullivan, Louis 136
Sun Oil Company 68
Sunoco 68
Swan, Joseph 103, 109
Swayne, Circuit Justice 35
Swissvale (Pennsylvania) 58–59

Tacitus 181
Tainter, Charles 102
Tate, Daniel 30
Taylor, R.I. 86–87
Telluride (Colorado) 146–147
Terry, Charles A. 176
Tesla (car) 89
Tesla, Daniel 90
Tesla, Milutin 90
Tesla, Nikola 1, 63, 81, 89, 90–98, 99, 101, 111, 134, 142–143, 144, 151, 154–156, 157, 158, 175–177, 178, 209, 238
Tesla Electric Company 92, 94
Tesla Electric Light and Manufacturing Company 92, 95
Tesla Motor Company 95
Texas Pacific Railroad 39
Thaw, Henry Kendall 63
Thaw, William 63
Thomas, Lowell 216
Thomas Edison National Historical Park 110
Thomson, J. Edgar 74, 215
Thomson, Elihu 110, 173
Thomson, Sir William 64, 145, 148, 202, 203; *see also* Kelvin, Lord
Thomson-Houston Electric Company 110, 117, 118, 121, 123, 134, 138, 173, 175, 242
Tiberius, Emperor 181
Tivoli (Italy) 78
Tonawanda (New York) 156
Tonnaleuka Club 49
Topsy 44, 122
Toshiba 219
Toucey, J.M. 55–56
Towne, A.N. 24
Townsend, William Kneeland 175–177, 242
track circuit 54–55
Tracy, Frances "Fanny" Louisa 115
Trafford (Pennsylvania) 163
Train Line 37
transformer 78–80, 82, 83–85, 87, 98, 233–234, 235–237
Transformer Building 155–156
transmission (gas) 73
transmission (power) 76, 77, 81–82, 84, 87, 98, 146–149, 156, 157, 214, 220, 231–234
transmission (radio) 219
triple valve 37, 41–42, 225–228
trolley 171–178; *see also* streetcar
trolley pole 174–175
Troy (New York) 14, 20, 22
Trust Company of America 182
tuberculosis 115
Tucson (Arizona) 200
tunnel: Mt. Cenis 25; Niagara Falls 144; Port Huron 179; Solitude 62, 64
turbine 144, 148, 168, 193; *see also* steam turbine
Turbinia (ship) 165, 167, 168
turbo-generator 168
Turin (Italy) 78

Index

Turtle Creek (Pennsylvania) 44, 67, 158–161, 214–216, 221
Tyler, John 62

Union Carbide Suit 85
Union College (Schenectady) 18, 205
Union Electric Signal Company 55–56
Union Pacific Railroad 32
Union Passenger Railway 171
Union Switch and Signal Company (US&S) 3, 56–59, 77, 85–86, 151, 184, 213, 217
Union Trust Company 182
United Copper Company 182
United States Board on Geographic Names 8
United States Court for Receivers 184
United States Treasury 182
University of Pittsburgh (Pitt) 3, 5, 8, 171
University of Turin 77, 96
University of Gottingen (Germany) 113
University of Pittsburgh 3, 5
Uptegraff, W.D. 131
usury 181
Utah 146

vacuum 34, 43, 103
Vail, Benjamin 92
valid (patent) 109, 175, 242
Vancouver Island (British Columbia) 200
Van Depoele, Charles Joseph 173, 175–178, 242
Van Depoele Electric Manufacturing Company 173
Vanderbilt, Alfred Gwynne 193
Vanderbilt, Cornelius 27, 32, 109, 116, 135, 192
Vanderbilt, Emily (wife of William Sloane) 192
Vanderbilt, William H. 104, 109
Vedder, Albert Isaac (grandfather) 11
Vedder, Emeline (mother) 11; see also Westinghouse, Emeline
Vedder, Mary Stilson (grandmother) 11
Venus 157
Vermont 11, 158
Vienna (Ontario) 99
Villard, Henry 104–105, 108, 112, 134
volt 106–107, 149, 179
voltage 55, 77–78, 81–82, 86, 98, 118–121, 143, 147, 149–150, 156, 157, 173, 221, 229–234, 235

WABCO Holdings, Inc. 217
Wabtec Corporation 217
Walker, Marguerite Erskine (wife) 21; see also Westinghouse, Marguerite
Walker Company 176

Wallace, William James 177, 242
War of Spanish Succession 12
Warner, Charles Dudley 113
Warner Theater (Pittsburgh) 177
Washington, George 159–160, 220
Washington, D.C. 135, 193–194, 195, 205, 207, 211
Washington (Pennsylvania) 177
water wheel 147
Ways and Means Committee 195
Welker, Martin 35
West Mifflin (Pennsylvania) 3, 220
West Orange (New Jersey) 109–110, 111, 122, 124
West Orange laboratory
West Virginia 29, 76
Western Electric 139
Western Express (train) 195
Western Pennsylvania Humane Society 206
Western Union Telegraph Company 92, 100
Westinghouse 175, 190, 221
Westinghouse, Agnes Sylvia (granddaughter) 200
Westinghouse, Albert (brother) 14–15, 17–18
Westinghouse, Aubrey Harold (grandson) 200
Westinghouse, Catherine (sister) 15
Westinghouse, Elizabeth (sister) 15
Westinghouse, Emeline Vedder 189; see also Vedder, Emeline
Westinghouse, Evelyn (daughter-in-law) 199–200, 223; see also Brocklebank, Evelyn Violet
Westinghouse, George (father) 11–12, 14, 19, 30, 31
Westinghouse, George III (son) 65–66, 69, 132, 136, 193–194, 196–201, 204, 208
Westinghouse, George Thomas (grandson) 199–200
Westinghouse, Henry (brother) 15
Westinghouse, Henry Herman (brother) 15, 31, 78, 204, 207, 208, 216, 217
Westinghouse, Herman (brother) 15
Westinghouse, Jay (brother) 14–15
Westinghouse, John (brother) 14–15, 17–18
Westinghouse, Margaret Virginia (granddaughter) 200
Westinghouse, Marguerite 5, 26, 30, 32, 33–34, 42, 60–66, 68–69, 77, 96, 132, 136, 157, 186, 188–194, 196–198, 200, 204, 206, 207, 208, 214, 223; see also Walker, Marguerite Erskine
Westinghouse, Mary (sister) 15
Westinghouse, Richard Lawrence (grandson) 200

Westinghouse, Violet Louisa (granddaughter) 200
Westinghouse & Co. 14, 16, 17
Westinghouse Air Brake Company 31–36, 37–39, 41–43, 44–52, 59, 130, 140, 158, 161, 166, 168, 184, 197, 202, 204, 206, 207, 214, 217, 221, 225–228
Westinghouse Air Spring Company 204, 208
Westinghouse-Baldwin 179
Westinghouse Bridge 214–216
Westinghouse Broadcasting 3, 220
Westinghouse Education Center 3
Westinghouse Electric and Manufacturing Company 133–134, 139–143, 145–157, 158–164, 166–168, 175–177, 180, 183–187, 189, 199, 202, 204, 213, 214, 216, 218–219, 220, 239, 242
Westinghouse Electric Company 1, 57, 80, 85–88, 93–95, 106, 117, 131–132, 218
Westinghouse Electric Corporation 219
Westinghouse Lighting Corporation 219
Westinghouse Machine Company 166–167, 169, 204, 213, 239
Westinghouse Memorial Bridge 3
Westinghouse Memorial (Pittsburgh) 5, 211–213, 216
Westinghouse Memorial Society 211
"Westinghouse Ordinance" 73
Westinghouse Park 5, 190, 221
Westinghouse Park 2nd Century Coalition (WP2CC) 190, 221
"Westinghouse-Parsons Steam Turbine" 239
Westinghouse Research Center 3
Westinghouse Small Appliances India 219
Westinghouse Solar Lights 219
Westmoreland County (Pennsylvania) 60
Wetmore, Lila Courtney (wife of William Stanley) 86
Whitcomb, G.D. 31
White, Stanford 63
White, Thomas H. 218
"White City" 135–143, 192
White Consolidated Industries (WCI) 218
White House 65, 195
White-Westinghouse 218
Whittlesey, F.P. 87
WikiTree 223
Wilder, Gene 138
Wilkins, William 62
Williams, Edwin 28, 31
Wilmerding 44–52, 158, 160–161, 163, 166, 168, 188, 195, 198, 206, 217, 221

Wilmerding News (newspaper) 46, 50
Wilmerding Renewed 52
Wilmerding Times (newspaper) 51
Wilson, August 3
Winchester Avenue Railroad Company 175, 242
Wistenhaus, John Ferdinand (grandfather) 11
Wistinhausen, John Hendrik (great grandfather) 11, 158
"Wizard of Menlo Park" 101, 103
Wood, Eric Fisher 211
Wood Street (Pittsburgh) 177
Woodlawn Cemetery (New York City) 207–208
Woodvale (Pennsylvania) 127
Woodyard, W.H. 196
Woolson, Albert 163
World Series 5
World War I 121, 170, 211
World War II 76
World's Columbian Exposition 46, 125, 134, 135–137, 139, 142–143, 148, 158; *see also* Chicago World's Fair and Columbian Exposition

Xcel Energy 147

Yale Corinthian Yacht Club 197
Yale Daily News 197
Yale Military Drill Company 198
Yale University 78, 197–198
Yale University Tennis Club 197
YMCA 52, 177, 195

ZBD transformer 233
ZF Friedrichshafen AG 217
Ziegler, Matilda "Tillie" 123
Zipernowsky, Karoly 233

www.ingramcontent.com/pod-product-compliance
Ingram Content Group UK Ltd.
Pitfield, Milton Keynes, MK11 3LW, UK
UKHW050702160426
5217IPUK00038B/2011